CAST AND WROUGHT
ALUMINIUM BRONZES
PROPERTIES, PROCESSES
AND STRUCTURE

This book is written as a tribute to

Pierre G. Durville
for his pioneering work in the
development of wrought aluminium bronze

and to

Charles H. Meigh
for his poineering work in the
development of cast aluminium bronze

CAST AND WROUGHT

ALUMINIUM

BRONZES

PROPERTIES, PROCESSES AND STRUCTURE

Harry J Meigh CEng MIMech E

CRC Press
Taylor & Francis Group
Boca Raton London New York

CRC Press is an imprint of the
Taylor & Francis Group, an **informa** business

First published in 2000 by
IOM Communications Ltd

IOM Communicaions Ltd
is a wholly-owned subsidiary of
The Institute of Materials

This paperback edition first published in 2008 by
Maney Publishing

Published 2017 by CRC Press
Taylor & Francis Group
6000 Broken Sound Parkway NW, Suite 300
Boca Raton, FL 33487-2742

ISBN 13: 978-1-861250-62-9 (hbk)
ISBN 13: 978-1-906540-20-3 (pbk)

**Visit the Taylor & Francis Web site at
http://www.taylorandfrancis.com**

**and the CRC Press Web site at
http://www.crcpress.com**

The right of Harry J Meigh
to be identified as the author of this book
has been asserted in accordance with the
Copyright, Designs and Patents Act 1988
Sections 77 & 78

Typeset in the UK by
Dorwyn Ltd, Rowlands Castle, Hants

CONTENTS

Part 1   Cast and Wrought Aluminium Bronzes: Properties and production processes

v

FOREWORD

This book has been written at the request of the Copper Development Association of Great Britain to bring up to date the information contained in the excellent book by P J Macken and AA Smith, published in 1966, which has hitherto been the standard reference book on aluminium bronze throughout the industrial world.

Considerable research has been done since 1966 in the metallurgy of these alloys which has allowed guidelines to be established regarding the composition and manufacturing conditions required to ensure reliable corrosion resistance. This book brings this knowledge together in a form which aims to be readily understandable to engineers and designers whose knowledge of metallurgy may not be extensive.

It has been divided into two parts to make it easier for the reader to home-in on the information in which he/she is interested:

Part 1 seeks to meet the needs of people who are responsible for the selection of materials: designers, engineering consultants, metallurgists, architects, civil engineers etc. It provides information on the compositions and corresponding properties of the cast and wrought alloys available, as well as on the types of components obtainable in these alloys. It includes two chapters on corrosion. It also provides information, for the benefit of manufacturers, on the various manufacturing processes: casting, hot and cold working and joining. It does not seek to provide detail technical guidance for particular cases, but gives general principles that have to be observed.

Part 2 deals with the microstructure of the main aluminium bronzes and is for the benefit of those who wish to obtain a deeper knowledge of this range of alloys.

Additional information is provided as appendices, including recommendations on machining. An extensive list of references is also given at the end of the book.

The ISO(International)/CEN(European) type of alloy designation is used throughout this book as it indicates the nominal composition of the alloy (e.g. CuAl10Ni5Fe4). The alloying element are shown in bold type for clarity (particularly since the 'l' of Al is easily mistaken for a '1'). American equivalent alloy designations are indicated in tables in which compositions and properties are given.

ACKNOWLEDGEMENTS

The author wishes to thank the following, without whose help and expert knowledge on various aspects of aluminium bronzes, this book would not have been as comprehensive as it aimed to be.

Mr Vin Calcutt of the Copper Development Association (UK). It was at his suggestion that this book was written. His constant support and encouragement and the information he provided was invaluable.

Mr Dominic Meigh, Consultant (son of the author). His knowledge of the metallurgy of aluminium bronze was of particular importance as was his very thorough reading and comments on all chapters. His help with computer technology and, in particular, his guidance in producing illustrations was much appreciated.

Professor G W Lorimer, Head of Materials Science Centre, University of Manchester and UMIST, who supplied unpublished reports on the microstructure of aluminium bronze alloys which complemented articles published by his department. This work represent the most comprehensive treatment of the metallurgy of aluminium bronzes.

Mr Arthur Cohen of Copper Development Association Inc., who provided a large number of technical references.

Monsieur Pierre Neil, whose articles are published under the name Pierre Weill-Couly, for is valuable comments on the chapters dealing with the microstructure, the corrosion resistance and the welding of aluminium bronzes.

Monsieur Christian Durville, grandson of Pierre Durville, who supplied interesting information on the early production of aluminium bronze billets for subsequent working.

Dr Roger Francis of Weir Materials whose expert comments on the corrosion resistance of aluminium bronze have been much appreciated.

Mr Simon Gregory of Alfred Ellis & Sons Ltd, and

Mr J C Bailey, Delta (Manganese Bronze) Ltd for their valuable information on wrought aluminium bronze processes and products.

Mr Alan Eklid of Willow Metallurgy, Consultants, for his knowledgeable and helpful comments on wrought processes and products and on continuous casting.

Mr Richard Dawson of Columbia Metals for his expert advice on the welding of aluminium bronzes.

Mr Dave Medley, of Scotforge, USA, for the valuable information he supplied on the wear performance of aluminium bronzes.

Dr I M Hutchings, Reader in Tribology, Cambridge University, for reading through the draft chapter on wear resistance and for the valuable comments that he made.

Mr M Sahoo and colleagues of CANMET for their unpublished work on the effects of impurities.

Dr G S Murgatroyd AIM, AMIBF, formerly of Sandwell College, for the loan of his (unpublished) doctorate thesis on aluminium bronze.

Mrs S Inada-Kim of the Imperial College of Science Library, for the help she provided to **Sonia Busto Alarcia,** a Spanish student who sifted through the College's references on aluminium bronze and carefully collated information from these references.

Mrs Maureen Clutterbuck of the Cheltenham College of Technology for her guidance in the production of graphs.

HISTORICAL NOTES

Earliest aluminium bronze

Although aluminium, which is present in clay, is the most common metallic element in the earth's crust,[92] it was not before 1855 that it was first produced by a Frenchman, Henri Sainte-Claire Deville (1818–1881), by a sodium reducing process.[126] This was a very expensive process but the high resultant cost of aluminium did not deter metallurgists from carrying out experiments to alloy it with every known metal. Soon a metallurgist by the name of John Percy reported that 'a small proportion of aluminium increases the hardness of copper, does not injure its malleability, makes it susceptible of a beautiful polish and varies its colour from red-gold to pale yellow'. The Tissier brothers in Rouen, who were assistants to Sainte-Claire Deville, brought the attention of the French Academy in 1856 to the properties of aluminium bronze and a week later a paper by Debray described the work done on this alloy by the Rousseau brothers at their Glassière Works in the suburbs of Paris.

The high cost of the alloy and the fact that its performance did not always match the claims of its advocates meant that there was little interest in using it. An alpine mountain howitzer was cast in aluminium bronze for the French artillery in 1860 and, although it successfully passed every test it was subjected to, it was too expensive to be used for gun manufacture. It seems however that the alloy was used, despite its cost, for making some ships propellers.

In 1885, Cowles Bros in America successfully produced aluminium bronze at a much lower cost. The process consisted in reducing corundum, a mineral containing aluminium oxide, by melting it with granulated copper and coarse charcoal in an early form of electric furnace. Aluminium was thus refined and alloyed to copper in one operation. Controlling the aluminium content must have been difficult and, since corundum may also contain other oxides, such as those of iron, magnesium and silicon, the presence of these other elements may, with the exception of iron, have had a deleterious and unpredictable effect on properties. The Cowles Company set up a subsidiary in Stoke-on-Trent in England and the two companies produced six grades of aluminium bronze ranging from 1.25% to 11% aluminium.

A further breakthrough occurred in 1886, when Charles M. Hall and Paul L. T. Héroult, working independently, first successfully produced aluminium at an economically viable price by an electrolysis process, for which Héroult took out a patent. For reasons that are not clear, instead of adding pure aluminium directly to copper in a furnace to produce aluminium bronze, it was produced by a variant of the Héroult electrolytic process. This consisted in melting pure alumina, by a powerful electric current, over a molten bath of copper and electrolysing the whole melt with alumina as the anode and copper as the cathode. Aluminium ions thus released alloyed with the copper cathode to form aluminium bronze. Aluminium-containing alloys, including aluminium bronze, started to be produced by the Héroult process in 1888 by the Société Métallurgique Swiss in Switzerland and in Germany by its associated company Allgemeine Elektrizität Gesellschaft of Berlin. The American company Wilson Aluminium Company of Brooklyn, New-York also produced 3 to 18% Al aluminium bronze by an indirect electrolysis process using copper and corundum. The demands for aluminium bronze being still fairly modest, the tonnage produced was low.

This phase lasted only a short time. As the demand for aluminium rose and its price fell, there was no advantage in producing aluminium bronze by the indirect electrolytic method and most users began to make their own alloy from the component metals.

First systematic research into copper-aluminium alloys

In 1905, Dr L. Guillet[82] published his research into the whole range of combinations of copper and aluminium and concluded that the only alloys that could be used industrially were those which contained less than 11% or more than 94% of aluminium. He produced what was probably the first equilibrium diagram of copper and aluminium as well as many photomicrographs of great theoretical value. A similar but more detailed and extensive investigation was published in 1907 by Professors H. C. H. Carpenter and Mr C. A. Edwards of the National Physical Laboratory, Teddington, England. They came to the same conclusion regarding the useful range of alloys and their equilibrium diagram (Fig. H1) closely resembled that of Dr Guillet. Fig. H2a gives an enlarged view of the aluminium bronze section of this diagram which it is interesting to compare with the more recent binary diagram shown in Fig. H2b. They were aware that the freezing range of the useful copper-rich alloys was very narrow but, because of the limitations of the research instrumentation available at the time, it was not possible to determine accurately the temperatures at which solidification began (the 'liquidus' line) and ended (the 'solidus' line). The other interesting point is that they were aware that, if a 10% aluminium alloy was cooled slowly between 600°C and 500°C, a structural change occurred: namely, a 'needle-like' structure was, at least in part, changed into a 'lamellar' structure; but they did not label the lamellar structure (later called 'gamma 2') nor did they realise its detrimental effect on corrosion resistance, probably because the transformation was only partial, due to too fast a rate of cooling.

Their report, published by the Institution of Mechanical Engineers,[44] gave however a lot of interesting information on tensile, hardness, torsion and alternating stress properties as well as on micro-structure and corrosion resistance.

It is clear from this report that, since the cost of aluminium had dropped dramatically thirty years previously, an increasing volume of both cast and wrought aluminium bronze was being produced by quite a number of companies, notably in

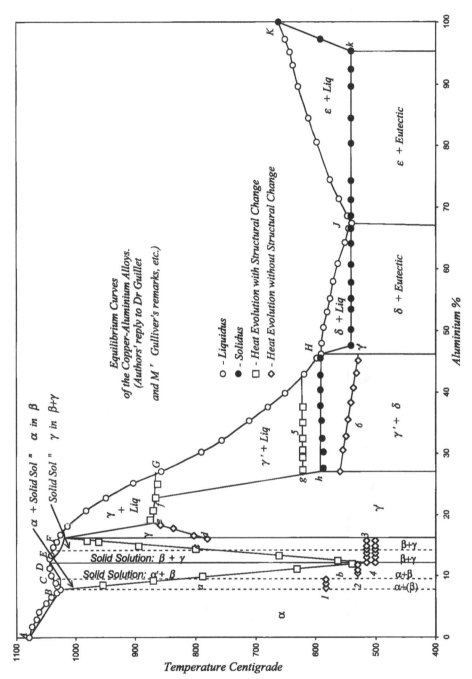

Fig. H.1 The copper–aluminium equilibrium diagram by Carpenter and Edwards.[44]

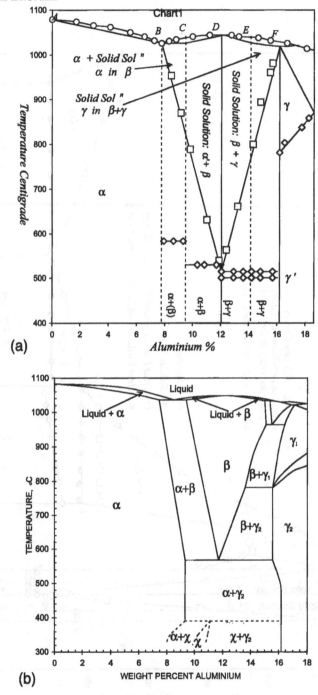

Fig. H2   (a) Enlarged aluminium bronze section of the copper–aluminium equilibrium diagram by Carpenter and Edwards;[44] (b) Latest copper–aluminium binary equilibrium diagram[127] for comparison.

the ship building industry. It seemed to have been used then mostly as a wrought material and its suitability in this connection was fully recognised. Rolled bars, sheets and even tubes were successfully produced. Included among the cast products however were large propeller castings. The growing use of this range of alloys is confirmed by a paper by E. S. Sperry[166] in an article in *Brass World* in 1910 which shows that, by that time, aluminium bronze, still usually consisting only of different combinations of aluminium and copper, had been tried by many firms. But there were problems and the author comments that

> no copper alloy held out more promise at the time it was produced commercially, and none has proved more disappointing than aluminium bronze'; but he adds: 'After much good and bad experience with it, I will frankly say that it is a bronze without a peer, and the early 'worshippers' of it did not over-rate it by any means.

What caused so much disappointment then as later, was the difficulty of producing sound billets and castings due to dross and shrinkage problems. It was recognised that it should be poured 'quietly' but it did not seemed to have occurred to any one at that time to pour it other than by the time-honoured 'bottom pouring' technique. This unshakeable adherence by so many founders to tradition was to discourage many designers in later years from specifying the alloy.

Another American, writing anonymously in *Brass World*[8] in 1911, gives interesting advice on how to cast aluminium bronze. It shows that much ingenuity and perseverance was being exercised in overcoming problems, including that of gas porosity.

Addition of other alloying elements

Although industrially produced aluminium bronze seemed to have consisted, at that time, only of copper and aluminium, the idea of adding other alloying elements had been considered. Already, by 1891 attempts were being made to add manganese to the basic copper-aluminium alloy. An American patent was taken out by Dr J. A. Jcacore at that date for the addition of 2 to 5% manganese.[182] But the adverse effects of some elements, present as impurities, proved a deterrent to progress in that direction. Carpenter and Edwards[44] report that

> very extended research was published by Professor Tetmajer in 1900. His alloys contained notable quantities of elements other than aluminium and copper. These impurities, principally silicon and iron, ranged from one to four per cent; and their influence on the properties of aluminium and copper has since been found to be so considerable, that his alloys are not comparable with the pure copper-aluminium alloys that can be prepared at the present time.

It seems, therefore, that by 1907 the wrought alloys still normally consisted only of copper and aluminium. Carpenter and Edwards[44] report that they contained 2%

aluminium for tubes, 5% for rods and 8–9% for propeller shafts. Castings were made in the 10% aluminium alloy.

By 1910, however, it was felt that the effects of adding some other alloying elements should be investigated. Lantsberry and Rosenhain, also of the National Physical Laboratory, thought there were three likely candidates: manganese, nickel and zinc. They realised, however, that it would take too long to investigate all three in one research programme and so they decided to concentrate firstly on manganese because of its de-oxidising effect (experience with other alloys had shown that the use of a de-oxidant had a beneficial effect on mechanical properties). They also knew that manganese, like aluminium, had a strengthening effect when alloyed to copper.

They decided to limit their investigation to the range of copper–aluminium alloys which had already been found to be commercially useful, namely up to 10% aluminium. After some preliminary trials with a range of alloys containing up to 10% manganese, they decided to concentrate their research on alloys with less than 5% manganese and later on three alloys containing 9–10% aluminium and 1–3% manganese. They concluded that such additions of manganese made no visible change to the micro-structure of the alloys, that it resulted in 'a higher "yield-point", a slightly higher ultimate stress and an undiminished ductility' and that 'taken as a group, the ternary alloys certainly attain a degree of combined strength and ductility decidedly superior to the best of the copper-aluminium alloys'. The ternary alloys were comparable to the corresponding binary alloys in dynamic test although slightly inferior in alternating stresses. They absorbed more energy on impact and 'their power of resisting repeated bending impact was very remarkable'. They also had significantly better resistance to abrasion: 'considerably above that of ordinary tool steel'. Finally, 'as regards resistance to corrosion, both in fresh and sea water, the ternary alloys which were investigated, appeared to be at least equal to the copper-aluminium alloys and, in some cases, show a slight superiority'.

It seems that, for the following ten years, the Alloys Research Committee of the Institution of Mechanical Engineers which had funded the above research by the National Physical Laboratory, concentrated their research on aluminium-rich alloys without investigating the effects of other alloying elements on aluminium bronzes.

Inventors of the Tilting Process

Pierre Gaston Durville

A French man, by the name of Pierre Gaston Durville (Fig. H3), was among the first to produce aluminium bronze on a commercial basis. He was born on the 13th March 1874, the son of Alexandre Durville, an architect in Paris. His interest in aluminium bronze began when he was working for the French motor manufacturer Delaunay Belleville, in Paris, during the period 1900–10, under the well-

Fig. H3 Pierre Gaston Durville
(1874–1959).

Fig. H4 Charles Harold Meigh
(1892–1968).

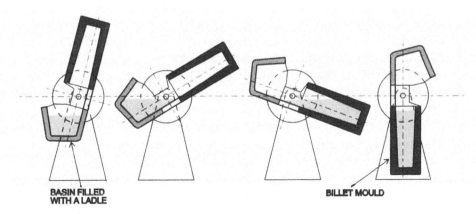

BASIN FILLED
WITH A LADLE

BILLET MOULD

Fig. H5 The principle of the Durville Process for pouring aluminium bronze
billets.[127]

known metallurgist Henri le Chatelier. Le Chatelier had a keen interest both in
aluminium and in aluminium bronze. As mentioned above, aluminium bronze
usually consisted, at that time, of only copper and aluminium, the most favoured
composition being 90% Cu and 10% Al. Le Chatelier had been a member of a
commission, set up by the French government in 1909, to recommend a suitable
alloy to replace the silver coinage then in circulation. The commission recommended, at le Chatelier's suggestion, that the possibility of using aluminium bronze

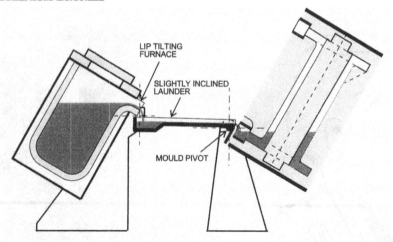

LIP TILTING
FURNACE

SLIGHTLY INCLINED
LAUNDER

MOULD PIVOT

Fig. H6 The Meigh Process for pouring aluminium bronze sand castings.

be studied. Difficulties in producing this alloy satisfactorily, however, resulted instead in pure nickel being introduced in 1912 for the 5 and 10 centimes pieces and in 1914 for the 25 centimes piece.

Meanwhile Durville had been developing a novel method of making aluminium bronze billets which would overcome the problems of oxide inclusions and shrinkage defects which were then being encountered. It came to be known as the 'Durville Process'. This process is illustrated in Fig. H5. The equipment consisted of an ordinary ingot mould connected by a short channel to a basin in such a way that the open ends of the ingot mould and of the basin faced each other. The ingot mould was inverted and the metal poured with a ladle into the basin. After carefully removing the dross on the surface of the metal, the equipment was slowly turned through 180° to transfer the metal without turbulence from the basin to the ingot mould. The avoidance of turbulence overcame the problem of oxide inclusion and, the fact that the hottest metal remained always on top, meant that the ideal condition was created for solidification to occur progressively from the bottom to the top of the mould, thereby overcoming the problem of shrinkage defects.

Le Chatelier encouraged Durville to set up his own business to produce billets commercially by this process. Accordingly, in 1913, Durville set up his company, 'Bronzes et Alliages Forgeables S.A.' with its office in Paris and its works in the little town of Mouy in the Oise department, sixty kilometres north of Paris.

The 90/10 copper–aluminium, which Durville manufactured, was intended almost exclusively as a wrought material and used for forgings, bars, stampings, etc. The work of converting the billets into wrought forms was subcontracted to a local steel mill.

With the problems of manufacturing aluminium bronze billets resolved, the French government decided in 1920 to replace the 50 centimes, 1 franc and 2 francs bank notes with aluminium bronze coins, due to its attractive gold-like appearance and technical suitability. The alloy used consisted of 8.5–9% aluminium with the balance in copper. This composition was a compromise between hardness for good wear property and ductility for the stamping process. The manufacture of this coinage then became, by far, the main item of production of the Durville company.

The company's success with coinage proved its undoing. The cash flow problems, resulting from high stock levels and delayed payments, forced the company out of business in 1924. It was bought by the Electro-Cable group and production was transferred to its works at Argenteuil, north-west of Paris. This too later went out of business. Pierre Durville had however retained the patent rights to his process and negotiated a five-year licence agreement in 1935 with 'Le Bronze Industriel' at Bobigny, north-east of Paris. His son Gilbert, who had been in charge of the laboratory at Mouy, joined Le Bronze Industriel together with other key personnel.

Pierre Durville died in 1959 at the age of 85.

Charles Harold Meigh MBE

In 1919, an Englishman by the name of Charles Harold Meigh (Fig. H4), who had served in the British Army during the war and who had recently married a close friend of Pierre Durville's daughter, joined the Durville company in Mouy. Charles Meigh was born on the 5th March 1892 at Ash Hall, near Hanley in Staffordshire, from a family which, for several generations, had been prominent in the Pottery industry. He had decided however to break with tradition and make his career in Engineering.

In 1923, four years after he had joined Durville's company, Charles Meigh left it to set up his own foundry near Rouen, called 'Forge et Fonderie d'Alliages de Haute Résistance'.

He was then able to fulfil his ambition to produce sand castings in aluminium bronze by a process, which made use of Durville's tilting principle, but significantly altered its application to suit the requirements and diversity of castings. This 'Meigh Process' is shown in Fig. H6. It comprised three important features:

- the casting was connected direct to the furnace by a short 'launder' or channel;
- a small basin, incorporated in the mould, received the metal from the launder; a small gate, connecting this basin to one of the risers, ensured that any dross was retained in the basin;
- tilting was through 90° only and began as soon as the small basin was full and continued until the mould was filled.

Small moulds were cast in a similar way but with hand ladles instead of a launder.

Fig. H7 High pressure centrifugal feed pump cast in an aluminium bronze containing 3% each of nickel, iron and manganese – Weight: 136 kg.[130]

By connecting the mould direct to the furnace, the turbulence involved in filling a large ladle and the higher melting temperature necessary to compensate for heat loss in transit, were avoided. The continuous process of filling, as the mould tilted, meant that hot metal, straight from the furnace, could compensate for the shrinkage of the metal as it began to solidify in the mould during pouring, thereby creating ideal conditions for directional solidification and limiting the amount of 'feeding' required after casting.

The Meigh Process was later introduced into England at Birkett Billington and Newton of Stoke-on-Trent who used it under licence to produce aluminium bronze castings. Charles Meigh collaborated with the French Admiralty in developing the use of other alloying elements and perfected an alloy containing nominally 3% each of nickel, iron and manganese and 9–10% aluminium. Fig. H7 shows a centrifugal pump body casting made in this alloy at that time. This was an early form of nickel–aluminium bronze.

Charles Meigh returned to England in 1937 and set up a new company in Cheltenham. He played a part in the growing interest shown by the British Admiralty in the use of aluminium bronze and set up the Meigh Process at the Chatham Naval Dockyard. One interesting casting that he designed and produced during the 1939–45 war was an aerial torpedo tail-fin in aluminium bronze (Fig. H8) which, unlike the previous tail-fins fabricated in steel, did not distort on impact with the

Fig. H8   Aerial torpedo fin, 1939–45.[130]

sea. This greatly improved the accuracy of aerial torpedoes and Charles Meigh was awarded the MBE after the war in recognition of his contribution to the war effort. He died in 1968 at the age of 76.

Leading contributors to the metallurgy of aluminium bronze

Many researchers have made valuable contributions over the years to the metallurgy of aluminium bronze, as is evident from the list of references at the end of this book – a list which does not claim to be fully comprehensive. Certain names do stand out however, if only by the frequency of the references to their work in articles by subsequent researchers. The following are among these.

Equilibrium diagrams and structure

Mention has already been made of the work done Dr L Guillet in France and published in 1905 and by H. Carpenter and C. Edwards[44] of the National Physical Laboratory in Teddington England in 1907, on the equilibrium diagram of the copper–aluminium binary alloys (Figs H1 and H2). The equilibrium diagram shown in Fig. 11.4 (Chapter 11) is based on the work of Stockdale (1922–4), modified by Smith and Lindlief[164] (1933), Hisatsune[95] (1934) and Dowson[66] (1937).

Equilibrium diagrams of ternary alloys seem to have been first produced by the following:

- Copper–aluminium–nickel by W. Alexander in 1938.
- Copper–aluminium–iron by A. Yutaka in 1941.
- Copper–aluminium–silicon by F. Wilson in 1948.
- Copper–aluminium–manganese by D. West and D. Thomas in 1956.

Most of the basic work on the structure of complex nickel–iron–aluminium bronzes was carried out in the early nineteen fifties by Cook, Fentiman and Davis of the Metal Division of Imperial Chemical Industries of Birmingham, England and most other authors refer to their work.

The equilibrium diagram for the high manganese (12%) complex alloy with 8% Al would seem to have been first produced by O. Knotek of Bern in Switzerland in 1968. This type of alloy was principally developed by Stone Propellers of Charlton, London.

Identification of kappa phases – Mechanism of corrosion

Cook, Fentiman and Davis would appear to have been the first to have designated 'kappa' (κ) a phase that arose as a result of the breakdown of the beta (β) phase in the complex copper–aluminium–nickel–iron system (see Chapter 13). Previously, W Alexander had designated Fe(δ) a κ-related phase in the copper–aluminium–iron system and A. Yutaka had designated NiAl, another κ-related phase, in the copper–aluminium–nickel system.

Following Cook, Fentiman and Davis's research, other researchers began to differentiate between various κ phases and this work went on in parallel with research into the mechanism of corrosion. The names of the Frenchmen F Gaillard, Pierre Weill-Couly (Forge et Fonderie d'Alliages de Haute Résistance) and Dominic Arnaud (Centre Technique des Industries de la Fonderie) came to prominence in this connection. P Weill-Couly established that there was an important relationship of aluminium to nickel content which must be respected if corrosion is to be avoided (see Chapters 12–13).

Another important name, frequently quoted, is that of the Swiss metallurgist P. Brezina of Esher Wyss who collaborated closely with the above researchers and who produced a key paper in 1982 on the heat treatment of complex aluminium bronzes.

Between 1978 and 1982, British Ministry of Defence (Naval) metallurgists, E. Culpan, J. Barnby, G. Rose, A. Foley and J. Rowlands published a number of papers which cast new light on selective phase corrosion of nickel–aluminium bronze. This was followed by the most comprehensive study to date of the structure and corrosion performance of the main types of aluminium bronzes carried out by the Materials Science Centre of the University of Manchester under Professor G. Lorimer and Dr N. Ridley. The researchers were F. Hasan, A. Jahanafrooz, J. Iqbal and D. Lloyd. The author is grateful to them for the wealth of information from their research which he has used in this book, including some work which has not yet been published.

Growing use of aluminium bronze

In spite of the attractive properties of aluminium bronze, the market for the alloy grew slowly. This was due in part to the reluctance to change of many users and, in the case of castings, to the difficulties experience by many founders in producing sound castings – often using traditional foundry techniques. It was not however until after the second world war that the demand for aluminium bronze began to grow sharply. Three main factors were responsible for this up-turn in the demand for both cast and wrought aluminium bronzes:

(a) the rapid growth in the oil industry, especially offshore extraction, and its impact on
(b) the demand for propellers for larger ships and for ships operating at higher speeds, and
(c) the need for a strong, shock- and corrosion-resisting alloy for submarines.

Rapid growth of the oil industry

Following the end of the 1939–45 war, the growth in the motor industry and in the use of oil for domestic and industrial heating, for power generation and for ships and railway engines, resulted in a rapid growth in the demand for oil. The demand for aluminium bronze in both the cast and wrought forms for pump, valves and heat exchangers grew in consequence.

The Suez Crisis (1956) created a boom in the construction of super-tankers to bring oil to Europe and America around the Cape. There was a corresponding boom in the demands for pumps and valves containing aluminium bronze – a demand which came to a temporary halt in 1976 due to over-construction of super-tankers.

The rise in the price of oil controlled by OPEC, made the development of offshore oil and gas extraction off Mexico and in the North Sea more economical and also worthwhile for the security of future oil supplies. North Sea gas began to flow ashore in 1967 and oil in 1973. This created new requirements for aluminium bronze, notably for fire pumps.

The prosperity which oil brought to the Middle East produced a demand for desalination plants which, at first, incorporated heat exchangers as well as pumps and valves, (the more recent process by osmosis no longer requires heat exchangers). This created a significant demand for both cast and wrought aluminium bronze.

Propellers

Prior to the 1939–45 war, the favourite material for ships propellers was manganese bronze (high tensile brass), but as the speed of ships increased so did the exposure of propellers to corrosion fatigue. As may be seen in Chapter 9, nickel–aluminium bronze is twice as resistant as manganese bronze and stainless steel to corrosion fatigue and the popularity of nickel–aluminium bronze has steadily grown to the point where it has become the favourite propeller material. The relative ease with which nickel–aluminium bronze propellers can be repaired by welding and straightened when damaged in service is another attractive feature of this alloy.

Nuclear Submarines

Navies were the first to appreciate the advantages of using aluminium bronze, although it took some time for this group of alloys to replace gun-metal.

The development of nuclear power made it possible to build submarines that could remain at sea for very long periods and therefore travel very long distances undetected. In the cold war situation that existed between the Soviet Union and the West, this ability to move undetected was of particular value for the nuclear deterrent on both side. Both conventionally-armed and nuclear-armed submarines were built, the first contract being placed on the Electric Boat Company in 1951.

The loss of the American nuclear submarine *Thresher* on 10 April 1963, which is thought to be have been due to the failure of a casting, proved to be a turning point for aluminium bronze. The excellent strength, and the shock- and corrosion-resisting properties of nickel–aluminium bronze made it very suitable for submarine castings. It also presented four significant advantages over gun-metal:

- the close-grain nature of aluminium bronze meant that defects could readily be seen when castings were subjected to radiography which had become a requirement for all critical submarine castings;
- the close-grain nature of the alloy also meant that aluminium bronze was inherently more pressure-tight than gun-metal;
- defective castings could be repaired by welding;
- size for size, aluminium bronze was 10% lighter than gun-metal – an important consideration for submarines where weight is a crucial design consideration.

The introduction of radiography, made it possible for the first time for founders to see the nature and exact location of defects as well as the effect of changes in techniques. It resulted in very significant improvement in the quality of aluminium bronze castings produced by founders involved in high quality naval work.

Development of alloys

It will be seen from Chapters 2 and 4 that a great variety of both cast and wrought alloy compositions, to suit different applications and different processes, have been developed since the early days. This makes aluminium bronze one of the most versatile family of alloys.

Part 1

CAST AND WROUGHT ALUMINIUM BRONZES

Properties and production processes

1
ALUMINIUM BRONZES AND THEIR ALLOYING ELEMENTS

The aluminium bronzes

Aluminium bronzes are copper-base alloys in which aluminium up to 14% is the main alloying element. Smaller additions of nickel, iron, manganese and silicon are made to create different types of alloys with properties designed to meet different requirements of strength, ductility, corrosion resistance, magnetic permeability etc.

The name 'Aluminium-Bronze', initially given to this range of alloys, is really a misnomer, since bronzes are alloys of copper and tin. For this reason, the term 'cupro-aluminium' has sometimes been used, notably in France, but other countries, and specially English-speaking countries, have generally retained the original term.

Both cast and wrought aluminium bronzes are widely used in equipment that operates in marine and other corrosive environments where, due to their superior strength, corrosion and erosion resistance, pressure tightness and weldability, they have increasingly supplanted other alloys in pumps, valves, propellers etc.

Properties of aluminium bronzes

The following attractive combinations of properties, offered by aluminium bronzes, make them suitable for a wide range of environments and applications.

- High strength – *some alloys are comparable to medium-carbon steel.*
- *Exceptional* resistance to corrosion – *in a wide range of corrosive agents. It should be stressed, however, that not all aluminium bronze alloys are resistant to corrosion. It is important therefore to select the appropriate alloys for corrosive environments.*
- *Excellent* resistance to cavitation erosion *in propellers and impellers.*
- Castable *by all the main processes: sand, centrifugal, die, investment, continuous casting.*
- Pressure tight, *when defect-free, due to close-grain structure.*
- Ductile and malleable – *can be cold or hot worked into plate, sheet, strip, rod, wire, various extruded sections, forgings and pressings.*
- Weldable – *fabrications can be made from both cast and wrought components and repairs and rectifications are possible.*
- *Good* machinability – *much easier and therefore cheaper than stainless steel.*
- *Good* shock resistance – *advantageous on warships, motor vehicles, railways etc.*
- *Exceptional* resistance to fatigue – *particularly suitable for propellers.*

- *Good* damping properties – *twice as effective as steel.*
- *Suitable at* high temperatures – *retains a high proportion of its strength up to 400° C and is exceptionally resistant (for a copper alloy) to oxidation at these temperatures.*
- *Suitable at* low temperatures – *suitable for cryogenic applications.*
- Good wear resistance – *notably for gears and bushes at low speed and at relatively high fluid velocities.*
- Low magnetic permeability – *especially cupro-aluminium-silicon alloy.*
- Non-sparking – *used in safety tools and explosive handling.*
- Attractive appearance – *used for ornamental purposes.*

Effects of alloying elements

In order to appreciate the significance of any particular combination of alloying elements, it is necessary to know their individual effects. This chapter gives only a general presentation of the effects of alloying elements: more detail information will be provided in subsequent chapters.

Aluminium

Aluminium has a marked effect on mechanical properties of aluminium bronzes. It is the element that has the most significant effect on resistance to corrosion. Most manufacturers control it to within ± 0.1%.

Mechanical properties
Figure 1.1 shows the effect of varying additions of aluminium to copper forming a range of alloys known as 'binary' alloys. Because of the high ductility of copper–aluminium alloys, proof strength is not a realistic concept. Only tensile strength and elongation are therefore shown. These alloys have exceptionally high elongation at about 6–7% Al which may reach 75%. This exceptionally high level of elongation is among the best attainable in any structural material. As we shall see, when we consider the wrought alloys in Chapter 5, the excellent ductility of this simpler type of alloys, with less than 8% aluminium content, means that they can be cold worked, although in practice cold working is normally only used as a final 'sizing' operation. Above 8% Al, elongation falls sharply down to zero around 13% when the alloy becomes brittle due to a progressive change of structure (see Chapter 11).

As may be seen, the tensile strength increases with aluminium content up to just over 10% Al. Thereafter, the change of structure, just mentioned, begins to occur and the tensile strength begins to fall.

Aluminium has a similar effect on wrought alloys as it has on cast alloys but its effect is accentuated by hot and cold working and by heat treatment.

The effect of iron and nickel additions can be seen in the difference between the curves shown in Figure 1.1, and will be discussed below. It can be seen however

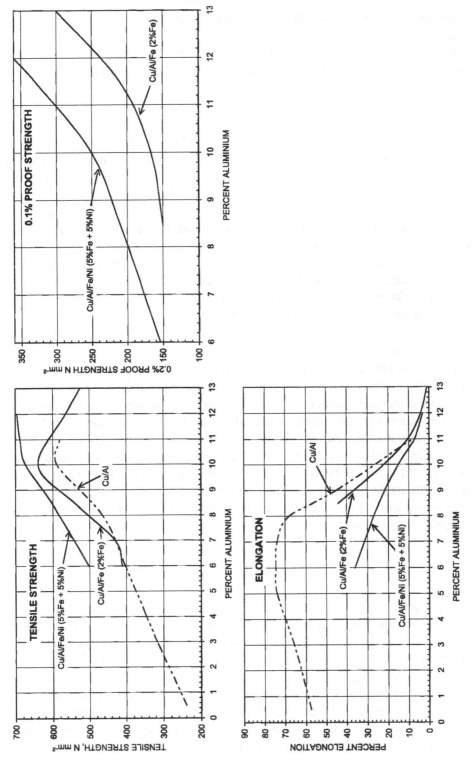

Fig. 1.1   Effect of aluminium, iron and nickel on mechanical properties.[127]

that aluminium has the most pronounced effect of all three alloying elements on mechanical properties.

In the case of Cu–Al alloys, fatigue and creep properties increase in proportion to the aluminium content, while impact strength remains at a fairly constant high level of 70–95 Joules.

Combination of properties

Alloys with aluminium contents within the range of 4.5–7.5% are used mainly in the wrought form. Even within the composition range of particular standard specifications, manufacturers are able to supply various alloys, some of which are noted for their forging properties or strength while others are more suitable for applications involving corrosion or shock resistance. In general, within the range of 8–11% aluminium, the hardness, strength, hot workability and, to a lesser extent, fatigue strength increase with aluminium content while ductility, creep at elevated temperatures and corrosion resistance tend to be adversely affected. Exceptional hardness values are obtained with aluminium contents of 11–13% and these alloys find application for wear resistant service where low ductility and impact strength and poor corrosion resistance are not a disadvantage.

Corrosion resistance

The resistance to corrosion attack in most environments is due to the tenacious protective film of aluminium oxide which forms on the surface of the alloy and which readily reforms if damaged. This oxide film is not however totally impenetrable and long term corrosion resistance is dependant on the sub-film structure of the alloy. As explained in Chapter 11, copper-aluminium alloys with less than 8.2% Al have excellent long term resistance to corrosion. As aluminium increases above 8.2%, however, the alloy structure becomes increasingly vulnerable to corrosion.

Alumina, the oxide of aluminium, is a very hard substance, used as an abrasive in shot blasting and other applications, and this accounts for the good erosion resisting properties of aluminium bronzes.

Iron

Mechanical properties

Figure 1.1 shows the effect on mechanical properties of a 2% iron addition. The trend is very similar to that of the binary copper-aluminium alloys with a slight increase in tensile strength and reduction in elongation. Between 3 to 5% Fe, tensile strength and proof strength tend to improve but elongation to reduce.[59] Increasing iron to 7% further increases tensile strength as well as elongation but causes no change in proof strength. Iron has also the effect of increasing the strength of the alloy at high temperature.[78] In practice, where iron is the only alloying element other than aluminium, the iron content seldom exceeds 4%.

The properties shown in Figure 1.1 for Cu–Al–Fe alloys with 2% Fe, are minimum mechanical properties achievable with a standard sand-cast test bar. In practice, the results of pulling different standard test bars gives a scatter of mechanical properties above the minimum shown in Figure 1.1. This applies to all aluminium bronze alloys and will be discussed at greater length in Chapter 3.

It will be seen that, as in the case of the binary Cu–Al alloys, the tensile strength of the Cu–Al–Fe alloys dips down above 10% Al for a similar reason of structural change (see Chapter 12). On slow cooling at high aluminium content these alloys can become very brittle.

Iron refines the crystalline structure of aluminium bronzes and this has the effect of increasing the toughness of the alloy, that is to say its ability to withstand shocks, as reflected in the Izod test. It causes grain refining only up to 3.5%, above which it has no further grain refining effect.[156] Iron improves hardness as well as fatigue and it also improves wear and corrosion resistance.[78] It also narrows the solidification range.

As in the case of Cu–Al alloys, fatigue and creep properties of Cu–Al–Fe alloys increase in proportion to the aluminium content, while impact strength remains at a fairly constant high level of 70–95 Joules.

Corrosion resistance

Copper–aluminium–iron alloys are not a good choice for corrosive environment. If care is taken in the choice of aluminium content and cooling rates the concentration and corrodible nature of certain structures can however be minimised. A full explanation of the effect of corrosive environment on this kind of alloys can be found in Chapter 12.

Nickel and iron

Mechanical properties

In conjunction with iron, with which it is always associated, nickel improves tensile strength and proof strength, as may be seen from Figure 1.1 in the case of Cu–Al–Fe–Ni alloys with 5% each of nickel and iron. Feest and Cook[70] have demonstrated that 4%–5% Fe in Cu–Al–Fe–Ni alloys has a refining action.

Figure 1.1 shows the variation of mechanical properties with aluminium content of this type of alloy. It will be seen that tensile properties are appreciably above those of the Cu–Al–Fe alloys with 2% Fe. Elongation, on the other hand, is significantly lower. It is evident, however, that the effect of aluminium on mechanical properties is much more significant than that of iron and nickel. In order to obtain a good combination of strength and elongation in this type of complex alloys, together with good corrosion resistance and workability, the aluminium content must be a compromise, and controlled to close limits. For example, the aluminium content of cast alloys containing iron and nickel is normally restricted to 9–10% in order to meet specified mechanical properties.

In most cast alloys, the nickel content usually lies in the range of 4.5–5.5%, whereas wrought alloys vary considerably in the nickel content that they specify: some alloys specify a range of 1–3% and others as much as 4–7%, depending on the combination of properties required for a given application.

The alloys containing approximately 5% each of iron and nickel, are the most popular cast and wrought aluminium bronzes because of their combination of high strength and excellent corrosion resistance (see below).

Nickel improves hardness but reduces elongation. The effect of nickel is in fact much more pronounced on elongation than on tensile properties, particularly at the lower range of aluminium values. According to Crofts,[58–59] increasing nickel to 7% further increases proof strength but reduces both tensile strength and elongation. The presence of nickel also improves resistance to creep. According to Thomson,[172] it reduces impact resistance.

Table 1.1 shows the effect on properties of varying the iron content while keeping the aluminium content constant. The figures have been arranged in ascending order of iron content and this shows that iron has the most marked effect on tensile strength and hardness whereas the variations in nickel content appear to have a less significant effect. The effect on proof strength and elongation is less clear. These figures indicate only a trend since, as we have seen above, the spread of mechanical properties obtained in practice makes it difficult to draw firm conclusions.

The effect on mechanical properties of varying the iron content in the presence of 5% nickel for a range of aluminium contents is shown in Table 1.2 where it can be seen that this effect is significant. These figures relate to die cast samples and are higher than they would be in sand cast samples (see 'Effect of cooling rate on mechanical properties' in Chapter 3).

We have seen that hardness is due mostly to the effect of aluminium and increases with aluminium content but the rate of increase is greater for the complex Cu–Al–Fe–Ni alloys.

Complex alloys with high aluminium content are ductile at high temperatures and are therefore hot worked. If the aluminium content of these alloys is increased

Table 1.1   Effects of variations in iron and nickel content on the mechanical properties of sand castings.[127]

COMPOSITION				MECHANICAL PROPERTIES			
Cu	Al %	Fe %	Ni %	Tensile Strength N/mm^2	0.2% Proof Strength N/mm^2	Elongation %	Hardness HB
Rem	9.4	2.7	5.2	602	263	20	149
Rem	9.4	3.2	3.1	618	247	25	143
Rem	9.4	4.1	3.8	641	231	23	152
Rem	9.4	4.6	3.7	657	247	25	156
Rem	9.4	4.8	5.1	649	278	19	163

Table 1.2 Effects of variations in aluminium and iron contents on the mechanical properties of a diecast alloy containing 5% nickel.[127]

COMPOSITION					MECHANICAL PROPERTIES		
Cu	Al %	Fe %	Ni %	Mn %	Tensile Strength N/mm²	Elongation %	Hardness HV
Rem	8.3	0.3	5.0	0.5	549	18	171
Rem	8.4	5.2	5.0	0.5	657	13	211
Rem	9.2	0.3	5.0	0.5	685	12	220
Rem	9.2	5.2	5.0	0.5	750	13	240
Rem	9.5	4.0	5.0	–	649	18	170
Rem	10.1	0.3	5.0	0.5	765	9	275
Rem	10.2	5.2	5.0	0.5	843	7	270
Rem	10.6	0.3	5.0	0.5	750	5	272
Rem	10.6	5.2	5.0	0.5	889	6	290

above 13%, they become brittle but very hard and therefore ideally suited for high wear resistance application provided the load is in compression.

Sarkar and Bates[158] report that a low nickel–iron ratio increases impact resistance. According to Thomson,[172] with a 6:3 nickel–iron ratio, slow cooling markedly reduces all properties including the general level of impact values, whereas with a 3:5 nickel–iron ratio, tensile and impact properties were only slightly affected by slow cooling. Edwards and Whitaker[69] report that increasing nickel reduces ductility which can be restored by subsequently increasing iron.

Corrosion resistance
As will be seen in Chapter 12, the main reason for the presence of nickel in some aluminium bronzes is to improve corrosion resistance, but it should be kept above the iron content for complete resistance in hot sea water. In the case of slowly cooled alloys, Weill-Couly and Arnaud[183] recommend the following relationship between aluminium and nickel content for an alloy to be corrosion resistant:

$$Al \leq 8.2 + Ni/2$$

It should be noted that, at the minimum nickel content allowed by some standard specifications, the maximum aluminium content allowed may be higher than the maximum corrosion-safe aluminium content given by the above formula.

Manganese

Mechanical properties
Alloys with high manganese content (8–15%) have been extensively used as a propeller material due to their high mechanical properties and good corrosion resistance. Manganese has a similar effect to aluminium on mechanical properties,

except that the manganese content needs to be six times greater than the aluminium content to have the same effect. Figures 1.2a and 1.2b show the effect of varying the aluminium content whilst keeping the manganese content constant at 12%. Figure 1.2c shows the effect of varying the manganese content whilst keeping the aluminium content constant at 8%. The trends of these two sets of curves are similar, though not identical. It is therefore only an approximation to say that 6% Mn is equivalent to 1% Al. In fact, plotting properties against 'equivalent' aluminium (Fig. 1.2d) shows a good correlation of tensile properties but a divergence of elongation at lower equivalent aluminium. This shows that a low actual aluminium content of 6% has an overriding influence on elongation.

If tensile properties shown in Figure 1.2d are compared with properties shown in Figure 1.1 for the Cu–Al–Fe–Ni type of alloy (with 5% each of nickel and iron), it will be seen that the high manganese alloy has higher strength properties. It also has better ductility and impact strength.

The combination of aluminium and manganese content significantly increases hardness.

Edward and Whittaker,[69] working mainly on high manganese alloys (6–8% and 11–14%), established the critical range of manganese content, in relation to aluminium content, below which tensile strength, yield strength and hardness and above which elongation will be less than optimum. These figures are given in Table 1.3.

Table 1.3   Critical range of Mn content in relation to Al content to achieve optimum mechanical properties by Edwards and Whitaker.[69]

Al Content %	Critical Mn Range %
7.0	16.5–24.0
7.5	13.5–20.0
8.0	10.5–16.0
8.5	8.0–12.0
9.0	6.0–8.0
9.5	4.0–4.0
10.0	2.0–0

Effect of nickel and iron on properties of high manganese alloys
The high manganese alloys also contain nickel (1.5–4.5%) and iron (2–4%). If the nickel content falls below 1%, the proof strength is reduced. Difficulty also arises if nickel significantly exceeds 2%, as a progressive drop in ductility then occurs. It should only be increased above 2% when better creep resistance is required at the expense of other properties. Iron is maintained at no less than 2.5% as a grain refiner, but mechanical and corrosion resisting properties are adversely affected if it exceeds 3%.

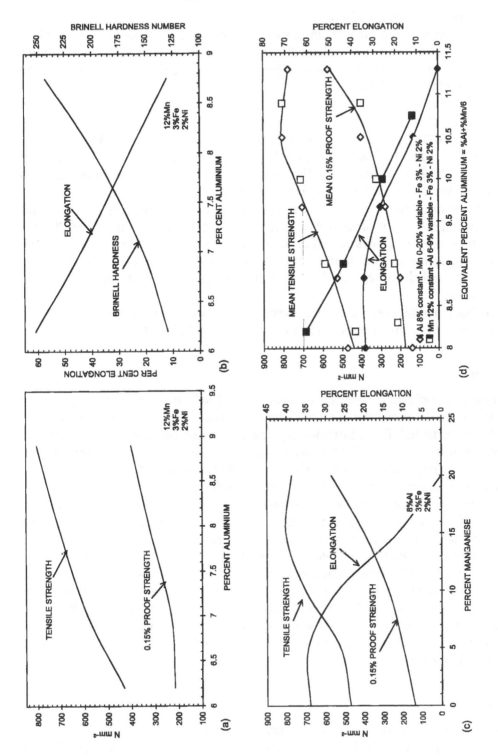

Fig. 1.2 Effect of manganese on mechanical properties.[127]

Effect of manganese on castability

The most common reason for adding small quantities of manganese to a Cu–Al–Fe or to a Cu–Al–Fe–Ni alloy is to deoxidise the copper prior to the addition of aluminium and to improve fluidity, thereby improving the quality of castings and the castability of thin sections.

Corrosion resistance

If a small addition of manganese is made to a Cu–Al–Fe or to a Cu–Al–Fe–Ni alloy to improve fluidity, it should be kept below 2% (see Chapters 12 and 13), since, at higher percentages, it encourages the formation of a corrosion-prone structure and it also renders the alloy more prone to crevice corrosion.

If, on the other hand, manganese is a principal alloying element (8–15%), it has beneficial effects on the structure with regard to corrosion resistance (see Chapter 14). There are two alloys in which manganese is used as one of the principal alloying elements in association with aluminium. One alloy contains 7.5–8.5% aluminium and the other 8.5–9.5% aluminium. See Chapters 3 and 14 for more information.

The high manganese alloys are, however, less resistant to stress corrosion fatigue in sea water than the nickel-aluminium bronze alloy CuAl10Fe5Ni5.

Silicon

As in the case of manganese, silicon acts as an aluminium substitute, the effect of 1% silicon on the properties of an alloy being equivalent to about 1.6% aluminium. If it is desired to add silicon intentionally, the aluminium content should be lowered at the same time. When silicon is present in an alloy of given aluminium content, the tensile strength and proof strength are raised with a marked drop in elongation. Silicon also improves machinability and, according to Goldspiel et al[78], it also improves hardness, and therefore bearing properties, but reduces impact resistance.

Up to 1% silicon acts as a grain refiner but silicon, present as an impurity in an alloy in excess of the minimum allowed by the specification, can however have a very detrimental effect on mechanical properties. For this reason, Goldspiel et al[78] recommend that silicon should not exceed 0.005% in propeller castings. The silicon bearing alloys all contain around 2% silicon and 7% aluminium. One alloy, CuAl7Si2, is used in the UK, in both cast and wrought forms, mainly for naval applications because of its low magnetic and high impact properties. In the USA, a similar alloy is used for its good machining and bearing properties and mainly in the wrought form.

Lead

Lead does not alloy in aluminium bronzes and, if present, takes the form of dispersed minute inclusions that weaken the alloy and have a detrimental effect on welding. For castings that are welded, the lead content should be kept to a minimum: below 0.1%, and preferably lower, as there is a danger of cracking adjacent to the weld.

In the USA, lead additions of over 1% have been made to improve the bearing properties of some aluminium bronzes under conditions of poor lubrication, but these materials have a much lower strength and elongation, as even small additions to improve machinability have a harmful effect on some mechanical properties.

Impurities

Zinc is perhaps the most common impurity in aluminium bronzes and may, on rare occasions, extend to 1% or even more. This is not considered to have a harmful effect on the mechanical properties or the corrosion resistance of the alloy unless considerably greater amounts are present.

The maximum permissible tin content is subject to some controversy. Generally small amounts up to approx. 0.2%[106] are not considered harmful. Magnesium has been recommended as a de-oxidant but even 0.01% has a harmful effect on ductility [106].

Phosphorus has the reputation of being a harmful impurity, but it does not affect mechanical properties unless more than 0.08% is present[106], although it may encourage hot shortness when more than 0.01% is present.

More information on the effects of impurities in aluminium bronzes is given in Chapter 3.

2
PHYSICAL PROPERTIES

The mechanical properties of cast and wrought aluminium bronze alloys are given in Chapters 3 and 5 respectively. This chapter deals specifically with other physical properties.

Being copper-based alloys, aluminium bronzes have certain physical properties similar to other copper alloys. For example they have good electrical and thermal properties by comparison with ferrous alloys although not as good as other copper alloys. Unlike most other copper alloys, they have a very short melting range. The presence of aluminium renders the alloy 10% lighter than other copper alloys and therefore comparable to steel. Aluminium bronzes have good elastic properties which is an advantage for shock resistance, but which render these alloys less rigid than steel from a structural point of view. Finally the strength of the more alloyed aluminium bronzes, coupled with their non-sparking properties makes these alloys well suited to explosive conditions.

Melting ranges

The melting ranges of aluminium bronzes relative to aluminium content are given in Table 2.1 and shown diagrammatically in Chapters 11, 12 and 13 in the form of Equilibrium Diagrams. It will be seen that aluminium bronzes have a characteristically narrow melting range.

Density

The density of aluminium (2.56 g cm^{-3}) is only 29% of that of copper (8.82 g cm^{-3}). It is not surprising therefore that the aluminium content has the most significant influence on alloy density. In fact, in the case of alloys containing only copper and aluminium, the density varies in direct proportion to the aluminium content, as illustrated in Figure 2.1. Nickel has almost the same density (8.80 g cm^{-3}) as copper and although iron (7.88 g cm^{-3}) is 11% lighter than copper, additions of iron and nickel do not appear to make a significant difference to alloy density for any given aluminium content. Manganese (7.42 g cm^{-3}) is 16% lighter than copper and consequently a high proportion of manganese results in slightly lower alloy density. Finally silicon (2.33 g cm^{-3}) is 9% lighter than aluminium and yet appears to have less effect than an equal proportion of aluminium in reducing alloy density.

Because of the small melting range mentioned above, aluminium bronze castings, provided they are free of porosity, solidify in a very compact form, as will be

Table 2.1   Density and melting range of aluminium bronzes.[127,173]

Alloys	Density g/cm³	Melting Range °C
Wrought alloys		
CuAl5	8.2	1050–1080
CuAl7	7.9	1040–1060
CuAl7Si2	7.8	980–1010
CuAl8	7.8	1035–1045
CuAl8Fe3	7.8	1045–1110
CuAl9Mn2	7.6	1045–1100
CuAl9Ni6Fe3	7.6	1050–1070
CuAl10Fe3	7.6	1060–1075
CuAl10Fe5Ni5	7.5	1060–1075
CuAl11Ni6Fe5	7.6	1045–1090
Cast alloys		
CuAl9Fe2	7.6	1040–1060
CuAl6Si2	7.8	980–1000
CuAl10Fe5Ni5	7.6	1050–1080
CuAl9Ni5Fe4Mn	7.6	1040–1060
CuMn13Al8Fe3Ni3	7.5	950–990

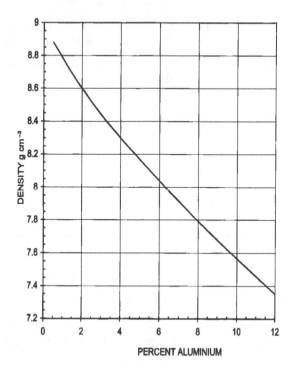

Fig. 2.1   Effect of aluminium content on the density of aluminium bronzes.[127]

explained in Chapter 4. Consequently, the as-cast condition is almost as compact as the wrought condition and there is therefore little difference in density between the cast and wrought forms of any given alloy.

Thermal properties

Coefficient of thermal linear expansion

Available figures for the thermal linear expansion of aluminium bronzes are given in Table 2.2. It will be seen that alloy composition makes little difference to the coefficient of thermal expansion, but that it increases with temperature range.

On solidification, a 4% volumetric contraction occurs in aluminium bronzes with a further 7% volumetric contraction on cooling to room temperature. This represents a linear contraction of 2 to 4% after solidification.

Table 2.2   Effect of composition and temperature range on linear coefficient of thermal expansion of wrought and cast alloys.[127-173]

Alloys	Coefficient of thermal linear expansion per K $\times$ 10^{-6}								
	−100 to 0 °C	−50 to 0 °C	0 to 100 °C	0 to 200 °C	0 to 300 °C	0 to 400 °C	0 to 500 °C	0 to 600 °C	0 to 700 °C
Wrought alloys									
CuAl5				17		18			
CuAl7						17			
CuAl7Si2				18		18			
CuAl8				16		17			
CuAl8Fe3				16		17			
CuAl9Mn2				16		17			
CuAl9Ni6Fe3				16		17			
CuAl10Fe3				15		17			
CuAl10Fe5Ni5						18			
CuAl11Ni6Fe5				16		17			
Cast alloys									
CuAl9Fe2	15.5	15.9	16.3	16.5	17.1	17.8	18.4	18.8	19.3
CuAl6Si2					16.2				
CuAl10Fe5Ni5	15.5	15.9	16.3	16.5	17.1	17.8	18.4	18.8	19.3
CuAl9Ni5Fe4Mn					16.2				

	−183 to 0 °C	0 to 100 °C	100 to 230 °C	230 to 325 °C
CuMn13Al8Fe3Ni3	15.17	17.7	18.56	21.34

Specific heat capacity

Specific heat capacity is given in Table 2.3. It is difficult to see any obvious relationship between the specific heat of various aluminium bronzes and their composition. The specific heat capacities in the cast form seem however to be higher than in the wrought form for any given alloy.

Table 2.3   Thermal properties of aluminium bronzes.[127,173]

Alloys	Specific Heat Capacity J/kg/ K	Thermal Conductivity J/sec/m/K	
	at 20° C to 100° C	at 20° C	at 200° C approx
Wrought alloys			
CuAl5	420	75–84	26
CuAl7	380	71	—
CuAl7Si2	380	45	–
CuAl8	420	63–71	20
CuAl8Fe3	420	59–71	20
CuAl9Mn2	420	59–67	–
CuAl9Ni6Fe3	420	38–46	13
CuAl10Fe3	420	38–46	13
CuAl10Fe5Ni5		33–46	–
CuAl11Ni6Fe5	420	59–67	20
Cast alloys			
CuAl9Fe2	434	42–63	
CuAl6Si2	419	45	
CuAl10Fe5Ni5	434	38–42	
CuAl9Ni5Fe4Mn	419	38–42	
		at 20° C	at 150° C
CuMn13Al8Fe3Ni3		12.14	12.98

Thermal conductivity

The thermal conductivity of aluminium bronzes is given in Table 2.3. It is influenced by a combination of composition and temperature.

The aluminium content has a marked influence on thermal conductivity, as is most clearly seen in the case of alloys containing only copper and aluminium (see Fig. 2.2). It will be noted that the thermal conductivity of these alloys drops from about 84 J $s^{-1}m^{-1}K^{-1}$ at 5% aluminium to about 59 J $s^{-1}m^{-1}K^{-1}$ at 12% aluminium. Table 2.3 gives an indication of the effect of other alloying elements. Iron appears to have little effect on thermal conductivity, but nickel, silicon and especially manganese significantly reduce thermal conductivity.

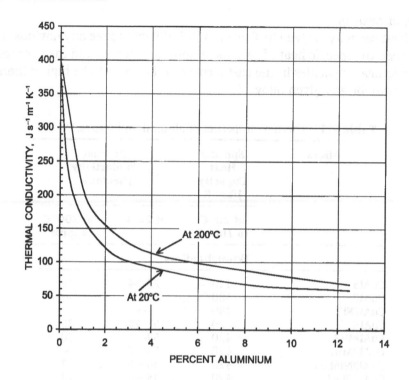

Fig. 2.2 Effect of aluminium content on the thermal conductivity of aluminium bronzes alloys.[127]

Electrical and magnetic properties

Electrical conductivity

As with thermal conductivity, the electrical conductivity of aluminium bronzes is influenced by a combination of composition and temperature and the effects of these are very similar for both forms of conductivity.

Aluminium content has the most marked effect, as may be seen most clearly in the case of alloys containing only copper and aluminium (see Fig 2.3). It will be noted that the electrical conductivity of these alloys drops from 17.5% of I.A.C.S. (International Annealed Copper Standard) at 5% aluminium to a minimum of about 10% of I.A.C.S. in the range of 12–14% aluminium (see Fig. 2.3). The effect of other alloying elements may be seen from the figures given in Table 2.4. As with thermal conductivity, iron has little effect on electrical conductivity, but again nickel, silicon and especially manganese have a marked influence.

Magnetic properties

Table 2.4 gives available magnetic permeability figures for aluminium bronzes. The alloy that contains nominally 2% silicon and less than 1% iron is ideally

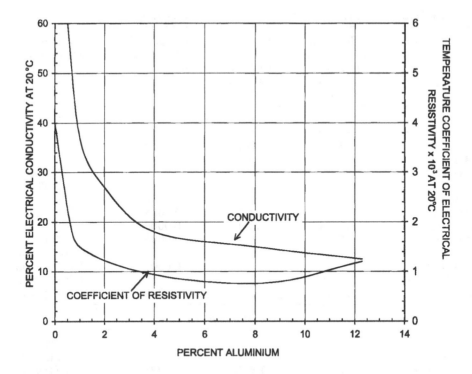

Fig. 2.3   Effect of aluminium content on the electrical conductivity and resistivity of aluminium bronzes.[127]

suited for non-magnetic applications and is the most suitable of copper alloys for extremely critical applications such as gyro compasses and other similar instruments. The magnetic properties of aluminium bronzes generally are largely dependent on the amount of iron which is precipitated in the structure, although other alloying additions may have some effect. Figure 2.4 shows graphically the effect of iron present on the magnetic permeability of an alloy containing 3.7% nickel. From this it can be seen that the iron content does not have a marked influence until it exceeds 0.5–1%. For the lowest possible magnetic susceptibility in nickel-free alloys, it has been found that the iron content should be less than 0.15% in the as-rolled condition, or less than 0.5% if it has been quenched from a high temperature (above 900°C). This is due to the change of solubility of iron.

Magnetic properties have been found to be related to the condition and heat treatment of the alloy and this is particularly so in the case of the high manganese alloys which have very low magnetic permeability when quenched from above 500° C but considerably increased permeability below this temperature, particularly on slow cooling (see Table 2.4).

Table 2.4   Electrical and magnetic properties of aluminium bronzes.[127–173]

Alloys	Electrical Conductivity (Volume) at 20°C % IACS	Electrical Resistivity (Volume) at 20°C 10^{-7} Ohm/m.	Temp. coeff. of electric. resist. per K at 20°C	Magnetic Permeability μ
		Wrought alloys		
CuAl5	15–18	1.0–1.1	0.0008–9	
CuAl7	15	1.0–1.5		
CuAl7Si2	7–8	1.9–2.2		1.05 (max) < 1.0001 (Typical)
CuAl8	13–15	1.1–1.3	0.0008	
CuAl8Fe3	12–14	1.2–1.4	0.0008	1.15 (Typical)
CuAl9Mn2	12–14	1.2–1.4	0.0008	
CuAl9Ni6Fe3	7–9	1.9–2.5	0.0005	1.0002 < 1.0001 Heat treated
CuAl10Fe3	7–9	1.9–2.5	0.0005	1.15 (Typical)
CuAl10Fe5Ni5	7–10			1.50 (Typical)
CuAl11Ni6Fe5	12–14	1.2–1.4	0.0008	1.50 (Typical)
		Cast alloys		
CuAl9Fe2	8–14	1.3–1.4		< 1.30 (Typical)
CuAl6Si2	8–9			1.04 (max.)
CuAl10Fe5Ni5	7–8	1.9–2.2	< 0.0001	1.40 (Typical)
CuAl9Ni5Fe4Mn	7–8	2.2		1.40 (Typical)
CuMn13Al8Fe3Ni3	3	5.5		1.03 (quenched) 2–10 (sand cast) 15 (slowly cooled)

Elastic properties

Moduli of elasticity and of rigidity

Table 2.5 gives figures for the elastic properties of both wrought and cast alloys. It will be seen that elastic properties of wrought alloys are higher than those of cast alloys. In the case of wrought alloys, light cold working reduces elasticity and heat treatment increases it.

Elasticity is closely related to the composition and structure of the alloy concerned. A study of Chapters 11 to 14 is therefore necessary for a fuller understanding of elasticity. The aluminium content has a marked effect, as is most clearly shown in the case of alloys containing only copper and aluminium (see Fig. 2.5). Above 8.5% aluminium the modulus of elasticity of these alloys falls sharply with increases in aluminium in rapidly cooled castings, whereas prolonged annealing of the wrought alloy below 565° C has the opposite effect. This is because, as explained in Chapter 11, the structure of these alloys changes at higher aluminium. In rapidly cooled castings the resultant structure lowers the modulus, whereas

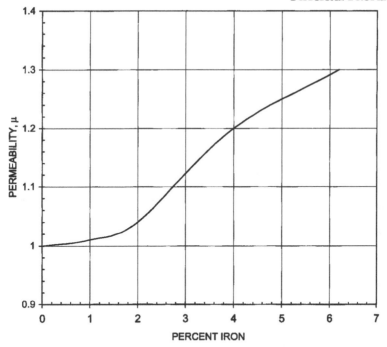

Fig. 2.4 Effect of iron content on the magnetic permeability of an alloy containing 10% Al and 3.7% Ni.[127]

Table 2.5 Elastic properties of aluminium bronzes.[127-173]

Alloys	Modulus of Elasticity (Tension) at 20° C kN mm⁻²			Modulus of Rigidity (Torsion) at 20° C kN mm⁻²			Pois-* son's Ratio
	Annealed	Lightly cold worked	Heat Treated	Annealed	Lightly Cold Worked	Heat Treated	
Wrought Alloys							
CuAl5	123–128.5	118		47.5	43.5		
CuAl7	108–120			45.0			
CuAl7Si2	110–125			41			
CuAl8	121–126	113.5		46.5	42.0		
CuAl8Fe3	120–122.5			45.5			
CuAl9Mn2	105			39.0			
CuAl9Ni6Fe3	130–133.5	128–131	134–140	48–49.5	47.5–48.5	49.5–52.0	0.3
CuAl10Fe3	130–133.5	128–131	134–140	48–49.5	47.5–48.5	49.5–52.0	0.3
CuAl10Fe5Ni5	124–130						0.3
CuAl11Ni6Fe5	120			44.5			0.3
Cast Alloys							
CuAl9Fe2		100–120			41–44		
CuAl6Si2		100–111					
CuAl10Fe5Ni5		116–124			45–48		0.3
CuAl9Ni5Fe4Mn		115–121			42		0.3
CuMn13Al8Fe3Ni3		117			44.4 (cast) 46.2 (forged)		0.34

* Poisson's ratio = $\dfrac{\text{lateral strain}}{\text{longitudinal strain}}$

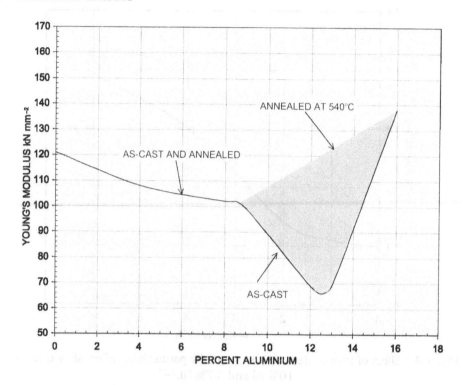

Fig. 2.5 Effect of aluminium content and heat treatment on modulus of elasticity of copper-aluminium.[127]

prolonged annealing below 565° C raises the value considerably by allowing time for the structure to change to a more elastic state. Above 12.5% Al (or even below if slowly cooled), another change of structure occurs which renders the alloy more brittle. It remains elastic however almost to the point of fracture at a higher stress value.

A similar trend occurs with other alloying elements which tend to increase the modulus of elasticity in both the as-cast and hot-worked conditions, although the changes of structure associated with higher aluminium content at high cooling rates, will offset this effect.

Figures given in Table 2.5 apply to a 20° C ambient condition. An increase in temperature up to 200° C causes no drop in modulus of elasticity, although at higher temperatures this falls off fairly rapidly. A typical value at 300° C being approximately 90 kN mm^{-2}.

Damping capacity

This property is closely related to the modulus of elasticity of the alloy: the higher the modulus, the lower the damping capacity. Thus the alloys having the highest modulus, have a poor damping capacity. On the other hand, those with a structure containing a high proportion of the 'beta phase' (see Chapters 11–14) have a high damping capacity ($1–2 \times 10^{-3}$) which remains constant over a wide frequency range. The damping capacity of alloys with low aluminium content is dependent on pre-treatment: (i.e.) it increases as the quenching temperature is raised and decreases with aluminium content.

Non-sparking properties

Sparks are tiny particles that are detached from their parent object by the force of impact of a harder instrument or object in air. Elements like iron, when finely divided and hot, can ignite spontaneously as they oxidise, becoming even hotter. This results in dull red particles rapidly becoming bright white at a much higher temperature. At this temperature the particle is visible as a spark and can cause fire or explosion in a combustible environment. In common with most other copper-base alloys, the particles detached from an aluminium bronze object due to impact against a ferrous or other harder objects, do not attain a dangerous temperature and are not therefore visible as a spark. In view of their high strength, these alloys are among the most favoured for applications where this is important. They may therefore be safely selected for non-sparking tools and equipment for handling combustible mixtures such as explosives.

3

CAST ALUMINIUM BRONZES

A – Cast alloys and their properties

Standard cast alloys

Table 3.1 gives the compositions and mechanical properties of cast aluminium bronzes to CEN (European) specifications, together with their former British designations and nearest American (ASTM) equivalents. Details of the latter are given in Appendix 1. These alloys are the most commonly used commercial cast alloys. Table 3.2 gives details of two other alloys of special interests (see below) which are to British Naval specifications. Since specifications are subject to occasional review, it is advisable to consult the latest issue of the relevant specification.

Cast aluminium bronzes may be grouped into three categories:

- High strength alloys
- Medium strength alloys
- Non-magnetic alloys

High strength alloys

The most widely used is the high strength alloy, CuAl10Fe5Ni5 which, in addition to high strength, has excellent corrosion/erosion resisting properties and impact values. It also has the highest hardness values of the aluminium bronzes. It is used in a great variety of equipment such as pumps, valves, propellers, turbines, and heat exchangers.

A slight variant of this alloy, with a more restricted composition, is designated CuAl9Fe4Ni5Mn (see Table 3.2). It is not a European standard. It is normally heat treated and, as a consequence, has enhanced mechanical and corrosion resisting properties. It is used in the same kind of equipment as the previous alloy but in applications requiring particularly good corrosion resisting properties, such as naval applications. The high aluminium, high nickel and high iron alloy CuAl11Fe6Ni6 has high hardness properties (at the expense of elongation) and is mainly used for its excellent wear resisting properties (see Chapter 10).

The high manganese containing alloy, CuMn11Al8Fe3Ni3-C, has higher mechanical properties than the above alloys and has been used extensively for marine propellers. It has also better ductility and impact strength than the first named alloy, CuAl10Fe5Ni5, but is less resistant to stress corrosion fatigue in sea water, as will be seen below. For this reason it is being increasingly superseded by CuAl10Fe5Ni5.

Table 3.1   Composition and minimum mechanical properties of cast aluminium bronzes to CEN specifications.

DESIGNATION			CEN COMPOSITION (%)				
International ISO 1338 European CEN/TC 133	Former British equivalent	Current American ASTM equivalent	Al	Fe	Ni	Mn	Cu
MEDIUM STRENGTH ALLOYS							
CuAl9-C	–	–	8.0–10.5	1.2 max	1.0 max	0.50 max	88.0–92.0
CuAl10Fe2-C	BS1400 AB1	C 95200	8.5–10.5	1.5–3.5	1.5 max	1.0 max	83.0–89.5
CuAl10Ni3Fe2-C	(French alloy)	–	8.5–10.5	1.0–3.0	1.5–4.0	2.0 max	80.0–86.0
HIGH STRENGTH ALLOYS							
CuAl10Fe5Ni5-C	BS1400 AB2	C 95800	8.5–10.5	4.0–5-5	4.0–6.0	3.0 max.	76.0–83.0
CuAl11Fe6Ni6-C	–		10.0–12.0	4.0–7.0	4.0–7.5	2.5 max	72.0–78.0
CuMn11Al8Fe3 Ni3-C	BS1400 CMA1	C 95700	7.0–9.0	2.0–4.0	1.5–4.5	8.0–15.0	68.0–77.0

See specifications for allowable impurities

DESIGNATION			MINIMUM CEN MECHANICAL PROPERTIES				
European CEN/TC 133 Designation (Number)	Former British equivalent	Current American ASTM equivalent	Mode of casting	Tensile Strength N mm^{-2}	0.2% Proof Strength N mm^{-2}	Elongation %	Hardness Brinell
MEDIUM STRENGTH ALLOYS							
CuAl9-C (CC330G)	–	–	Die cast	500	180	20	100
			Centrifugal	450	160	15	100
CuAl10Fe2-C (CC331G)	BS1400 AB1	C 95300	Sand	500	180	18	100
			Die cast	600	250	20	130
			Centrifugal	550	200	18	130
			Continuous	550	200	15	130
CuAl10Ni3Fe2-C (CC332G)	(French alloy)	–	Sand	500	180	18	100
			Die cast	600	250	20	130
			Centrifugal	550	220	20	120
			Continuous	550	220	20	120
HIGH STRENGTH ALLOYS							
CuAl10Fe5Ni5-C (CC333G)	BS1400 AB2	C 95800	Sand	600	250	13	140
			Die cast	650	280	7	150
			Centrifugal	650	280	13	150
			Continuous	650	280	13	150
CuAl11Fe6Ni6-C (CC334G)	(French alloy)	–	Sand	680	320	5	170
			Die cast	750	380	5	185
			Centrifugal	750	380	5	185
CuMn11Al8Fe3 Ni3-C (CC212E)	BS1400 CMA1	C 95700	Sand	630	275	18	150

Table 3.2　Composition and minimum mechanical properties of cast aluminium bronzes of special interest, to British Naval Standards with ASTM equivalents.

DESIGNATION			CEN COMPOSITION (%)					
ISO TYPE Designation	British specification	Current American ASTM equivalent	Al	Fe	Ni	Mn	Si	Cu
CuAl9Ni5Fe4Mn	NES 747 Pt 2	–	8.8–9.5	4.0–5.0	4.5–5.5	0.75–1.30	0.1 max	bal.
CuAl6Si2	NES 834 Pt 3	C 95600	6.1–6.5	0.5–0.7	0.1 max	0.5 max	2.0–2.4	bal.

			MINIMUM MECHANICAL PROPERTIES				
			Tensile Strength N mm^{-2}	0.2% Proof Strength N mm^{-2}	Elongation %	Hardness Brinell	Impact Strength Joules
CuAl9Ni5Fe4Mn	NES 747 Pt 2		620	250	15	160	23
CuAl6Si2	NES 834 Pt 3	C 95600	460	175	20	–	–

Medium strength alloy

The medium strength alloy CuAl9-C is used only in die casting and centrifugal casting where the chilling effect of the mould enhances the mechanical properties. Although the relatively rapid cooling rate will ensure that a highly corrodible structure will not occur, this alloy may be susceptible to some 'de-aluminification' corrosion as explained in Chapters 8, 9 and 11, but the attack may not penetrate significantly.

Although cast by all processes, CuAl10Fe3 is used principally in die casting and in continuous casting for subsequent rework. Its excellent ductility makes it resistant to cracking on rapid cooling. It also has very good impact properties, the latter being of great importance in such applications as die-cast selector forks for motor vehicle gear boxes. It is not advisable, however, to sand-cast this alloy for use in corrosive applications, as the slower cooling rate is liable to give rise to a very corrodible structure (see Chapter 12). In the faster cooling conditions of die casting and centrifugal casting, the alloy may be susceptible to 'de-aluminification' as the CuAl9-C alloy above.

Alloy CuAl10Ni3Fe2-C is an alloy of French origin. It is a compromise between the high strength nickel-aluminium bronze CuAl10Ni5Fe5-C, and the nickel-free alloys. It has good corrosion resisting properties and is the most weldable aluminium bronze alloy (see Chapter 7).

Low magnetic alloy

The low magnetic alloy is the silicon-containing aluminium bronze CuAl7Si2 (see Table 3.2). Its principal attractions are its low magnetic permeability combined with excellent corrosion resisting and impact properties. It also has good ductility and machinability. One of its main uses is in equipment for naval mine counter-measure vessels.

Factors affecting the properties of castings

Effect of alloy composition on properties

Mechanical properties
As explained in Chapter 1, aluminium has a pronounced effect on mechanical properties. It will be seen from Figures 3.1 to 3.3 that tensile properties increase with aluminium content whereas elongation reduces. These graphs therefore provide useful guidance to the foundry in selecting an aluminium content that will ensure that all the specified properties are achieved. Manganese and silicon have similar effects to aluminium: 6% manganese being approximately equivalent to 1% aluminium and 1% silicon being approximately equivalent to 1.6% aluminium. It will be seen that the low iron alloys, CuAl10Fe2-C (Fig. 3.1) and CuAl7Si2 (Fig. 3.2), have comparable properties, the silicon alloy having an equivalent aluminium content of around 10%.

Iron, on its own, has some effect on mechanical properties, as may be seen from Figure 3.1 which shows the effect of iron contents of up to 4.95% when compared with the lower iron content of the CuAl10Fe2-C alloy. But its effect appears unpredictable. In association with nickel, iron has a significant effect on mechanical properties, as may be seen by comparing the more complex CuAl10Ni5Fe5-C alloy (Fig. 3.3) with the low iron alloys CuAl10Fe2-C (Fig. 3.1) and the silicon containing alloy CuAl7Si2 (Fig. 3.2). Iron and nickel appear, however, to have no discernible effect on mechanical properties within the limits of composition of alloy CuAl10Ni5Fe5-C. Figure 3.3 also highlights the effect of higher aluminium contents by comparing the higher aluminium-containing ASTM alloy C95500 with the CuAl10Ni5Fe5-C alloy.

The properties shown in Figures 3.1 to 3.3 should be compared with minimum mechanical properties specified in Tables 3.1 and 3.2 above. They are tensile test on standard sand-cast test bars. As may be seen, the mechanical properties of a standard test bar may be significantly higher than the minimum called for by specifications. The resultant spread of properties shown on the graphs may be due in part to differences in the pouring temperature, the temperature of the mould and the speed of pouring. But it is also likely to be an inherent feature of any alloy, namely that the crystal structure of a test bar is similar but not identical throughout, resulting in differences in properties at various cross-sections along its length. The breaking point of one defect-free test bar may therefore show better properties than the breaking point of another defect-free test bar of the same composition. One could argue that a mean line through the spread of test results, may be more representative of the overall strength of a casting than the minimum test result. This is because neighbouring parts of a cross-section of a casting lend

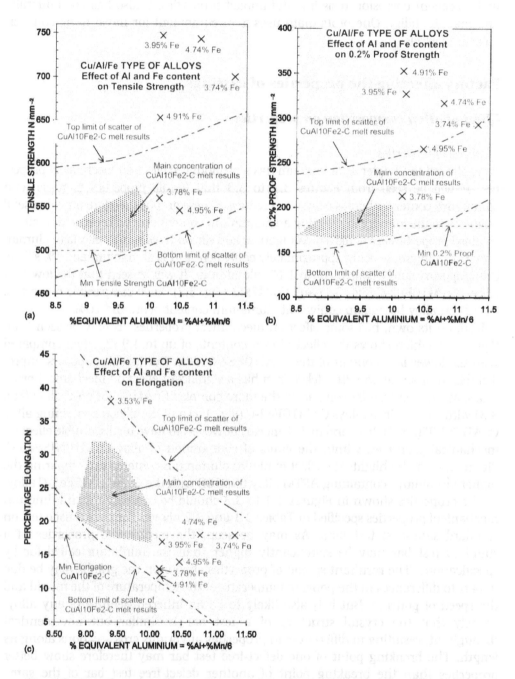

Fig. 3.1   Tensile test results on 40 melts of Cu–Al–Fe alloys, showing effect of Al and Fe on mechanical properties. By courtesy of Meighs Ltd.

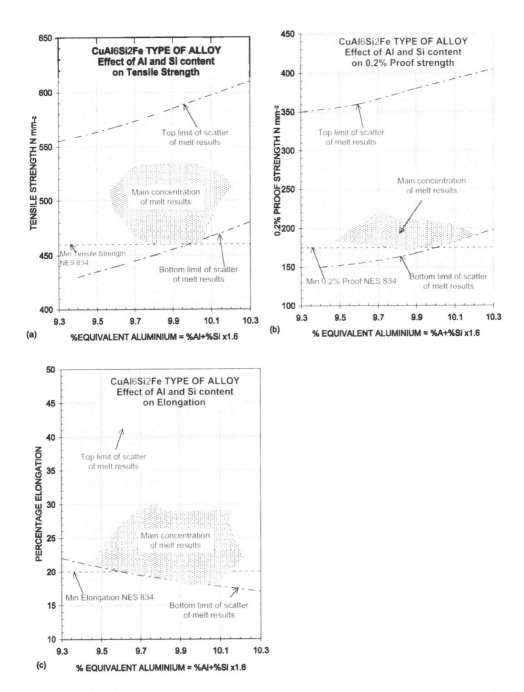

Fig. 3.2 Tensile test results on 200 melts of a CuAl6Si2Fe-C alloy, showing effect of Al and Si on mechanical properties. By courtesy of Meighs Ltd.

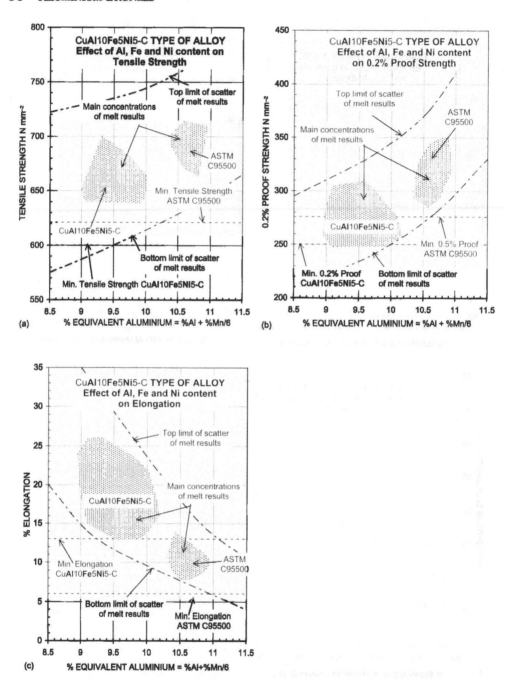

Fig. 3.3 Tensile test results on 193 melts of **CuAl10Fe5Ni5-C** and ASTM 95500 alloys, showing effect of Al, Fe and Ni on mechanical properties. By courtesy of Meighs Ltd.

strength to each other. On the other hand, designing a casting to the minimum properties specified for the alloy, provides an inherent margin of safety.

We shall see below that, being aware of the spread of mechanical properties, illustrated in Figures 3.1 to 3.3, is important for an understanding of the effects of impurities.

Fatigue properties

The fatigue properties of four cast aluminium bronzes are given in Table 3.3. In the case of manganese-aluminium bronze there is a significant reduction in fatigue strength in salt spray as compared to fatigue strength in air, as may also be seen in Figure 3.4.

Table 3.3   Fatigue strength of cast aluminium bronzes.[127-173]

Alloy	Condition	Tensile Strength N mm^{-2}	Fatigue limit 10^8 cycles N mm^{-2}
CuAl9Fe2	As cast	551	200
	As cast	552	150
CuAl10Fe5Ni5	As cast	551	220
	As cast	690	190
CuAl9Ni5Fe4Mn	As cast	655	210
	Heat treated	827	260
CuMn11Al8Fe3Ni3	As cast	649–727	232–247 in air 131 in sea water or salt spray

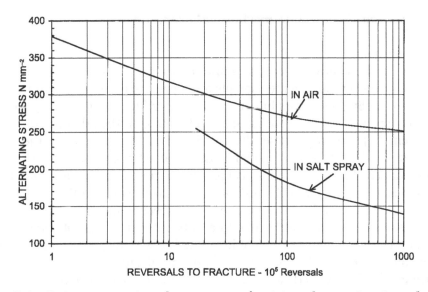

Fig. 3.4   Fatigue properties of manganese-aluminium bronze in air and salt spray.[127]

Effect of impurities on mechanical properties

M. Sadayappan *et al.*[157] of CANMET have carried out experiments on the effects of a variety of impurities on mechanical properties of nickel-aluminium bronze CuAl10Fe5Ni5. Their findings are shown plotted on Figure 3.5 together with the top and bottom range of the scatter of tensile test results, shown on Figure 3.3 above, for the same alloy produced under normal production conditions. The properties of the 'base' sample, free of impurities, are shown on each graph. It will be seen that mechanical test results in the case of most impurities fall within the limits of a normal scatter of results. The lowest figures for tensile strength and proof strength are for samples containing lead although the samples with the highest lead content do not have the lowest strength. On the other hand, lead would appear to have a beneficial effect on elongation, which is very surprising and unlikely to be indicative of a general tendency. In fact, the mechanical test results of the samples containing lead are more likely to be a function of the low aluminium content. The two samples containing beryllium show a distinct improvement in tensile and proof strength and worsening of elongation. It will also be seen that the base sample, free of impurities, shows low strength but high elongation, which reflects its low aluminium content.

Generally speaking, it is difficult to draw conclusions on the effects of the level of impurities tested to date. It is however in the nature of some impurities to have unpredictable effects. For example, silicon above the minimum allowed by specifications, can have very detrimental effects on mechanical properties. See also comments on the effects of impurities at the end of Chapter 1.

Effect of section thickness on mechanical properties

Effect of cooling rate
The mechanical properties of cast aluminium bronzes can vary considerably with variations in cooling rate from the solidification point to room temperature. A fairly rapid rate of cooling, as occurs in continuous, centrifugal or die casting, enhances mechanical properties. Slow cooling in a sand mould, on the other hand, results in lower strength properties. These changes in properties are due to the effect of cooling rate on the structure of the alloy, as explained in Chapters 11–13.

As mentioned above, the mechanical properties quoted for any alloy are those of a standard test bar. A 25 mm dia. test bar is a relatively small casting and its rate of cooling in a sand mould is relatively fast. The mechanical properties of a large-sectioned sand casting are likely therefore to be inferior to that of a standard test bar, if the casting is allowed to cool at its normal slow rate in the mould. Accelerating the rate of cooling of a casting by 'knocking it out' of the mould early will enhance mechanical properties but will result in built-in stresses that are likely to give problems in machining and may cause distortion or even cracking in some cases. It may also have adverse effects on corrosion resistance. It is therefore

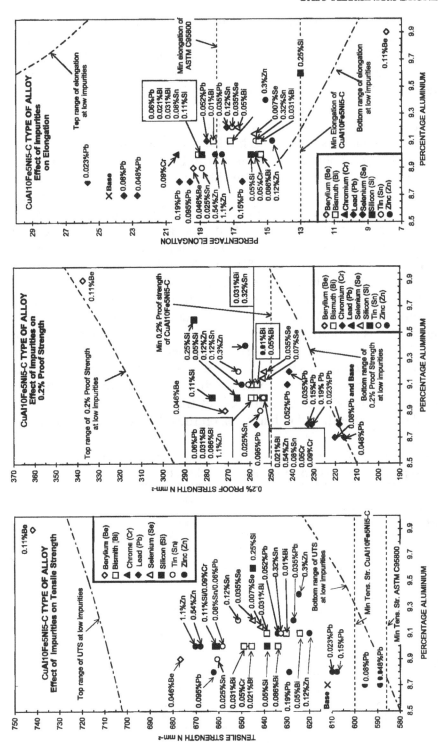

Fig. 1.1 Effect of various impurities on the mechanical properties of nickel–aluminium bronze, by M. Sadayappan *et al.*[157]

normally preferable, when designing a casting, to take account of variations in mechanical properties with section thickness than rely on rapid knock-out which is, in any case, difficult to regulate. It is also possible to improve the properties of selected parts of a casting by the use of metal chills or chilling sand, provided these can be incorporated in the overall 'methoding' of the casting.

Effect of section thickness on mechanical properties
The effect of section thickness on the mechanical and fatigue properties of sand-cast ship propellers, in nickel-aluminium bronze CuAl10Fe5Ni5, has been very thoroughly investigated by P Wenschot.[187] He obtained data from 117 castings varying in weight from 6 kg to 60 tonnes, having cast sections varying from 25 mm to 450 mm. Table 3.4 gives average values of mechanical and fatigue properties for a range of cast thicknesses. It will be seen that properties tend to deteriorate fairly rapidly as section thickness increases to 150 mm but the rate of deterioration is less as the thickness rises from 250 mm to 450 mm. Generally speaking, all mechanical properties, including elongation, reduce with section thickness.

By plotting values of mechanical properties on a linear scale against section thicknesses on a log scale, Wenschot[187] found that there was a relationship that was broadly linear as is illustrated on Figures 3.6a to 3.6d. As encountered above in Figures 3.1 to 3.3, mechanical properties obtained from standard size test bars can show a scatter of test results for the same alloy composition. It will be seen that there is a significant scatter of test results, which is most pronounced in the case of elongation (Fig. 3.6c) and of tensile strength at the heaviest sections (Fig. 3.6a). The spread of proof strength and of hardness, on the other hand, is relatively small

Table 3.4 Effect of casting thicknesses on mechanical and fatigue properties of a ship's propeller to CuAl10Fe5Ni5 type alloy, by P. Wenschot.[187]

		Average Values of Properties				
Range of cast section thickness mm	Number of castings	Tensile Strength N mm^{-2}	0.2% Proof Strength N mm^{-2}	Elongation %	Brinell Hardness H_B	Corrosion fatigue life* to failure 10^6 Cycles
20–30	33	679	262	22.3	163	100
30–60	4	636	252	18.3	160	90.3
60–75	3	613	241	18.9	160	–
75–110	4	589	230	19.3	149	–
150–160	3	582	210	20.7	136	–
250–280	12	503	201	14.0	129	33.3
280–320	5	511	199	15.0	128	33.9
320–360	12	487	196	13.8	131	29.8
360–380	17	496	197	15.0	128	29.0
380–420	8	478	195	15.6	126	22.9
420–450	16	489	189	15.9	129	26.3

* at 127.5 N mm^{-2} stress amplitude and zero mean stress

and constant. There are also likely to have been some differences in the aluminium content of the 117 castings tested which would also contribute to the spread of mechanical test results.

Test bars machined from a given area of a casting, only represent the properties of a very small section of that area of the casting. As previously mentioned, it can be reasonably argued that the average figures obtained from a number of test bars are more indicative of the resultant mechanical properties of a given casting section than the test figures of any one test bar.

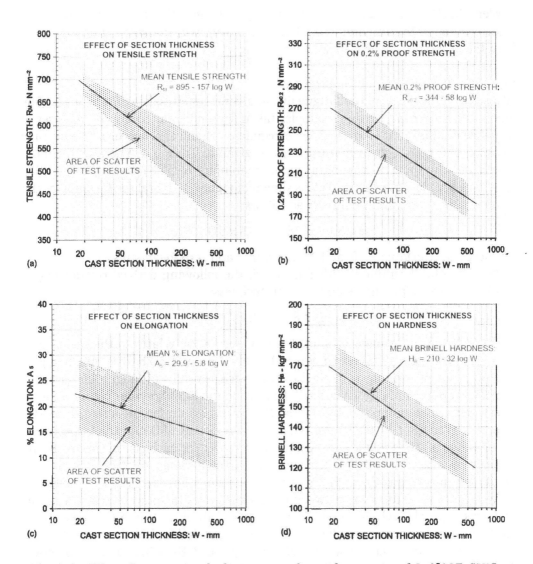

Fig. 3.6   Effect of cast section thickness on mechanical properties of CuAl10Fe5Ni5 type alloy, by P. Wenschot.[187]

The formulae for the lines running through the middle of the scatter of values are as follows:

for Tensile Strength: R_M = 895 – 157 log W   (1)
for 0.2% Proof Strength: $R_{p0.2}$ = 344 – 58 log W   (2)
for Elongation: A_5 = 29.9 – 5.8 log W   (3)
for Hardness: H_B = 210 – 32 log W   (4)

where W is the section thickness

Wenschot[187] found that mechanical properties near the surface of a propeller blade casting, at 90 mm and at 250 mm thickness, was not significantly different from the properties at the centre.

Effect of section thickness on fatigue properties
Wenschot[187] determined the fatigue life, at a constant stress amplitude of 127.5 N mm^{-2} and zero mean stress, of samples taken from different section thicknesses of the same castings that were used to determine mechanical properties. The average value obtained for each section thickness is given in Table 3.4. There was again a scatter of values and it was found more meaningful to plot fatigue life (on a log scale) against tensile strength. This is shown of Figure 3.7b and the formula for the mean number of cycles to failure is as follows:

$$\log N_f = 5.78 + 3.33 \times 10^{-3} R_M \qquad (5)$$

where N_f is the fatigue life at 127.5 N mm^{-2} and zero mean stress.

By combining formula 1 with formula 5, the following derived relationship is obtained between fatigue life and section thickness:

$$\log N_f = 8.76 - 0.52 \log W \qquad (6)$$

This relationship, shown graphically on Figure 3.7a, relates to the mean lines through the scatter of values shown in Figures 3.6a and 3.7b. It is interesting to note that, if this formula is applied to the average values of fatigue life given in Table 3.4, the points, thus calculated, lie close to the line shown in Figure 3.7a.

In order to determine the effect of section thickness on fatigue strength at 10^8 reversals, Wenschot[187] tested a number of 25 mm and 450 mm thick specimens over a range of stress amplitudes. Figure 3.7c shows, for each size of specimen, the number of cycles at which failure occurred over a range of stress amplitudes. Drawing a mean line through the scatter of values for each section thickness, showed that the relationship between number of cycles to failure and stress amplitude was approximately linear when they were both plotted on a log scale. Since the two mean lines were parallel, Wenschot[187] concluded that the mean line for intermediate section thicknesses would also be parallel. By using equation 6 above, the fatigue life at a stress amplitude of 127.5 N mm^{-2} could be calculated for a number of section thicknesses and lines for each section thickness drawn parallel to

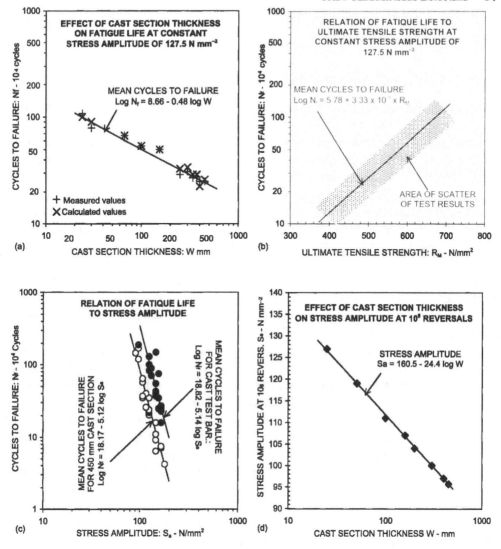

Fig. 3.7 Corrosion fatigue properties of nickel–aluminium bronze in seawater, by P. Wenschot.[187]

the two lines shown on Figure 3.7c. It was then possible from these lines to obtain the relationship between fatigue strength (S_a) and section thickness at 10^8 reversals which is shown on Figure 3.7d. The formula for this relationship is as follows:

$$S_a = 160.5 + 24.4 \log W \qquad (7)$$

Effect of mean stress
The above research on fatigue properties of nickel-aluminium bronze was carried out at zero mean stress. In practice, a propeller blade has a relatively high tensile

mean stress and this reduces the alloy's ability to endure fluctuating stresses. This effect must therefore be taken into account in the design of propeller blades.[187]

Effect of heat treatment on mechanical properties

The most common heat treatment applied to aluminium bronze castings is a 'stress relief anneal' applied to welded castings in order to achieve an homogeneous metallurgical structure and corrosion resistance throughout the heat-affected zone. For the CuAl9Ni5Fe4Mn alloy, to British Naval Specification NES 747 Pt 2 (see Table 3.2), the heat treatment consists in soaking at 400–700° C for one hour per 25 mm of section thickness followed by cooling in air. This heat treatment is also used to relieve internal stresses in a casting which has been rapidly cooled during the pouring process, in order to minimise distortion during subsequent machining. For the CuAl9Fe2 type of alloy, a lower soaking temperature of 350–400° C is adequate.

In addition to improving corrosion resistance, heat treatment generally improves mechanical properties. This will be seen from Figure 3.8 which gives mechanical test results of 884 melts. By comparing this graph with Figure 3.3 for the non-heat treated CuAl10Fe5Ni5 alloy, it will be seen that heat treatment has a beneficial effect on tensile properties, though little effect on elongation. One effect of heat treatment is to nullify the differences in pouring conditions of the test bars. The fact that there still is a significant spread of test results confirms that this spread is inherent to the nature of an alloy as suggested above.

Apart from the above stress relief anneal heat treatment, aluminium bronze castings are not normally heat treated as the required properties can usually be obtained by careful selection of alloying elements. There is also the possibility of distortion particularly in the case of propeller castings. However, heat treatment is applied on occasions for special purposes when exceptional combinations of properties are required.

Generally, the simple alloy CuAl9Fe2 is only heat treated when maximum resistance to wear is required at the expense of ductility. In this case the component should be quenched from 900° C and re-heated to 400° C for 1–2 hours when the maximum hardness should be obtained.

The complex alloy CuAl10Fe5Ni5 may be heat treated to improve proof strength, tensile strength and hardness with some reduction in ductility. Water quenching after 1 hour at 900–950° C and subsequent reheating for about 2 hours at 600–650° C results in the change in properties as shown in Table 3.5.

While these two heat treatments correspond with those most commonly adopted in practice, there is a wide variety of treatments available which will modify the properties of the casting. These are further discussed in Part 2.

With large castings, it may not be possible to carry out any heat treatment, as the necessary equipment required for heating and quenching may not be available. This factor alone imposes limitations on the heat treatment of castings but, in

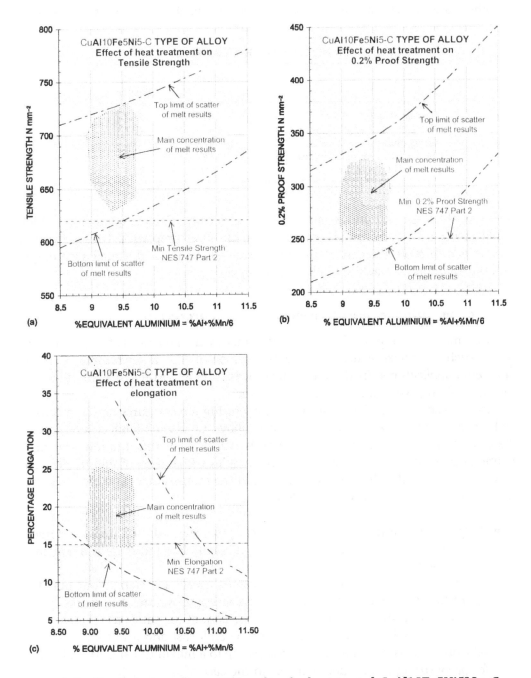

Fig. 3.8 Tensile test results on 844 melts of a heat treated CuAl10Fe5Ni5Mn–C alloy, showing the effect of heat treatment on mechanical properties. By courtesy of Meighs Ltd.

Table 3.5 Effect of heat treatment on mechanical properties of alloy CuAl10Fe5Ni5.[127]

MECHANICAL PROPERTIES		
	As cast	Heat Treated
Tensile Strength, N mm^{-2}	664	757
0.1% Proof Strength, N mm^{-2}	293	432
Elongation, %	17	18
Izod Impact Strength, Joules	18	21
Brinell Hardness, HB	170	210

addition, thickness variations may prevent uniformly effective quenching which could result in a considerable variation in the final properties. Because of the danger of distortion, the casting to be heat treated should be designed accordingly and the appropriate machining allowances made.

Effect of operating temperature on mechanical properties

Castings may be required to be used in equipment which operates at high temperatures, such as power generating machinery, or at low temperatures, such as cryogenic applications. In these circumstances, it is important to know how the mechanical properties of the alloys are affected.

Table 3.6 shows the effect of a range of operating temperatures on the mechanical properties of the two most common cast alloys, CuAl9Fe2 (chill cast or gravity die cast) and CuAl10Fe5Ni5 (sand cast). It will be seen that the low and high temperature tests were done with different samples which were of slightly different compositions. But the effect of the difference in the temperatures at which the tests were done is nevertheless clearly evident:

- At low temperature, the tensile strength, proof strength and impact strength are all increased. The effect on elongation is not clear but there is a tendency for it to be reduced.
- At high temperatures the effect is reversed except for elongation, but it is interesting to note that the effect on proof strength is much less marked than that on tensile strength. Elongation in the case of the sand cast sample (d) is significantly reduced, yet in the case of the chill cast and heat treated sample (e) elongation is very markedly increased up to a maximum at 400° C. Figures for proof strength were not available in the case of sample (e).

Effect of temperature on mechanical properties of manganese-aluminium bronze
The effect of operating temperature on the mechanical properties of CuMn11Al8Fe3Ni3 (not included in Table 3.6) is shown in Figure 3.9.

Table 3.6   Effect of operating temperature on mechanical properties of two standard alloys.[173]

DESIGNATION	CASTING CONDITION AND COMPOSITION (Remainder Cu)	TYPICAL MECHANICAL PROPERTIES				
		Testing temp. °C	Tensile Strength N mm^{-2}	0.2% Proof Strength N mm^{-2}	Elongat. %	Impact Strength Joules
CuAl9Fe2	chill cast	−196	793	334	17	32
Sample a	9.91% Al,	−100	752	329	24	38
	1.98% Fe,	−70	710	325	23	39
		−40	710	289	27	39
		20	676	301	24	44
Sample b	**9.8 mm section**	20	669	293	29	−
	gravity die cast	150	626	271	30	−
	10.05% Al,	200	615	281	29	−
	2.9% Fe	250	595	274	29	−
		300	545	294	28	−
		350	465	271	25	−
		400	313	281	50	−
CuAl10Fe5Ni5	**Sand cast**	−196	749	407	10.5	10
Sample c	**6 mm machined**	−100	719	367	8.2	12
	bar	−70	719	430	10.5	13
	10.05% Al,	−40	700	355	10.7	14
	5.1% Ni	20	686	319	12.6	14
	4.1% Fe,					
	1.12% Mn					
Sample d	**Sand cast**	20	673	301	8	−
	50 mm dia bar	204	567	279	7	−
	10.37% Al,	316	505	270	5	−
	5.77% Ni					
	4.46% Fe					
Sample e	**Chill cast and**	20	880	−	10.8	−
	heat treated	100	853	−	12.9	−
	10.28% Al,	200	774	−	13.7	−
	4.97% Ni	250	715	−	10.0	−
	4.75% Fe	300	634	−	13.3	−
		350	431	−	28.1	−
		400	296	−	55.3	−
		450	217	−	44.9	−
		500	168	−	32.5	−

Tests at temperatures down to −183° C have shown that proof strength and tensile strength increase progressively and, although there is a gradual reduction in elongation, the figure never falls to a dangerous level. Work by Lismer[117] confirmed this but revealed a change in fracture characteristics and ductility between −150° and −196° C. At −196° C the Charpy impact strength was 16.3 Joules compared to 21.7 Joules at −150° C and 42 Joules at room temperature.

Tensile data for short-term exposure to elevated temperatures, shown in Figure 3.9, indicate that a useful degree of strength is maintained at temperatures up to 350° C. A particular point of interest is the absence of any reduction in elongation

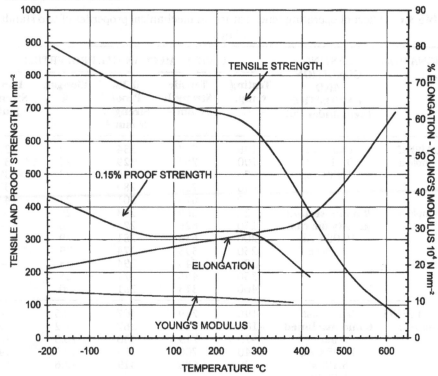

Fig. 3.9 Effect of operating temperature on the mechanical properties of CuMn11Al8Fe3Ni3.[127]

due to a ductility dip in the range 300–600° C, as occurs with many other aluminium bronzes (see Fig. 7.1, Chapter 7).

Creep resistance is reported to be excellent at temperatures up to 175° C and in this range it is possibly superior to any other copper-base casting alloy. At higher temperature the manganese-aluminium bronzes are less suitable than other aluminium bronzes and are not favoured for prolonged use above ~280° C.

The effect of operating temperatures of 204° C and 260° C on stress-rupture is given in Table 3.7.

Table 3.7   Stress/rupture data on manganese-aluminium bronze.[127]

Test temperature °C	Stress to cause rupture in specified time N mm⁻²	
	1,000 hrs	100,000 hrs
204	538	464*
260	374	232*

* Extrapolated values

B – Casting processes

Processes

Aluminium bronze castings are made by all the main foundry processes:

- Sand casting
- Shell mould casting
- Ceramic mould casting
- Investment casting
- Die or permanent mould casting
- Centrifugal casting
- Continuous and semi-continuous casting

The principles governing the manufacture of sound castings are dealt with in the next Chapter. Although explained in terms of sand castings, they apply to any casting process.

Sand casting

Making castings from sand moulds is the most versatile method of producing components of a great variety of sizes and complexity.

There are two categories of sand moulded castings: floor moulded and mechanised moulded.

Castings that are required in relatively small quantities are normally floor moulded, using pattern equipment usually made of wood, although resin patterns and, occasionally, metal patterns are also used . They range in size from a fraction of a kilogram to several tonnes. Probably the largest aluminium bronze castings made are propellers for super tankers which can weigh in excess of 70 tonnes.

Castings that are relatively small (typically less than 45 kg), and which are required in batches of 50 or more, are normally more economical to produce in a mechanised sand foundry using metal patterns. Cores may, however, be made from wooden core boxes.

Components of all shapes, sizes and configurations are sand-cast in aluminium bronze for a variety of equipment. They include pumps, valves, propellers, heat exchangers, turbines, bearings, strainers, filters, compressors, water meters, paper making machinery, pickling equipment, slippers for rolling mills, seal housings, pipe fittings, glass moulds, ships fittings and a great variety of miscellaneous machinery. Most sand moulded aluminium bronze castings are made in the high strength alloy, $CuAl10Fe5Ni5$, or its equivalents for reason of strength and resistance to corrosion.

Shell mould casting

In shell moulding, a metal pattern is heated and sprayed with a specially bonded sand which rapidly set on contact with the hot pattern, forming a thin shell of hardened sand. The shell mould is normally made in two parts and, together with a similarly made shell core or cores, is assembled and cast. The relative fragility of the mould limits the size of castings which can be made by this process. Shell moulding produces castings with a better finish and greater dimensional accuracy than hand-moulded sand castings, but comparable to machined moulded sand castings. Shell cores are sometimes used in conjunction with ordinary sand moulds to produce castings with better internal finish, for example in the case of pump impeller cores.

The pattern equipment for shell moulding is relatively expensive and renders this process uneconomical unless the number of castings produced justify the outlay.

Shell moulded castings are used for smaller components in much the same applications as are listed above under 'sand castings' and are also principally made in the high strength aluminium bronzes.

Ceramic mould casting

Ceramic material is used in this process to produce moulds which are similar in construction to a sand mould. Because of the high cost of ceramic, this process (sometimes referred to as the 'Shaw Process') is normally used only for relatively small moulds. Castings of excellent finish and of a high degree of accuracy are made by this process which obviates the need for some machining operations. The use of

Fig. 3.10 Large stern tube made in aluminium bronze for naval use (Westley Brothers).

ceramic cores, in conjunction with sand moulds, for castings of even relatively large turbine rotors and pump impellers, can save a lot of time-consuming hand-dressing of the vanes and may result in a net saving in manufacturing cost, as compared with the use of sand cores.

Ceramic moulded castings are used extensively in the aircraft industry for precision castings required in numbers not large enough to justify the alternative process of die casting or investment casting (see below). These castings are made in either the high or medium strength aluminium bronzes to suit each requirement.

Investment casting

Investment casting is a process for producing large quantities of intricate tiny parts which do not require coring. In this process, a replica of the casting is made in wax and is dipped in a ceramic slurry. It is allowed to dry and then stoved and this causes the ceramic material to set into a hard shell and the wax to melt away. The result is a one piece mould in which the metal is later poured. This process is highly automated and the small moulds are made in sets arranged like a Christmas tree.

Die casting or permanent mould casting

Die casting consists in pouring metal by gravity or under pressure into a permanent mould made from a special heat-resisting metal. Most aluminium bronze die castings are gravity poured. Pressure die casting of aluminium bronze has been tried but is not considered economic due to the short life of the die at the high operating temperatures and high rate of production involved.

Die casting is the ideal process for small and fairly intricate components which need tight dimensional accuracy and consistency together with excellent surface finish and which are required in large quantities. Cores have to be retractable unless they are made in sand, shell or ceramic.

Medium strength aluminium bronze, CuAl10Fe3, is probably the most widely used copper-base alloy for gravity die casting. Its fluidity in the molten state, good reproduction of details, excellent surface finish and relatively slight attack on the material of the die (notwithstanding its fairly high melting point), make it a most suitable die casting alloy. This alloy has outstanding impact, wear and fatigue properties and is therefore a most appropriate choice for components subjected to repeated shock loading such as gear selector forks in motor vehicles. The high-strength aluminium bronze, CuAl10Fe5Ni5, is less suitable in view of its higher melting point and somewhat inferior fluidity in the molten state. It is also prone to hot tearing on rapid cooling if its shrinkage is hindered.

More complex shapes can be produced by die casting than by forging and a wide range of components can be satisfactorily produced as gravity die castings in aluminium bronze. Castings produced in this way vary in weight from a few grams to several kilograms. Castings of up to 20 kilos have been made, but the higher the weight of the component, the more restricted is the variety of shapes achievable.

The degree of dimensional accuracy which may be obtained in die cast components is normally ± 0.25 mm on all parts of the casting, although on certain dimensions in one half of the die, this may be reduced to ± 0.125 mm. Die casting can therefore result in a reduction of machining cost as compared with forging or hot pressing. The properties and dimensional tolerances of a die casting can be further improved by a subsequent coining operation; the resultant surface hardening is of particular value in increasing the wear resistance of critical faces such as those of gear teeth.

Centrifugal casting

All cylindrical aluminium bronze products, including bushes and gear blanks, are ideally suited to the centrifugal casting process, and the properties are superior in many respects to both sand and chill castings.

The principle of centrifugal casting is essentially simple: molten metal is introduced quietly into a rapidly rotating mould and is retained by centrifugal force against the circumference where it solidifies. Thus the exterior surface of the casting takes the form of the inside of the mould. With cast iron or steel moulds there is rapid chilling of the metal, so that fine-grained structures are obtained with the maximum chill occurring at the outer face. This is of particular advantage for gear wheels and similar products, whose exterior surfaces suffer heavy wear and occasionally impact loading. In addition to this grain refining effect, there are further structural advantages. As the solidification takes place almost entirely from the outside, a form of directional solidification occurs which concentrates porosity and impurities in the metal last to freeze along the bore of the cylinder. Subsequent machining of the bore removes this unsound metal. Centrifugal castings, made in chill moulds, may therefore have slightly greater density than normal chill castings. Higher speeds of rotation are required for aluminium bronze than for tin bronzes and gunmetals, and very high rates of chilling should be avoided with aluminium bronze as this can cause surface cracking.

Centrifugal castings in sand moulds are also frequently produced although, inevitably, the principal advantages of chill cast centrifugal products with regards to good mechanical properties do not apply. It is, however, more suitable for certain cylindrical and other similar shapes.

From the dynamics of the process, it is clear that all castings should be symmetrical around the axis of rotation. Apart from plain cylinders, flanged castings and gears with hub and web faces of different diameter and unsymmetrical shapes can be made with the aid of shaped sand cores inserted into the permanent moulds. Castings as small as 50 mm dia. and up to 2000 mm dia. can be made by the process. The speed of rotation reduces as the diameter increases and the peripheral speed maintains the optimum load of 60 times gravity, applied by centrifugal force. The upper limits are governed by the equipment available, rather than by any other fundamental factors.

Continuous and semi-continuous casting

There are two related casting processes for producing stock lengths of uniform solid or hollow cross-sections:

- The *continuous casting* process is used to produce a variety of both solid and hollow sections which may be either regular or irregular. When the process is properly applied and controlled, the product has a good surface finish which needs little machining (some faces may not require any further machining).
- The *semi-continuous process* is primarily used to produce simple standard cross-sections: e.g. round, square or rectangular. They are generally intended for subsequent hot working and are consequently cast in standard lengths. The surface finish is quite good for this purpose but some proof machining might be required for more demanding processes such as forging.

Continuous casting

The vertical continuous casting process, is illustrated in Figure 3.11 (which shows the Delta Encon Process).[64] In this process, the alloy is melted in an induction furnace (A) and then transferred to a holding furnace (B) which has a controlled nitrogen atmosphere. At the base of this furnace is a graphite die, housed in a water-cooled copper jacket (C).

At the point of entry of the liquid metal into the holding furnace, the metal flows, free of turbulence, beneath a weir which allows any dross caused on pouring to float to the surface. This ensures that no dross is carried along in the flow of metal to the die. A plunger (not shown) controls the flow of metal from the holding furnace to the die. This allows a die to be replaced by another one whilst the holding furnace contains liquid metal.

The casting process is started using a 'starter bar' of the same size and configuration as the intended product. The cast bar is lowered by a set of rollers (D) which controls the speed of lowering to match the rate of heat withdrawal within the die. This synchronisation of the lowering speed with the rate of heat withdrawal is critical for the successful operation of the process: too rapid a lowering speed will lead to spillage of metal and too slow a lowering speed to the premature freezing of the metal in the die. The process ensures that the metal is fed progressively as solidification occurs. It creates ideal conditions for directional solidification. The operational controls, as well as the design and composition of the die, also have a bearing on dimensional accuracy which can be to ± 0.1 mm.

A sliding clamp grips the moving bar and an abrasive cutting wheel (E) cuts it to the desired length as casting proceeds. The cut pieces are then transferred to a conveyor system (F).

The process is truly *continuous* since bars of any lengths can theoretically be produced. In practice, a continuous casting installation is designed to meet the demand for certain types and sizes of bars and this determines (a) the capacity of the

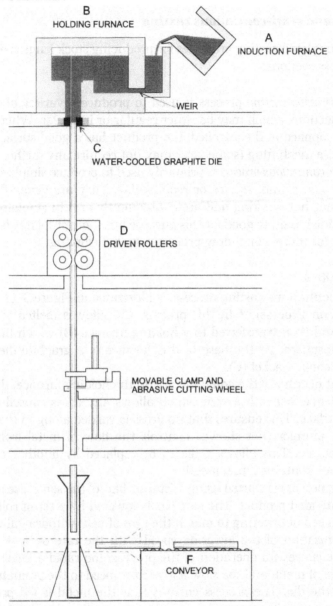

Fig. 3.11 A continuous casting installation.[64]

furnace and (b) the height of the installation. Production runs of any given size of product rarely exceeds 3 tonnes.

The process may be either vertical or horizontal. In a horizontal installation, the die is mounted in the side of the holding furnace near its base. In this case, the flow of metal is less easy to control but the furnace is more readily topped-up and longer bars may be produced.

Graphite dies are quite fragile and rarely last for more than one cast, but they are less expensive than the copper dies used in semi-continuous casting. More accurate and intricate shapes can be produced with graphite dies. Hollow bars are produced using a two-piece die and a tapered graphite post, known as a mandrel, which forms the bore. Carbon pick-up is not considered to be a problem with aluminium bronzes.

The range of sizes that can be produced is being continually extended. In general, round bars of less than 25 mm are unusual. Hollow bars of 40 mm outside diameter are possible but difficult to produce with bore of less than 18 mm diameter.

Round rods and tubes are mechanically straightened after casting by a process known as 'reeling' and subsequently checked for dimensional accuracy, concentricity, and straightness. Other sections are straightened by stretching.

Semi-continuous casting

Semi-continuous casting is a simpler process than continuous casting although the principles of operation are the same. It consists in pouring molten metal at a controlled rate from a holding furnace, via a launder, into a water-cooled copper die. Care is taken to avoid turbulence in pouring. The bottom part of the die is lowered slowly as solidification proceeds. The rate of pouring and the speed of lowering of the casting are synchronised to correspond to the rate of heat withdrawal from the die. As the casting emerges from the bottom of the die, it is cooled by a water spray. Generally speaking, the quantity of metal in the holding furnaces determines the cast length of the bar. The depth of the pit below the die must be such as to accommodate the longest length produced. There is therefore no need for a sliding cutting wheel as in the continuous casting process and this is the essential difference between the two processes.

The process, which is usually vertical, is called *semi*-continuous because it is designed to produce only a given length of bar per cast. The dies are usually made of copper and are repeatedly used. Although the cost of these dies is high, the regular demand for standard sections justifies the initial investment.

Billets of about 1.5 tonnes are regularly produced by this process. Sections of up to 450 mm dia. may be produced but sizes below 100 mm dia. are not usually cast by this method.

Advantages of the continuous and semi-continuous casting processes

- They are relatively simple methods of producing long lengths of bars of both uniform and variable cross-sections. Continuous casting can be more appropriate than wrought processes in some circumstances. As may be seen in Chapter 5, however, most wrought processes result in significantly higher mechanical properties.

- Both continuous and semi-continuous casting processes provide ideal conditions for solidification to occur directionally thereby resulting in shrinkage-free castings. See Chapter 4 for a more detailed explanation of directional solidification.
- Because the die is water-cooled, solidification is relatively rapid, producing a fine grain structure. This results in good hardness value and enhanced mechanical properties. The thinner the cross-section of the casting, the more rapid the rate of solidification and consequently the finer the grain structure.
- It is claimed that the quality of billets produced by semi-continuous casting is superior to that produced by the tilted mould process (see Chapter 4). Generally speaking, continuous casting is a more satisfactory way of producing stock billets for subsequent working. Near-net shape components are also readily cast by this method.

Choosing the most appropriate casting process

The choice of the most appropriate casting process depends on casting size, quantity, dimensional accuracy, appearance and cost. The following gives broad guidelines but, in many cases the choice of route may not be immediately obvious and will require discussion with various founders involving cost estimates:

- *very small castings in very large quantities*: investment casting,
- *small castings in relatively large quantities*: die casting for high precision but alloy choice restricted to low nickel alloys; otherwise: mechanised sand moulding or shell moulding; (cost and/or appearance of casting may determine the choice);
- *small castings of less than 45–50 kilograms in medium to large quantity*: mechanised sand moulding or shell moulding (cost and/or appearance of casting may determine the choice); for high precision, ceramic moulding would be best;
- *small castings of less than 45–50 kilograms in jobbing quantity*: floor moulding; the cost of shell moulding pattern equipment is likely to rule out this option; for high precision: ceramic moulds;
- *medium size castings up to 45–50 kilograms in large quantities*: mechanised sand moulding;
- *medium size castings in jobbing quantities and castings in excess of 45–50 kilograms in any quantities*: floor sand moulding; for cylindrical shapes, centrifugal castings may be best;
- *long lengths of bars of both uniform and variable cross-sections*: continuous or semi-continuous casting.

Applications and markets

Most sand castings are made in the high tensile aluminium bronzes for reasons of strength and resistance to corrosion. The largest proportion of these castings are used in ship building and other sea water applications where the properties of aluminium bronze are used to greatest advantage, but they are also to be found as components of equipment used in all the following industries:

Building, as structural components in 'prestige' buildings and as ornaments.

Coal mining, in various machinery fittings (non-sparking properties).

Cryogenics, mostly as pumps, valves, strainers and pipe fittings.

Explosives, as components of explosive handling equipment (non-sparking properties).

Glass, as glass moulds.

Oil and gas, as pumps, valves etc (non-sparking properties).

Paper making, as components of machinery.

Process plant, as components in a variety of chemical processes.

Power generation and transmission, as pumps, turbines etc.

Railways, for shock resisting fittings.

Steel manufacture, as rolling mills 'slippers', bearings and pickling hooks.

Water, as pumps, valves etc.

Chapter 9 gives details of aluminium bronze components used in corrosive environments (marine service, water supply, petro-chemical, chemical and building industries)

Fig. 3.12 Main and intermediate propeller shaft brackets for a mine counter-measure vessel cast in silicon–aluminium bronze (Meighs Ltd).

4
MANUFACTURE AND DESIGN OF ALUMINIUM BRONZE CASTINGS

A – Manufacture of castings

The making of sound castings

The great advantage of the casting process over other metal forming processes is its versatility. A wide variety of shapes can be produced by pouring molten metal into a mould. Nevertheless, certain physical conditions are necessary for this molten metal to solidify free of internal defects.

The danger of the following defects occurring need to be understood and taken into consideration in devising techniques and procedures for the making of sound aluminium bronze castings:

- Oxide inclusions
- Shrinkage defects
- Gas porosity

Oxide inclusions

The tenacious film of oxide that forms on aluminium bronze is mainly responsible for its excellent corrosion resistance. It creates, however, a problem for the foundry-man in that, as it forms on the molten metal, it is liable to get entrapped in the metal if there is turbulence in the pouring of a casting.

Shrinkage defects

Solidification of aluminium bronze
In common with other metals, as liquid aluminium bronze reaches its solidification temperature, tiny crystalline particles, known as 'nuclei', begin to form adjacent to the mould face where the metal has cooled most rapidly. These nuclei then grow into larger crystals with a Christmas tree-like form, known as a 'dendrites' (see Fig. 4.1). These dendrites grow away from the mould face as the temperature of the metal falls (see Fig. 4.2). Depending on casting thickness, other dendrites may later begin to form around nuclei that have formed at the centre of the section. The dendrites grow in thickness as the liquid metal, remaining between the arms of the dendrites, solidifies. Soon this sideways growth of dendrites is impeded by that of

other dendrites growing in the vicinity. Eventually, the outward growth of the dendrites is also prevented by the dendrites which have grown away from the opposite face of the mould or by dendrites which have nucleated at the centre of the section.

It will be seen from Figure 4.2 that this progressive solidification forms a **V**-shaped solidification front.

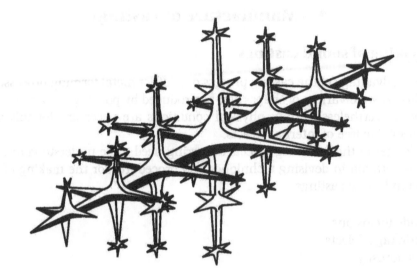

Fig. 4.1   A dendrite.[92]

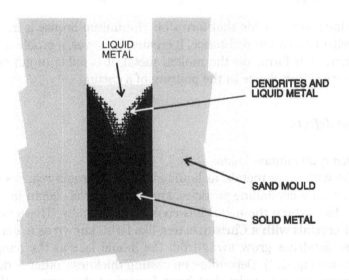

Fig. 4.2   Solidification process.

Solidification range

In the molten state, aluminium bronze consists of a substantially uniform solution of its alloying elements in each other (this solution may also contain compounds of some of the elements present). When, as the temperature falls, the first dendrites begin to appear, they initially consists of solid solutions that are richer in the higher melting point elements. Consequently, the liquid metal between the arms of the dendrites becomes richer in aluminium – the element with the lowest melting point – and its temperature of solidification is therefore lower. This is the reason why solidification occurs over a range of temperatures. This solidification range is significantly narrower for aluminium bronze than for other copper-based alloys. This, as we shall see, has an important bearing on the techniques that need to be used to produce sound aluminium bronze castings.

When the liquid metal between the arms of the dendrites finally solidifies, it creates a bond between them which gives the alloy its mechanical properties.

The effects of shrinkage

Aluminium bronze shrinks on solidification by approximately 4% volumetrically. This means that, as the liquid metal between the arms of the dendrites solidifies, it shrinks and various consequences may then ensue:

1. Liquid metal may percolate between the arms of the dendrites to compensate for this shrinkage. Alternatively, since the part-solidified metal has a pasty consistency, it will be compacted, by the combination of atmospheric pressure and the head of liquid metal above.
2. The level of the liquid metal above the part-solidified metal will fall correspondingly;
3. If the depth of the part-solidified metal is too great for shrinkage to be fully compensated, tiny cavities will form, known as shrinkage defects. Since they occur at the boundary between dendrites, they weaken the bond between them and therefore reduce the mechanical properties of the alloy.
4. If the liquid metal above the part-solidified metal is isolated from atmospheric pressure by a region of solid metal, due, for example, to a more rapidly solidified thinner section above (see Fig. 4.6a), shrinkage defects will occur as in (3).

Provided there is no gas dissolved in the metal (see below), these shrinkage defects are vacuum cavities, and the pressure difference between them and atmosphere is 10.13 N cm^{-2} which is equivalent to a 1.37 metres head of molten aluminium bronze. *This is a considerable force available to prevent shrinkage defects occurring, provided the metal in the process of solidification is exposed to atmospheric pressure.*

Gas porosity

Molten aluminium bronze has a great affinity for hydrogen which it may absorb from dampness in the atmosphere or, more markedly, from the combustion gasses of oil or gas-fired furnaces. As aluminium bronze approaches its solidification temperature, its solubility for hydrogen diminishes significantly. The dissolved gas is therefore liable to come out of solution, forming small bubbles in the liquid metal remaining between the arms of the dendrites. The longer a casting takes to solidify, the more dissolved gas will come out of solution. Furthermore, any tendency for shrinkage defects to occur, as explained above, will cause a reduction in pressure which further reduces the solubility of hydrogen in aluminium bronze and therefore causes more gas to come out of solution. Gas bubbles, in turn, create a back pressure which assists the formation of shrinkage defects. The presence of hydrogen gas is therefore doubly harmful.

Prevention of defects

Avoiding oxide inclusions

The Durville process of casting billets

As explained in the historical note at the beginning of this book, the need to pour aluminium bronze with the minimum of turbulence in order to avoid oxide inclusions, was realised by the French Foundryman, Pierre Durville, when he began to manufacture aluminium bronze billet at about the time of the first World War. His method of casting billets is illustrated in Figure 4.3. which shows the tilting motion of the mould and pouring basin during casting. The mould and basin are rigidly connected to one another. The metal is poured from a ladle into the pouring basin where it is allowed to rest for a short time to allow the lighter oxide to float to the

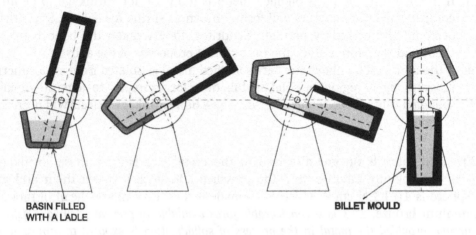

BASIN FILLED
WITH A LADLE

BILLET MOULD

Fig. 4.3   The 'Durville process' of casting billets.[127]

surface and be skimmed off. The whole assembly is then slowly inverted, thus transferring the metal from the pouring basin to the mould with little or no turbulence. Moulds are generally made of cast iron but copper has also been used.

The Meigh process of pouring sand castings
Charles Meigh, who had worked for Pierre Durville, set up his own foundry in France in 1924 and applied the tilting principle to the pouring of castings for the first time.

The process is illustrated in Figure 4.4 and the following are its main distinctive features:

- the mould is tilted through 90° only and pivots about the point of entry of the metal into the mould;
- the mould is connected directly to a tilting furnace by a launder;
- the end of the launder nearest the furnace is enlarged to form a pouring basin and the launder is only slightly inclined;

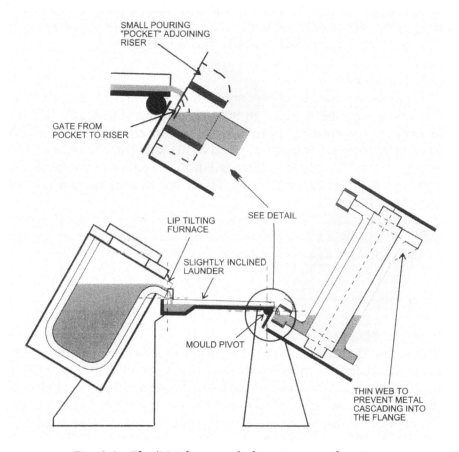

Fig. 4.4   The 'Meigh process' of pouring a sand casting.

- a small recess or 'pocket' is incorporated in the mould and is connected to an adjoining feeder by means of a narrow gate;
- the molten metal enters this small pocket and thence the adjoining feeder through this narrow gate;
- tilting begins as soon as the small pocket is full: the mould is thereafter tilted as it fills.

Any oxide forming in the pouring basin at the furnace-end of the launder is carefully skimmed off. As the molten metal flows down the launder, the oxide film that forms in contact with air creates a cover beneath which the metal flows protected from further oxidation. The small pocket at the point of entry into the mould allows the metal to settle before entering the adjoining feeder through the narrow gate. All these features are designed to avoid oxide inclusions in the casting.

It is important to visualise how the metal will flow through the mould as it tilts and fills in order to check whether the metal is likely to drop at any point within the mould. This would cause turbulence and oxide inclusions. To guard against this, thin webs are added to the pattern equipment to provide a path for the metal to flow without dropping. These webs are later removed from the casting unless they are desirable as a design feature. Such a web is shown in Figure 4.4.

Other tilting processes
Other tilting processes were later developed which are variants on the above two processes. Most involve tilting through 90° and pouring from a ladle rather than from the furnace. For example, there is a method of casting billets in which the mould is tilted through 90° only. It is known as the semi-Durville process. There is also a tilting method for pouring castings which is similar to the semi-Durville process and in which the mould incorporates a large pouring basin which is filled with a ladle prior to tilting.

Some castings, such as fixed-pitch marine propellers, are of a shape that does not lend itself to be cast by a tilting process. In such cases, others means are used to prevent oxide inclusions as explained below.

Directional solidification

Suitability of the tilting process
It is fortunate that the tilting principle, initially devised to overcome the problem of oxide inclusions, is also the most satisfactory way of avoiding shrinkage defects in billets and castings. This is because, as the metal flows into any part of the mould, it begins to solidify in contact with the mould face. As it does so, it shrinks as explained above, and the molten metal, flowing over it, compensates for this shrinkage. Being still liquid, this metal is able to transmit atmospheric pressure which, as also explained above, keeps the soft part-solidified metal compacted. In

this way shrinkage defects are avoided. In the Meigh process, the molten metal, coming straight from the furnace, is hotter and therefore accentuates the temperature gradient across the depth of the part-solidified metal.

Creating a temperature gradient which encourages the metal to solidify progressively from the mould face, is known as 'directional solidification'. The function of the feeders is to provide a reserve of molten metal which will remain liquid long enough to compensate for the shrinkage in the last part of the mould to solidify.

As previously mentioned, some castings, such as fixed-pitch propeller castings, are of a shape that does not lend itself to a tilting process. An alternative way of achieving directional solidification will be discussed below.

Unsuitability of bottom pouring
By contrast, in the traditional 'bottom pouring' method, the molten metal fills the mould cavity from the bottom.

At the end of the pour,

- solidification has begun to take place throughout the mould,
- the hottest metal is at the bottom of the mould,
- only the size of the feeders eventually creates a favourable temperature gradient,
- the depth of part-solidified metal is so great (i.e. the depth of the mould), that the head of liquid metal in the feeders together with atmospheric pressure may not be able to compact the metal. Only the metal initially 'chilled' in contact with the mould face is satisfactorily compacted.

In the case of alloys with a long solidification temperature range, such as gunmetal, bottom pouring results in a very diffused form of shrinkage porosity. Pressure tightness is often achieved by the addition of lead, which does not alloy, but which fills the micro-porosity and acts as a 'built-in impregnation'. In the case of alloys with a short solidification temperature range, such as aluminium bronze, bottom pouring tends to result in more concentrated shrinkage cavities.

For this reason, the following recommendations are made with the Meigh process in mind but would be applicable in principle to other tilting processes.

Sequence of solidification within a tilted mould
A casting is usually composed of sections of different thickness. For example, in the case of the valve body illustrated in Figure 4.5a, the flanges have the thickest sections and will therefore take longest to solidify. The main part of the valve body is next in thickness and the rings forming the valve seats are thinnest. The casting method must therefore be designed in such a way that solidification can proceed progressively from thinner to thicker sections, within an overall pattern of solidification for the whole casting.

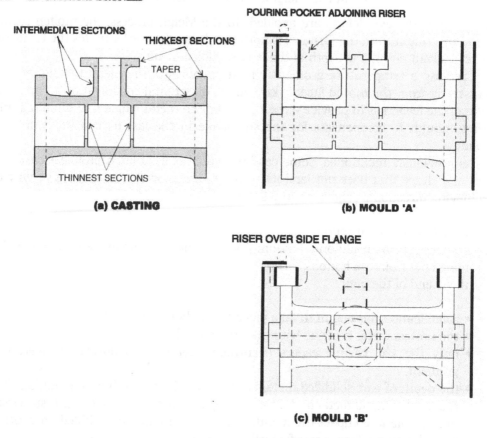

Fig. 4.5 A valve body casting illustrating progressively thickening sections and the cross-section of the corresponding sand moulds.[131]

This relatively simple casting illustrates the kind of choice which a 'methoding' technician has to make. There are two possible approaches shown on Figures 4.5b and 4.5c. Method 'B', shown on Figure 4.5c, represents the simplest and cheapest way of splitting the pattern and the simplest way of making the mould. For an ordinary commercial application, it is likely to be a successful method. The three flanges are lying in a vertical plain and are each surmounted by a riser. The pocket which receives the first metal from the launder, is located alongside the feeder which is above the left hand flange. This flange would therefore be first to fill. As the mould tilts, metal will start to flow from the flange into the main part of the valve body. Presently, it will fill the cavities forming the thin valve seats. Eventually it will reach the furthest flange – note the small web to prevent the metal cascading into that flange. The metal will then go on rising in the mould until it fills all three flanges. This mould filling sequence will ensure that thin sections will solidify before adjoining thicker sections which act as feeders. Finally, all three flanges will be last to solidify and will be fed by the feeders. Thus directional solidification is achieved throughout the casting.

In the case of a high integrity casting, however, which will be subjected to radiographic inspection, Method 'A', shown on Figure 4.5b, may be preferable. This is because, in Method 'B', the core of the side branch creates a barrier to the flow of metal on that side of the main cylinder. The metal has to flow beneath the core and rise again on the other side. This may result in too much heat going into the side core and adversely affecting directional solidification in that area of the casting. Method 'A' overcomes this problem, although it suffers from the bottom filling feature of the top branch (as cast). This can be remedied by introducing a temporary web (not shown) between the main flange to the left and the top branch.

In practice, casting shapes are often more complex and a number of conflicting factors have to be taken into consideration in order to establish the most favourable mould filling sequence.

Computer programs have been produced which predict the solidification sequence of the metal in a mould and therefore the areas where shrinkage defects would occur. The 'methoding' is accordingly modified to overcome the defects. Available programs are based on bottom pouring and therefore have to be correctly interpreted when applied to a tilting process. But they are nevertheless a very valuable tool for the methoding technician.

Feeders

Feeders need to be large enough to hold a supply of molten metal for the time required for the adjoining part of the casting to solidify. There is nothing gained in having tall feeders since atmospheric pressure is adequate to prevent shrinkage defects under conditions of directional solidification. The important thing is to keep the metal liquid in the feeders by the use of insulating sleeves, the application of exothermic powder and of vermiculite insulation. In the case of large castings that take a long time to solidify, fresh metal from the furnace is ladled into the risers at intervals of time to keep the metal in the risers liquid. The insulating vermiculite plays the vital role of preventing a solid cake of air-cooled metal forming on the top of the feeder which would reduce the effect of atmospheric pressure.

Casting features which may lead to shrinkage defects

Anything which may cause premature freezing ahead of the solidification front, will trap a pocket of molten metal behind it. When this metal eventually freezes, it will shrink and, since that area is cut off from a supply of molten metal, shrinkage cavities will form on freezing. This may happen in any of the following ways:

- *local thinning* of the section (see Fig. 4.6a);
- *an isolated mass* of localised heavier section (see Fig. 4.6b)
- *variations in the heat absorbing and conducting* characteristics of the mould material; the effect of this would be as shown in Figure 4.6c;
- a *hot spot* which is a point in the mould where the sand has been saturated with heat and is delaying the solidification of the adjoining metal. This usually

happens where a tongue or narrow promontory of sand is surrounded by metal on several sides. It can also occur where a thin wall of sand is sandwiched between two walls of metal or where a core is too small in relationship to the mass of metal surrounding it. The effect of a hot spot is shown in Figure 4.6d.

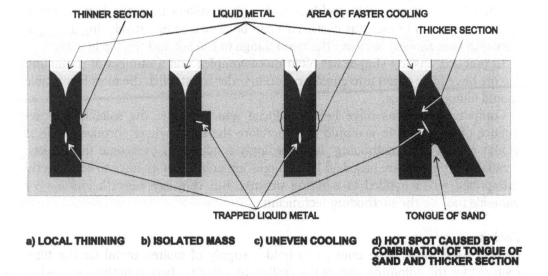

a) LOCAL THININING   b) ISOLATED MASS   c) UNEVEN COOLING   d) HOT SPOT CAUSED BY COMBINATION OF TONGUE OF SAND AND THICKER SECTION

Fig. 4.6   Casting features which may lead to shrinkage defects.[131]

Two or more of these effects may occur together in a given point in the mould. For example, at the inclined wall junction illustrated in Figure 4.6d, there is both a local thickening or isolated mass within the junction and a hot spot in the mould within the steep angle of the junction.

Some of these features may sometimes be avoided by better casting design, as we shall see later. If this is not possible, the following steps can be taken to minimise or eliminate the occurrence of shrinkage defects:

- *Local thinning*
  Local thinning may be due to pattern error or to local machining allowance or other design features. The former underlines the importance of careful dimensional checking of patterns; the latter can only be resolved by design modification, as will be explained later.
- *Isolated mass*
  Isolated masses are most conveniently dealt with by means of a metal 'chill'. If the isolated mass is a boss which is to be bored out by machining, an internal metal chill, which will leave a machining allowance, is more effective than a face chill. See below the effect of condensation on chills.

- *Variations in the heat absorbing and conducting characteristics of the mould material*
  This effect is inevitable in the case of sand mould, since sand and its binding agent are not a perfectly homogeneous material. It is most pronounced on thin wall sections. To ensure 100% soundness, it is advisable to introduce a slight taper (approximately 1/100) in the wall thickness in the direction in which solidification is intended to occur. This may require design approval. Variations in the heat absorbing and conducting characteristic of the mould material are liable to be accentuated if a mould dressing has been used. The use of mould dressing is therefore not advisable in the making of aluminium bronze castings. Furthermore, experience has shown that it is not necessary.
- *Hot spot*
  The effect of a hot spot can sometimes be overcome by the local use of chilling sand. A more satisfactory solution is a combination of good design practice (see below) and the use of chilling sand.

Directional solidification by a static process

As mentioned above, there are castings, such as fixed-pitch propeller castings, that do not lend themselves to a tilting process. Avoiding inclusions and shrinkage defects has to be achieved in a different way. Such a process, first developed by Charles Meigh, is illustrated schematically in Figure 4.7. Although similar in some respect to a bottom pouring process, it includes an important feature that creates the condition for directional solidification. This process consists of the following:

- *A pouring basin* into which the metal is poured by a ladle or, preferably, via a launder connected to a furnace.
- *An inclined sprue* of rectangular cross-section to prevent a vortex forming. The cross-sectional area of the sprue determines the rate of flow. The metal is poured into the basin at a rate that will ensure that, once the sprue is full, it remains so until the end of the pouring operation. This ensures that, once the sprue has been filled with metal, there is no further contact of the metal with air and therefore no more oxide formation.
- *A ceramic filter* at the bottom of the sprue which traps the oxide formed during the initial filling of the sprue and therefore protects the casting from inclusions.
- *A horizontal runner* to convey the metal to:
- *A vertical cylindrical shaped runner* connected to the mould cavity (e.g. a propeller hub) by *a thin continuous gate.* The vertical runner must have a generous cross-section in relation to the thickness of the continuous gate. There may be more than one set of vertical runner and continuous gate, depending on the size of the casting.

After passing through the filter, the metal rises up the vertical runner and through the continuous gate into the mould cavity. As the molten metal rises in the

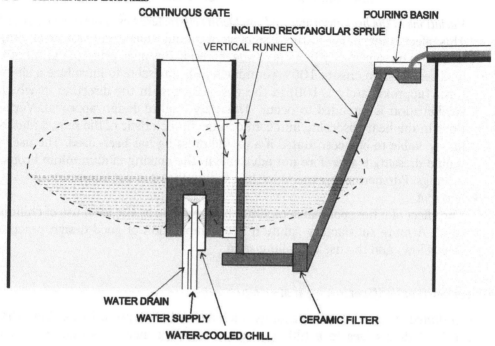

CONTINUOUS GATE

POURING BASIN

INCLINED RECTANGULAR SPRUE

VERTICAL RUNNER

WATER DRAIN

WATER SUPPLY

CERAMIC FILTER

WATER-COOLED CHILL

Fig. 4.7   A static method of obtaining directional solidification.

mould, it progressively solidifies in the continuous gate, thereby ensuring that the mould is filled from the top via the vertical runner. This creates the desired conditions for directional solidification.

One way, developed by Walter Meigh, of accentuating the temperature gradient in a propeller hub and further favouring directional solidification is to introduce a water-cooled chill in the bottom of the mould cavity. This consists of a pre-cast cylindrical receptacle, having an outside diameter less than the machined bore diameter of the hub. It is mounted as shown in Figure 4.7. Jets of water spray the inside face of the chill throughout the pouring operation and until the casting has completely solidified. The size of the chill must be such as not to prevent the molten metal in the hub acting as a feeder for the blades. Another advantage of such a water-cooled chill is that, by shortening the time of solidification, it results in a smaller grain casting with improved mechanical properties.

It has be said, however, that, if a very large propeller is bottom poured in the traditional way, directional solidification may nevertheless be achieved. This is because, due to its size, the metal in the hub remains liquid for a very long time and a favourable temperature gradient eventually forms from the bottom of the mould to the top allowing directional solidification to occur.

NB: The British Ministry of Defence issued in 1989 a Naval Engineering Standard (NES 747 Part 5) which gives guidance on making nickel–aluminium bronze castings and ingots both by the tilting method, recommended in this book, and by the traditional static bottom-pouring method, advised against in this book (see 'Unsuitability of bottom pouring' above).

Avoiding gas porosity

Electric melting of aluminium bronze very considerably reduces the danger of hydrogen absorption when compared with melting in oil or gas-fired furnaces. Thorough degassing with nitrogen is nevertheless very effective but needs to be checked with a gas tester. The gas tester works on the principle that the solubility of aluminium bronze for hydrogen falls significantly as pressure is reduced. Hence, if a molten sample of aluminium bronze, containing gas in solution, is placed in a near-vacuum, tiny bubbles of gas will be seen to escape (a glass window in the lid of the testing chamber makes monitoring of the test possible). In an extreme case of gas absorption, a 'mushroom' will form on the sample as the gas creates a large bubble inside the sample.

When the molten sample of aluminium bronze is placed in the tester, it is important that the pressure be reduced as rapidly as possible. One way of doing this is to have a valve-controlled connection between the testing chamber and a larger cylinder in which a near-vacuum has previously been created by a vacuum pump. As soon as the molten sample is placed in the chamber and the lid replaced, the connecting valve is opened, thus reducing the pressure very rapidly.

De-gassing should be the last operation immediately before pouring a large castings with large section thicknesses.

Blowing

Any form of condensation in a mould can cause blowing problems. As mentioned above, this is most likely to occur on chills which blow in contact with the molten metal, resulting in a mixture of gas porosity and slag. Dampness in other parts of a mould can have the same consequences. It is therefore important to make sure that a mould that has stood overnight in a damp atmosphere is dried before pouring. Blowing is also liable to occur when a core is made in two parts glued together. For this reason, it is advisable to make cores in one piece.

Differential contraction and distortion

At the time of solidification of a casting, there will inevitably be differences of temperatures between its various parts, due partly to differences of thickness and partly to the temperature gradient from the bottom to the top of the casting as cast. The parts of the casting at the higher temperature will want to contract more than the cooler parts. This differential contraction effect gives rise to internal stresses which are normally relieved if the casting is allowed to cool in its mould for an adequate period of time. In the case of a long and slender casting, however, this differential contraction can lead to distortion, particularly if the shape of the casting is asymmetrical. This can sometimes be overcome by introducing a slight opposite bend in the pattern equipment.

Another effect which may contribute to distortion is the differences in contraction, discussed below in the section on patterns.

It must be said however that cases of distortion are relatively rare since most castings are compact in shape.

Quality control, testing and inspection

Importance of quality control

The quality of a casting is crucial to its good performance in service. A defective casting will have some or all of its properties considerably reduced. It is therefore essential for the purchaser to specify his requirement at the outset and for the founder to have established and to observe agreed quality control procedures. It is also important for the purchaser to have an understanding of the problems facing the founder and to consider design modifications that will increase the likelihood of castings being defect-free.

Methoding records

One of the characteristics of aluminium bronze is that, if shrinkage defects have occurred in a casting, repeating the same method will result in the same defects. Similarly, a satisfactory method will give consistent good results provided nothing is changed. The method of producing a casting should therefore be carefully recorded for future reference, together with the corresponding inspection test results (photographs of DP tests, radiography, etc.). This also means that, when the method of making a particular casting is being perfected, modifications to the method can be done in the light of an analysis of previous stages in the development. This is particularly valuable in the case of high integrity castings required for naval service. In this connection, the use of computer-aided methoding can save much expensive and time-consuming trial and error.

Pre-cast quality control

Dimensional check of the pattern
Any new pattern should be carefully checked dimensionally. Subsequently, a check for damage or missing loose pieces should be made prior to issue to the foundry.

Analysis of composition
All good foundries analyse their melts prior to pouring to ensure that the composition conforms to specification. It is not sufficient merely to use previously analysed

ingots, since composition changes can occur on remelting, especially in the case of aluminium.

Gas vacuum test
This test has been explained above and is an essential quality assurance procedure.

Quality checks on castings

The type and extent of the quality checks required on a casting depends on two main considerations:

- The consequences of failure in service. This applies, for example, to castings used in submarine applications where failure can lead to considerable loss of life and of very costly equipment. Failure can also have grave operational consequences.
- The cost and consequences of scrapping a casting after machining.

In the former case, no expense should be spared to ensure that the casting is sound. In the latter case, it is advisable to have a testing procedure for the first casting to be produced and, possibly, a less costly procedure for subsequent castings. Much will depend on the size and complexity of the casting and on the proved reliability of the method of manufacture.

Dimensional check and visual inspection
It is particularly important to carry out a full dimensional check of the first casting to be made from a pattern. Thereafter checks for misplaced or distorted casting features, over-fettling etc. should be made. Some castings defects may also be visible with the naked eye, such as evidence of shrinkage defects at 'hot spots' or of dross on the surface of the casting.

Ultrasonic testing
Because of the variations in grain sizes, ultrasonic testing is not satisfactory for detecting defects in sand castings. It is however a useful tool for measuring wall thickness in sand casting (e.g. ensuring that there has been no movement in sand cores).

Ultrasonic testing is moreover a convenient technique for checking the quality of die castings, centrifugal castings and continuous cast products.

Dye-penetrant testing
This is a very effective method of detecting the kind of defects that may lead to a casting leaking on pressure test. Since it is not always practicable to pressure test a casting in the as-cast condition, dye-penetrant testing is normally a sufficient indicator that a casting will not leak when pressure tested by the purchaser after

machining. Experience is required to interpret correctly the results of a dye-penetrant test. Dye penetrant testing will sometimes reveal defects which have not been detected by radiography. It is therefore an important complementary test.

Dye penetrant testing should be a minimum test requirement once a satisfactory method of producing a casting has been established and found to be reliable.

Radiography
Radiographic inspection is a requirement for high integrity castings for naval applications. Although expensive to use, it is the most effective method of detecting castings defects. Because of the inherently compact nature of aluminium bronzes, the smallest cavity and inclusions show up clearly on films, provided correct radiographic techniques are used. Since it provides a record of the nature and extent of the defects, it is an invaluable tool for the experienced founder in developing his techniques on a given casting. In-house radiography makes it possible to take check shots on potential trouble areas of any castings. The cost involved is small in comparison to the cost to the foundry and to its customer of a rejected casting.

Proof machining
Proof machining is a necessary prelude to pressure testing and may uncover defects which can then be detected either visually or by dye penetrant testing.

Though generally less expensive than radiography, it can be a costly method of checking the soundness of a casting. It may however be justified in the light of the cost of scrapping a fully machined casting. It is nevertheless less effective than radiography.

One advantage of proof machining is that weld rectification (see Chapter 7) may be possible at that stage whereas it may be out of the question after final machining, due to distortion.

Pressure testing
Pressure testing of a proof machined casting is not an absolute guarantee that the casting will not fail after final machining. Used, however, in conjunction with dye penetrant testing, it provides a high degree of assurance of the pressure tightness of the finished casting.

Design of patterns

Pattern made to suit production method
A casting can only be as good as the pattern from which it is made. The method of producing a casting begins with the design of the pattern. It must be made to suit the best production techniques and thus avoid the need for costly modification or replacement at a later stage. It is vitally important therefore that, if the pattern is to be made by the purchaser of the casting, the foundry should be consulted

before patterns are made. Furthermore the methoding will involve the addition of certain features to the pattern equipment which may not be part of the finished casting.

In view of the high cost of pattern equipment, it is essential that it be made right first time. Pattern precision is vital, not only to achieve dimensional accuracy, but also because the directional solidification of the casting is critically dependant on section thicknesses being as planned.

The first decision to be made is which way up a pattern is to be in the mould. The cheapest and most convenient way of making a given pattern may be totally unsuited to the production of a sound casting. A typical example is the pump casing shown in Figure 4.9. The easiest way of making the pattern for this type of casting is to split it along the axis of symmetry, but this means that the potentially trouble-some joint flange, facing the camera in Figure 4.9, would lie vertically in the mould and it would be almost impossible to produce it and the adjoining parts of the casting free of shrinkage defects.

Fig. 4.8 A 900mm diameter propeller for a fast naval patrol boat and four 'butterfly valve' blades – all in nickel-aluminium bronze (Meighs Ltd).

Fig. 4.9   Centrifugal pump casing in nickel-aluminium bronze.[131]

Contraction allowance
Special attention needs to be applied to the correct choice of (linear) contraction allowances. Three factors will singly and jointly cause differences in contraction. They are:

- The production method used, because it affects the way the various parts of the mould are filled with metal and the speed of solidification.
- The restraint applied by the sand mould on the contraction of certain parts of the casting. This may be due particularly to sand cores but also to such features as flanges at opposite ends of a casting.
- The tendency of thin sections to contract less than thick sections. This is because thin sections solidify rapidly and some contraction occurs as the metal fills the section.

The larger the casting the greater the need to assess the likely effect of these factors on its contraction. Figure 4.10 gives linear contraction allowances for

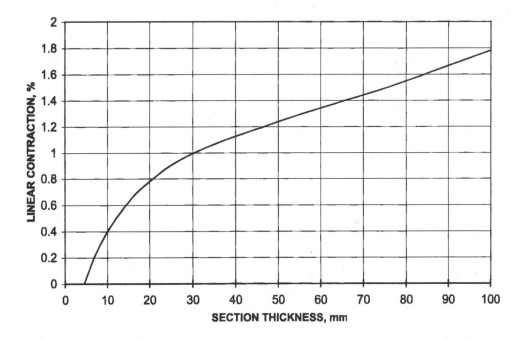

Fig. 4.10 Variations of percentage linear contraction with section thickness.[131]

different casting thicknesses which do not however take account of the possible restraining action of the mould material. Only experience can provide a guide of the likely contraction of a given casting. This is therefore another reason for close liaison with the foundry.

B – Design of castings

Introduction

Close co-operation between the designer and the founder is essential since the designer can not be expected to have acquired an intimate familiarity with every aspect of foundry technology. It is however desirable for designers to be aware of the basic principles involved and these have been explained in the first part of this chapter.

Of the three types of defects that are liable to occur in aluminium bronze castings, oxide inclusions and gas porosity are mainly dependent on good foundry practice. Shrinkage defects, on the other hand, are a factor of the combination of design features and of production methods.

Designing to avoid shrinkage defects

Applying design guidelines to avoid shrinkage defects need not over-restrict freedom of design since the founder has a choice of techniques to solve solidification problems. But it is clearly desirable to design so as to minimise problems as far as possible. It must also be recognised that there are occasions when there is no way, other than by a change of design, of preventing a shrinkage defect occurring. Limitations imposed by the laws of nature can sometimes be side-tracked but never ignored.

Simplicity of shapes

A casting should have as simple a shape as its function will allow and should not be designed as a 'cast fabrication' with numerous reinforcing webs to provide strength and rigidity. The resultant multiplicity of wall junctions are likely to create hot spots and consequent shrinkage defects. It is the main body of the casting which should be designed to have the necessary strength and rigidity. If weight considerations makes this unacceptable, the alternative of a weld fabrication of wrought and, possibly, some cast parts should be considered.

Occasionally it may be advantageous to attach troublesome features, such as mounting brackets, to the main body by welding. This may have the added advantage that a given basic design of, say, a pump or valve, could be used in a variety of applications requiring different installation arrangements. A further advantage is the consequent saving in pattern outlay and storage space requirement.

Taper

As previously explained, it is desirable for wall thicknesses to be slightly tapered in the direction in which solidification is planned to take place. This can often be achieved by merely tapering the machining allowances and thus leaving the basic design unaffected.

Relationship of thin to thick sections

Where the wall thickness of the casting changes, care must be taken to ensure that solidification can proceed progressively from thinner to thicker sections, within an overall pattern of solidification for the whole casting. It is desirable to design the casting with this in mind. Thus, in the case of the flanged valve casting previously shown on Figure 4.5a, there is a gradual transition from thin to thick sections. The thicker flanges, which will be surmounted by feeder heads, can then act as 'feeders' to the body. It should be noted that the practice of tapering the wall of the casting towards the flanges must not result in excessive thickness at the root of the flanges where it could give rise to shrinkage defects.

The more complex the shape of the casting, the greater the need for consultation with the experienced aluminium bronze founder, if a pattern of directional solidification is to be achieved throughout the casting.

Wall junctions and fillet radii

Wall junctions fall into five categories of shapes: **L**, **T**, **V**, **X** and **Y** or variants of these. In each case, an increase in mass results at the centre of the junction, as is illustrated by the inscribed circle method (see Fig. 4.11). It will be seen that the biggest increase in mass occurs with **V**, **X** and **Y** junctions. If the walls are of unequal thickness, as in Figure 4.11b, the resultant increase in mass at the centre of the junction is smaller. Furthermore, as explained earlier, unequal wall junctions can be advantageous in an overall pattern of directional solidification.

Wall junctions may also cause some parts of the mould to form a promontory with molten metal on two sides. Thus, in the case of **V**, **X** and **Y** junctions, tongues of sand are formed which are liable to give rise to hot spots.

For all these reasons it is preferable to avoid **V**, **X** and **Y** junctions. Wherever possible an **L** junction should be converted into a curved wall of constant thickness or gently tapering from one thickness to the other (see Fig. 4.12).

The fillet radii of junctions should be large enough to prevent the creation of hot spots, but not so large as to increase unduly the mass at the junctions. As a general guide, fillet radii should be equal to half the wall thickness of the thinner wall, in

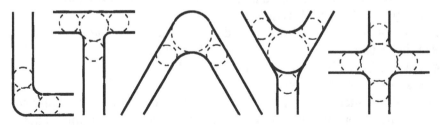

(a) CIRCLES SHOW INCREASE OF MASS AT JUNCTIONS OF WALLS OF EQUAL THICKNESS

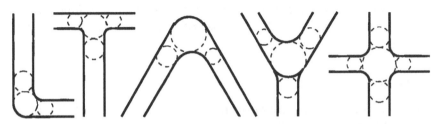

(b) CIRCLES SHOW SMALLER INCREASE OF MASS AT JUNCTIONS OF WALLS OF UNEQUAL THICKNESSES

Fig. 4.11 Wall junctions.[131]

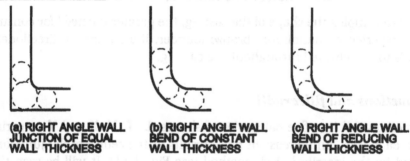

(a) RIGHT ANGLE WALL JUNCTION OF EQUAL WALL THICKNESS

(b) RIGHT ANGLE WALL BEND OF CONSTANT WALL THICKNESS

(c) RIGHT ANGLE WALL BEND OF REDUCING WALL THICKNESS

Fig. 4.12   Replacing sharp L junction by curved wall.[131]

the case of a junction of unequal wall thicknesses. When in doubt, the founder should be consulted.

The founder will need to ensure that the thicker mass, created by a junction, can solidify directionally and may therefore request a tapering of that mass towards the feeder.

Isolated masses

The size, shape and location of isolated masses such as bosses may be critical. If they can be located in such a way that they can conveniently be connected to a feeder, so much the better. Otherwise they must be of a size that can be effectively chilled. This is normally done by a piece of metal inserted in the mould at the point where a faster cooling rate is required.

Care must be taken in locating isolated masses, such as bosses, to ensure that the effects of the isolated mass is not aggravated by tongues of sand which will give rise to hot spots. This can happen, for example, if a boss is located too close to a flange unless it is merged with the flange.

Some castings, such as pump casings, may have to incorporate certain shapes in order to ensure a non-turbulent flow of the fluid through the casings. These shapes may give rise to isolated thicker sections which may be liable to shrinkage defects.

Another typical case of an isolated mass is the spindle guide of a valve, shown in Figure 4.13. The size of the central boss and the thickness of the supporting web, in relation to the wall thickness of the body, is critical. In all such cases, consultation with the founder is strongly recommended.

Webs and ribs

As previously mentioned, the use of strengthening webs and stiffening ribs should, as far as possible, be avoided. They may sometimes be advantageously replaced by curved wall sections of uniform thickness, as shown in Figure 4.14. However, the relatively low elastic modulus of aluminium bronze (see Chapter 2) makes the use of webs and ribs occasionally essential to achieve rigidity in certain applications.

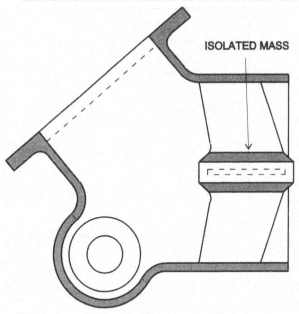

Fig. 4.13   Example of an isolated mass: the spindle guide of a valve.[131]

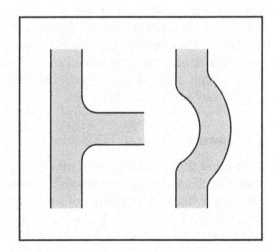

Fig. 4.14   Replacing a rib by a curved uniform section.[131]

If it is necessary to incorporate ribs and webs in the design, they should be thinner than the parts to which they are connected and be normal to them in order to avoid sharp tongues of sand in the mould. A hot spot can be avoided by means of a cut-away as shown in Figure 4.15. Alternatively, consideration might be given to welding on the ribs and webs.

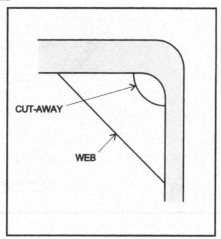

Fig. 4.15 Cut away in a web to avoid a hot spot.[131]

Cored holes

Since cores are almost completely surrounded by metal, care must be taken to ensure that they do not create hot spots. This is liable to happen if the thickness or diameter of a core is too small in relationship to the thickness of the metal surrounding it. In such a case, it may be necessary to machine out the core if at all possible. Consulting the founder may be necessary in such cases. There are, in any case, other aspects of core design, such as core supports and adjustment to shape to favour solidification which may have to be discussed.

Effect of machining allowance

It is important to bear in mind that the addition of machining allowances may, in some cases, have an adverse effect on directional solidification. Figure 4.16a illustrates the case of a casting where the addition of a machining allowance has resulted in undesirable changes in cast wall thickness. This can be remedied by thickening the non-machined parts to produce a constant thickness (Fig. 4.16b).

Other design considerations

Fluidity and minimum wall thickness

The presence of its surface film of aluminium oxide partly restricts the fluidity of molten aluminium bronze. If the molten metal momentarily ceases to flow in any part of a mould during the pouring operation, the oxide film may prevent it from resuming its flow. The metal will tend to flow around or over the affected part but

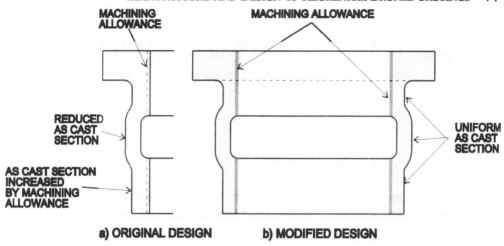

Fig. 4.16   Design modified for uniform section as cast.[131]

may not merge with it because of the presence of the oxide film. This will leave a crack-like defect in the casting, known as a 'cold shut'.

The minimum wall thickness that can be cast without the risk of a cold-shut is dependant on the pouring temperature and on the distance that the metal has to flow from its point of entry into the mould. It also depends on the size and length of runners bringing the metal to a particular area of the casting and to the wall thickness of the parts of the casting through which the metal has had to flow to reach the part of the casting under consideration.

The minimum allowable wall thickness is therefore a function of the size and complexity of the casting, of the running method used and of casting temperature. Table 4.1 gives recommended minimum wall thicknesses for cylindrical shaped castings of various diameters and lengths. This may serve as a guide for other

Table 4.1   Minimum castable wall thicknesses (mm) for sand cast cylinders.[131]

DIAMETER mm	LENGTH mm							
	80	150	300	600	1200	1800	2400	3000
80	6	8	10	12	–	–	–	–
150	8	10	10	12	14	–	–	–
300	10	10	12	14	16	16	18	22
600	14	14	14	14	16	18	20	22
1200	14	14	16	16	18	18	20	24
1800	14	16	16	18	20	20	22	24
2400	–	20	20	20	20	22	24	25
3000	–	20	20	22	22	24	24	25

shapes of castings. It must, however, be borne in mind that, in more complex castings, the metal may have to flow further to fill internal features, such as vanes and partition walls, and the casting may, therefore, have to be generally thicker to allow for this. This is an area where the designer needs to acquire experience of what is practicable through consultation with an experienced founder.

Weight saving

Advantage should be taken of the strength properties of aluminium bronze by reducing section thicknesses in order to save weight. One must however bear in mind the limitations imposed by the fluidity of the alloy and the need to achieve directional solidification throughout the casting. This may mean that weight has to be added to certain parts of the casting and kept to a minimum elsewhere.

Effects of thickness on strength

The tensile properties given in specifications relate to a standard test bar of 25 mm diameter and is a function of its speed of solidification and subsequent cooling rate. Parts of a casting which solidify and cool faster than the standard test bar will have improved tensile properties, whereas parts which take longer to freeze and cool will have lower tensile properties. Table 4.2 illustrates this point in the case of a standard nickel-aluminium bronze sand casting.

The designer therefore needs to bear in mind in his calculations the effect of cooling rate on tensile strength.

Table 4.2   Effect of section thickness on mechanical properties.[131]

Wall Thickness mm	Tensile Strength N mm^{-2}	Elongation %
5	708	29
8	662	26
9,5	646	24
19	631	21
38	585	18.5
76	569	18
152	538	18

Hot tears

As will be explained in Chapter 7, aluminium bronze, in common with some other alloys, experiences a drop in ductility as it cools from high temperature. Nevertheless, the relationship of the strength and ductility of the alloy to the crushing resistance of the sand, as occurs in the case of a sand core, is such that aluminium bronze, unlike cast steel alloys, shows little tendency to hot tears, provided the casting is sound. In practice, this means that, if the wall thickness of a casting is insufficient to give it strength to crush the sand, it will normally stretch without tearing, provided there are no defects in the casting which could lead to cracking under stress. This applies whether the sand which is restraining contraction is a core or a part of the mould. The heavier the section, the greater its strength and therefore the greater its ability to crush the sand and undergo its full contraction. The thinner the section on the other hand, the faster its rate of cooling with the result that solidification occurs so rapidly that shrinkage is partly compensated for during the pouring process. This means that there is less overall shrinkage and crushing of the sand after the mould has been filled.

The relatively slow pouring speed associated with the Meigh tilting process, combined with the directional solidification which it produces, have a significant effect in reducing the risk of hot tears.

Because aluminium bronze sand castings are not normally prone to hot tears or contraction cracking, fillet radii do not need to be as large as in the case of steel castings, where large radii are recommended to avoid stress concentrations. Nor is it so necessary to taper the junction of the body of a casting with a flange, as is recommended for steel castings to avoid hot tears, although a slight taper is nevertheless beneficial.

Composite castings

In spite of the above mentioned drop in ductility of aluminium bronze at high temperature, it is possible to cast aluminium bronze around an other metallic component, to form a composite casting. An example of this is given in Figure 4.17, where a nickel-aluminium bronze rotor has been cast on to a steel shaft. The rotor, being chill-cast, has enhanced strength, wear and corrosion resisting properties and the combination benefits from the higher strength of the wrought steel. In the latter stages of cooling, the aluminium bronze rotor increases in tensile strength and exercises a powerful grip on the shaft. This principle can be applied to a variety of applications in order to take advantage of the respective properties of two metals: aluminium bronze providing resistance to corrosion or non-sparking properties and steel giving strength and saving cost. This principle can also be used instead of the conventional shrinking-on process.

Design of castings for processes other than sand casting

The preceding notes and design recommendations apply mostly to sand castings, although the general principles outlined are valid whatever the mould material. The better the heat conductivity of the mould material, as in die casting, the more favourable the conditions for directional solidification and also the more predictable the conditions that lead to shrinkage defects. The greater strength of metal and ceramic moulds however imposes severe constraint on contraction and may lead to

ALUMINIUM BRONZE CAST ROTOR   STEEL SHAFT

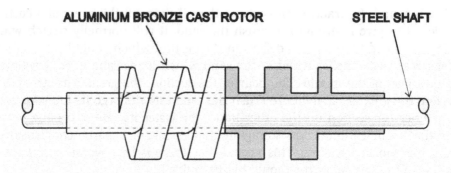

Fig. 4.17   Half-sectioned drawing of a composite screw pump rotor.[131]

hot tears, particularly in die casting. The low-nickel aluminium bronze is less prone to this problem and is therefore the popular die casting alloy.

Each process therefore imposes its own constraints on casting design and consultation with a founder who specialises in the chosen process is essential.

5
WROUGHT ALUMINIUM BRONZES

Wrought processes and products

Due to their excellent ductility, aluminium bronzes are among the few alloys that can be successfully wrought as well as cast. Before considering the available wrought alloys it is advantageous to consider the working processes to which wrought aluminium bronzes can be subjected and the resultant products since processes and products have a bearing on the choice of alloys.

Wrought aluminium bronze components begin life as a solid cast ingot, billet or slab which is then normally hot worked into a desired shape and section. This may subsequently be cold worked or machined to final dimensions. Wrought products include sheets, strips, plates, rods, bars, tubes, rings, wire, various sections and forgings.

The following processes are used in producing a variety of wrought aluminium bronze products:

- Forging
- Rolling
- Extruding
- Drawing
- Miscellaneous processes using sheet metal as raw material (bending, stamping, coining, pressing, deep drawing, spinning etc)

Some products can only be manufactured by one of these processes but, in many cases, a choice of manufacturing route is possible involving either forging, rolling or extrusion or a combination of these processes. The choice is governed mainly by cost and by the availability of suitable plant and equipment. In some cases there is also an alternative casting route, as was seen in Chapter 3, but the wrought process is able to offer significantly better mechanical properties than the casting process.

Many forgemasters, rolling mill owners and extruders, who normally work in high strength materials, have acquired the know-how to work aluminium bronzes. This means that there is potentially a great variety of shapes and sizes of wrought products that can be manufactured in aluminium bronzes.

The choice of wrought alloys, their properties and hot-working temperatures will be discussed later.

Forging

Forging is the most flexible method of hot forming metal: it permits a wide variety of shapes and sizes of components to be manufactured ranging from very small drop forgings, produced in large quantities, to very large hand forged components weighing several tonnes. It is the only wrought process that can be used for manufacturing components of non-uniform cross-section. It is also the only way to produce bars of uniform round, square, rectangular or hexagonal cross section which are too large to be rolled (typically over 200 mm width or dia). Whereas rolling is dependent on the availability of rollers of the desired size, forging is much more versatile in the sizes it can produce.

In hammer or 'open' forging, the hot billet is worked into the desired shape whereas in hot pressing or 'closed die forging' a hydraulic driven ram squeezes the billet into shape. Both processes progressively work the billet into a required shape. Open forging has the effect of removing all traces of cast structure from billet sections much more thoroughly than is done in the closed die forging. The shaping process in hammer forging is either controlled visually by a skilled operator (hand forging), or by the dimensions of a two-part die (drop forging) – see Figure 5.1. Hammer forging, however, cannot be controlled to the same degree as modern press forging in which the rate and amount of reduction can be pre-set for a given size and weight of billet. The settings are determined by the strain rate which the particular alloy can tolerate. There are however skilled hammer forging operators who have demonstrated that a five tonne hammer can be controlled to crack and egg without breaking it!

For uniform sections, a method of forging, known as GFN forging, is particularly efficient. It consists of four hydraulically powered hammers centred in a cruciform assembly mounted vertically. The heated billet (which is held horizontally) is passed through this hammer assembly and is worked into shape as it is fed backwards and forwards. Long lengths of uniform sections are produced in this way. This is done so accurately, that only a single machining operation, known as 'peeling', is needed to finish the product within a tolerance of 0 to +0.1 mm. The

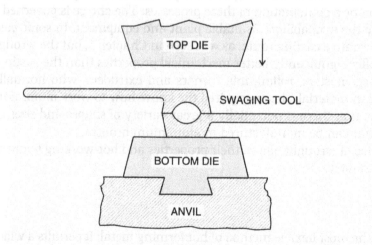

Fig. 5.1 The forging process.[92]

same process may also be used to reduce a billet to a size that can be rolled or extruded for the manufacture of smaller sections. One of the big advantages of GFN forging is that, since the work-piece is 'hammered' in opposite directions simultaneously, all the energy goes into the work-piece instead of some being absorbed by the anvil, as occurs with other forms of forging. This helps to maintain the work-piece at the hot-working temperature. The following are sizes of uniform sections made by GFN forging; other sizes may be produced if no suitable rolling or extrusion facilities are available:

Round bars up to 480 mm dia.

Square bars up to 400 mm across.

Flat bars in a wide range of combinations of thickness and width within a maximum width (400 mm typically).

Hollow sleeve type forgings of selected inner and outer diameters (with or without inside or outside flanged ends) can be produced in lengths of up to 10 metres.

Manufacturers usually state a maximum weight of forging which is governed by their melting capacity. Within this limiting weight, other cross-sections can be forged. Solid squares, rectangles, rounds and hexagons of up to 25 tonnes weight can be produced as open die forgings. Forged components are usually offered either 'black' (as forged) or 'bright' (peeled, bar turned or ground).

Small circular tube plates for heat exchangers, may be more economically manufactured by forging than by cutting out from plate. The diameter can be worked into shape within limits that require only a single machining run to bring it within tolerance.

In addition to tube plates, blocks for valve manifolds and for other purposes, rings, discs, hollow bars, stub shafts, stepped pump shafts, pipe flanges, etc. are hand-forged and a great variety of shapes are produced by drop forging. One special application of drop forging is known as 'heading' or 'up-set forging' and consists in heating one end of a bar which is then formed by a two-part die to produce a bolt or rivet head. This generates a much stronger head than one produced by machining the bolt or rivet from bar.

Extruding

Figure 5.2a shows diagrammatically the conventional extrusion process. The hydraulic ram to the right forces the heated cylindrical shaped billet through a die of the desired cross section to the left (a process similar in principle to making icing for a cake). The size of dies and billets, which the extrusion machine can accommodate, determines the cross-section and length respectively of the extrusion. Extruding is mostly used to manufacture small sizes of round, square, rectangular and hexagonal sections and is the only way to produce irregular cross-sections that cannot be rolled.

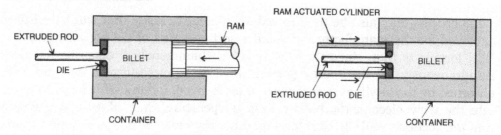

Fig. 5.2   The extrusion process.[92]

The following is the range of sections that are typically extruded but larger sizes could be produced if a larger extrusion plant is available:

- Round rods and bars of 9.5 to 76 mm dia.
- Hexagon rods and bars of 8.25 to 85 mm across flats.
- Square rods and bars of 12.75 to 63.5 mm across flats.
- Rectangular flats of about 3 mm minimum thickness and 120 mm maximum width depending on availability of dies.
- Shaped sections. These require special dies and the design of shaped sections needs to be discussed with the manufacturer to establish what can be achieved. Extruders who have successfully produced **I**, **T**, **L**, **U** and other sections in high strength materials such as stainless steels and titanium are likely to be able to produce these same sections in aluminium bronzes. Sections of up to 300 mm across are possible.

The conventional length in which extruded sections are sold is 3 metres. It is possible to produce longer lengths to meet special requirements, subject to the limitation of billet size. Transportation must be taken into account although small diameters can be coiled.

One problem associated with extrusion is the formation of a 'cornet' of oxide within the rear section of the extruded length. This is due to the fact that, as the billet is forced through the die, the film of oxide on the outside of the billet is drawn by friction into the centre of the rear of the billet and hence into the back of the extruded section. Up to 50% of the length of the extrusion may be affected and has to be cropped off, resulting in very low yield. A method of overcoming this problem, known as indirect extrusion, is illustrated in Figure 5.2b. In this process, the cylinder to the left is powered by a hydraulic ram and forces the die against the hot metal, causing the metal to be forced backward through the die and through the short length of cylinder.

Hollow sections require the use of a mandrel to form the inside diameter of the tube. The billet used for this purpose is a solid bar or continuous cast tube which has been bored out. Extruders who have experience in extruding tubes in high strength metals are able to extrude the high strength aluminium bronzes, even though the grip of the alloy on the mandrel is considerable. Typical sizes of extruded tubes range from 25 mm to 100 mm OD with 5 mm minimum wall thickness but larger tubes of, say, 380 mm OD with 30 mm wall thickness could be extruded if a large enough extrusion plant is available. The alternatives to extruded tubes are spun or continuously cast tubes, although these alternatives are likely to have lower mechanical properties. Short lengths of hollow sections can be made by forging techniques.

It is also possible to extrude other types of hollow sections provided the quantity required justifies the cost of the special dies. The manufacturer would need to be consulted on the design of such a hollow section.

The extrusion process has many advantages, viz.: its accuracy, the variety of sections it can produce and the relatively fast transformation of a billet into the required section. It extrudes a section almost to size in one operation. The disadvantages are: a) the cost of the die which need to be frequently refurbished or replaced and b) the low yield of direct – as opposed to indirect – extrusion. Nevertheless the speed of the process, by comparison with rolling, makes it more economical for smaller sections. Being a more precise process than rolling, it is a more suitable process for the production of hexagonal sections.

Rolling

Rolling is a convenient way of producing large plates and sheets and long lengths of uniform sections. It is the only process for producing uniform sections that are too large to be extruded or too small to be economically forged. It consists in passing a bar or slab of metal between a succession of pairs of shaped or plain rolls which reduce the thickness in stages by a squeezing action as illustrated diagrammatically in Figure 5.3.

In the case of plates and sheets a thick slab is used as the initial work-piece. It normally only needs to be slightly wider than the plate or sheet to be produced since

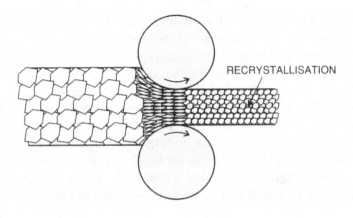

Fig. 5.3   The hot rolling process.[19]

the rolls prevent sideways expansion of the work-piece. As the thickness is reduced, the resulting extension is all lengthwise. But it is possible, and frequently good practice, to produce material significantly wider than the slab or 'cake'. The first few passes in the hot breaking down mill can be cross-rolled which has the double advantage of facilitating wide plate production and of reducing directionality (the tendency for the 'grain' of the metal to be all in one direction as in timber). Subsequent rolling is, as previously mentioned, all lengthwise. This practice may not be possible where casting and rolling facilities are integrated.

In the case of rods, bars and other sections, the billet used is cylindrical or square shaped (with rounded edges). It may be either continuously or individually cast and sometimes forged. Grooved rolls of the desired contour are used in the production of round, square, rectangular or other sections; each set of successive rolls having smaller grooves to reduce the work-piece to size progressively. In some installations, one reversible mill has rolls with different size grooves along its length and the work-piece is passed to and fro along the width of the rolls to reduce it in size.

Rolling of aluminium bronze is done by hot-working using conventional mill techniques. Working temperatures depend on the alloy as will be discussed later. The following are typical sizes of components produced by rolling but other sizes could be produced if the necessary size of rolling mill is available:

- Round Rods and Bars of 25 to 430 mm diameter, in lengths of up to 10 metres.
- Square and rectangular bars of 25 to 150 mm thickness.
- Plates and Sheets of 3 to 100 mm thickness and of up to 3.5 × 6 metres in size.

Component length is limited by many factors, including billet weight, handling, furnace size and transportation.

Hot rolling is not so accurate as extruding but, if it is followed by cold rolling, it is possible to obtain much tighter tolerances than by extrusion. Rolling is more restricted however in the shapes it can produce. The main advantage of rolling is that, although the initial cost of rolls is high, they have a much longer life than extrusion dies and the yield is almost 100%. The disadvantage is that the process requires a lot of space for the successive sets of rollers and is more labour intensive. This makes it generally less economical for smaller sections. Sections can be supplied either 'black' (as-rolled) or 'bright' (pealed, bar turned or ground).

Rings of up to 500 mm width and ranging in size from about 350 mm OD by 25 mm wall thickness to 3 metres OD by 100 mm wall thickness are commonly produced by a process known as Ring Rolling. Even larger sizes of up to 6 metres dia by 1.25 metres thick and weighing up to 25 tonnes, can be produced by this process. The process begins hot forming a disc at the plastic deformation temperature of the alloy. A hole is then punched through the centre of the disc and the resulting ring is known as a 'preform'. This preform is then re-heated to the hot working temperature and threaded on to a free turning roll on which it hangs

Fig. 5.4 Proof machined tubeplates for heat exchangers produced by hot rolling from octagonal cast slabs. Testing includes analysis, full tensile testing together with dye penetrant non destructive testing for porosity after machining. Alloy C63000 to ASME code SB171 with 1rs release to BS EN 10204. (Alfred Ellis & Sons Ltd).[92]

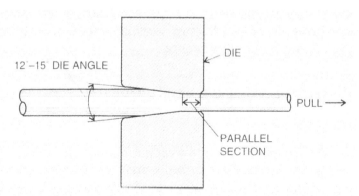

Fig. 5.5 The wire drawing process.

vertically. A driven roll exerts pressure on the face of the preform and this gradually 'pinches' the work piece increasing both its inside and outside diameters and reducing the wall thickness, until the desired dimensions are achieved. The width of the ring remains constant due to the pressure of the inner and outer rolls.

Drawing

Drawing is a cold working process which consists in pulling a previously extruded or rolled section through a die of fractionally smaller dimension. Many manufacturers cold draw their extrusions to bring them to final size, to improve surface finish and to enhance mechanical properties. Cold drawing induces stresses in the work piece which need to be relieved by subsequent heat treatment (see Chapter 6).

Cold drawing is the only way to produce sections that are too small to be extruded to size, as in the case of wire used for welding and metal spraying. Wire can be drawn down to 0.8 mm dia. See Fig 5.5. In the case of the harder aluminium bronzes, the reduction in size at each pass is very small and many cycles are necessary. The die loading tends to be high, especially in the case of the higher strength nickel-bearing alloys, and efficient lubrication is vital to minimise die wear. Frequent annealing is also necessary (see Chapter 6).

Miscellaneous processes

There is a variety of cold working processes which use sheet as raw material. Thus aluminium bronze sheet can be readily bent to form cylinders as well as **L** and **U** sections with radii equal to the thickness of the sheet material. Other cold-working operations are generally confined to the softer alloys, those containing approximately 8% or less aluminium, although a limited amount of cold work can be performed on alloys with up to 10% aluminium. Typical cold-working alloys with aluminium contents of 5%–8% are quite ductile and can be rolled, drawn, pressed, bent, and coined. They are, however, stronger than most other copper alloys, and require more frequent annealing. A rough working guide is that the workability is slightly less than that of a phosphor bronze.

Wrought alloys: properties and applications

Composition and properties

Table 5.1 gives the composition of the wrought aluminium bronze alloys to CEN standards and Tables 5.2 to 5.4 the forms and mechanical properties of these alloys. American (ASTM) wrought alloy specifications are given in Appendix 1. This information is provided only as a guide since specifications are subject to review from time to time. It is advisable therefore to consult the latest specification issue.

As will be seen from these Tables and discussed at greater length below, mechanical properties of wrought alloys are affected, not only by the composition of the alloy, but also by the hot or cold working process, the size of the product and any heat treatment applied after working. Wrought alloy specifications therefore specify the min/max mechanical properties to be achieved for specified limits of composition, for the particular form and for a given range of sizes. CEN specifications also specify the 'material condition' (sometimes referred to as the 'temper') which, in the case of aluminium bronze alloys, is one of the following:

Table 5.1 Composition of wrought aluminium bronze alloys to CEN standards.[51]

DESIGNATION				COMPOSITION Wt % (Remainder Copper)				
CEN Designation	Number	Former BS	Nearest ASTM	Al	Fe	Ni	Other specified elements	Total impurities
CuAl5As	CW300G		C 60800	4.0–6.5	–	–	As: 0.1–0.4	0.3
CuAl6Si2Fe	CW301G	CA107	–	6.0–6.4	0.5–0.7	–	Si: 2.0–2.4	0.2
CuAl7Si2	CW302G		C 64200	7.3–7.6	–	–	Si: 1.5–2.2	0.2
CuAl8Fe3	CW303G	CA106	–	6.5–8.5	1.5–3.5	–	–	0.2
CuAl9Ni3Fe2	CW304G	CA105	–	8.0–9.5	1.0–3.0	2.0–4.0	–	0.2
CuAl10Fe1	CW305G	–	C 61800	9.0–10.0	0.5–1.5	–	–	0.3
CuAl10Fe3Mn2	CW306G		–	9.0–11.0	2.0–4.0	–	Mn: 1.5–3.5	0.2
CuAl10Ni5Fe4	CW307G	CA104	C 63200	8.5–11.0	3.0–5.0	4.0–6.0	–	0.2
CuAl11Fe6Ni6	CW308G	–	–	10.5–12.5	5.0–7.0	5.0–7.0	–	0.2

- M (on its own) which means: as manufactured, without specified mechanical properties.
- R, followed by a three digit number, which is the mandatory minimum tensile strength in N mm^{-2}.
- H, followed by a three digit number, which is the mandatory minimum hardness in HB, except in the case of tubes and of plates, sheets and circles (Table 5.3) where it is the mandatory minimum hardness in HV.

American (ASTM) specifications specify the minimum properties for different 'tempers' of the finished product. The 'temper' is the degree of softness/hardness of the metal resulting from hot or cold working or heat treatment (see paragraph on Temper below and Chapter 6).

In order to explain the development of the microstructure of aluminium bronzes, from those with the simplest to those with the most complex composition, the various alloying systems are categorised in Part 2 as follows, according to the number of alloying elements that they contain:

- *Binary systems*, which contain only two elements, namely copper and aluminium,
- *Ternary systems*, which contain copper and aluminium plus one other element (iron, nickel, manganese, silicon etc.),
- *Complex systems* which contain more than three elements (copper, aluminium plus two or more other elements).

Since it is the structure of an alloy which determines its hot and cold working characteristics, wrought aluminium bronzes are also classified into the following categories, according to the number of different constituents or 'phases' of their microstructure. A 'phase' is a constituent of an alloy which has a given characteristic appearance under the microscope and has certain specific properties which affect the properties of the alloy as a whole.

Table 5.2   Mechanical properties of wrought rod, bar and profiles to CEN standards.

RODS, BARS AND PROFILES

Rod product specification: prEN 12163 (rod for general purposes)
Bar and Profiles product specification: prEN 12167 (profiles, rectangular bars)

Designation	Material† condition	Nominal dia or width across-flats		Tensile Strength R_m	0.2% Proof Strength $R_{p0.2}$	Elongation		Hardness			
Symbol (Number)		mm		N mm⁻²	N mm⁻²	%		HB		HV	
		from	up to and including	min	approx	$A_{11.3}$ % min	A % min	min	max	min	max
CuAl6Si2Fe (CW301G)	M	5	80	*	*	*	*	*	*	*	*
CuAl7Si2 (CW302G)	R500	5	80	500	(250)	18	20	–	–	–	–
	H120	5	80	–	–	–	–	120	150	125	155
	R600	5	40	600	(350)	10	12	–	–		
	H140	5	40	–	–	–	–	140	–	145	
CuAl10Fe1 (CW305G)	M	10	80	*	*	*	*	*	*	*	*
	R420	10	80	420	(210)	–	20	–	–	–	–
	H105	10	80	–	–	–	–	105	145	110	150
	R530	10	80	530	(420)	–	10	–	–	–	–
	H130	10	80	–	–	–	–	130	170	135	175
	R630	10	30	630	(480)	–	5	–	–	–	–
	H155	10	30	–	–	–	–	155	–	165	–
CuAl10Fe3Mn2 (CW306G)	M	10	80	*	*	*	*	*	*	*	*
	R590	10	80	590	(330)	–	12	–	–	–	–
	H140	10	80	–	–	–	–	140	180	145	185
	R690	10	50	690	(510)	–	6	–	–	–	–
	H170	10	50	–	–	–	–	170	–	180	–
CuAl10Ni5Fe4 (CW307G)	M	10	80	*	*	*	*	*	*	*	*
	R680	10	80	680	(480)	–	10	–	–	–	–
	H170	10	80	–				170	210	180	220
	R740	10	80	740	(530)	–	8	–	–	–	–
	H200	10	80	–	–			200	–	210	–
CuAl11Fe6Ni6 (CW308G)	M	10	80	*	*	*	*	*	*	*	*
	R750	10	80	750	(450)	–	10	–			–
	H190	10	80	–	–			190	235	200	245
	R830	10	80	830	(680)	–	–	–	–	–	–
	H230	10	80	–	–			230	–	240	–

†Material condition:
M and* mean: as manufactured without specified mechanical properties
R, followed by a three digit number, is the mandatory minimum tensile strength in N mm⁻²
H, followed by a three digit number, is the mandarory minimum hardness in HB

Table 5.3 Mechanical properties of rolled flat products and tubes to CEN standards.

PLATE, SHEET AND CIRCLES
for general purposes

Product specification: prEN 1652

Designation		Thickness mm		Tensile Strength R_m N mm^{-2}		0.2% Proof Strength $R_{p0.2}$ N mm^{-2}	Elongation %		Hardness HV	
Symbol (Number)	Material condition	from	up to and including	min	max	min	A over 2.5 mm	min	min	max
CuAl8Fe3	R480	3	15	480	–	210	30	–	–	
(CW303G)	H110	3	15	–	–	–	–	110	–	

PLATE, SHEET AND CIRCLES
for boilers, pressure vessels and heat exchangers

Product specification: prEN 1653

Designation		Nominal thickness mm			Tensile Strength R_m N mm^{-2}	0.2% Proof Strength $R_{p0.2}$ N mm^{-2}	Elongation A %	Hardness HV
Symbol (Number)	Material condition	from	over	up to and including	min	min	min	approx
CuAl8Fe3	R450	–	50	–	450	200	30	(130)
(CW303G)	R480	2.5	–	50	480	210	30	(140)
CuAl9Ni3Fe2	R490	10	–	100	490	180	20	(125)
(CW304G)								
CuAl10Ni5Fe4	R590	–	50	–	590	230	14	(160)
(CW307G)	R620	2.5	–	50	620	250	14	(180)

TUBE

Product specification: prEN 12451

Designation		Tensile Strength R_m N mm^{-2}	0.2% Proof Strength $R_{p0.2}$ N mm^{-2}	Elongation A %	Drift expansion %	Hardness HV	
Symbol (Number)	Material† condition						
	Annealed	min	min	min	min	min	max
CuAl5As)	R350	350	110	50	30	–	–
(CW300G	H075	–	–	–	30	75	110

†Material condition:
R, followed by a three digit number, is the mandatory minimum tensile strength in N mm^{-2}
H, followed by a three digit number, is the mandarory minimum hardness in HV

Table 5.4 Mechanical properties of forgings to CEN standards.

		Thickness		Tensile Strength	0.2% Proof Strength	Elongation	Hardness	
Designation		mm		R_m	$R_{p0.2}$	A		
Symbol (Number)	Material† condition	up to and including 80	over 80	N mm⁻² min	N mm⁻² min	% min	HB min	HV min
		die- and hand forgings	hand forgings					
CuAl8Fe3	M	X	X	*	*	*	*	*
(CW303G)	H110	X	X	(460)	(180)	(30)	110	115
CuAl10Fe3Mn2	M	X	X	*	*	*	*	*
(CW306G)	H120	–	X	(560)	(200)	(12)	120	125
	H125	X	–	(590)	(250)	(10)	125	130
CuAl10Ni5Fe4	M	X	X	*	*	*	*	*
(CW307G)	H170	–	X	(700)	(330)	(15)	170	185
	H175	X	–	(720)	(360)	(12)	175	190
CuAl11Fe6Ni6	M	X	X	*	*	*	*	*
(CW308G)	H200	X	X	(740)	(410)	(4)	200	210

Figures in brackets are not mandatory and are for information only
†Material condition:
M and * mean: as manufactured without specified mechanical properties
H, followed by a three digit number, is the mandarory minimum hardness in HB

- *single-phase alloys*, that is alloys consisting mainly of a single copper-rich solid solution, known as the α (alpha) phase.
- *duplex (twin-phase) alloys*, that is alloys consisting mainly of a mixture of two solid solutions: α+β: a copper-rich solid solution α (alpha) and a solid solution β (beta) richer in aluminium than α.
- *Multi-phase alloys*, that is alloys consisting of a mixture of several solid solutions and/or compounds which have precipitated out of solution.

As will be seen in Chapters 11 to 14, these different types of structures are to be found in different alloying systems.

Single-phase alloys

Nature and working characteristics

Single-phase alloys contain approximately 8% or less aluminium and consist of a copper-rich solid solution, known as the α (alpha) phase. This phase is very ductile at room temperature and, as with the alpha brasses, is amenable to extensive cold rolling or drawing before annealing becomes necessary. The following alloys come into this category:

- CEN alloys (Tables 5.1, 5. 3 and 5.4): **CuAl5As** and **CuAl8Fe3**
- ASTM alloys (Appendix 1): CuAl5 (C 60800), CuAl7Sn0.3 (C 61300), CuAl8 (C 61000), **CuAl7Fe2** (C 61400) and **CuAl2.8Si1.8Co0.4** (C 63800).

Alloys **CuAl6Si2Fe** and **CuAl7Si2**, on the other hand, although having less than 8% aluminium, are equivalent to alloys with higher aluminium contents due to the effect of silicon which, as explained in Chapters 1 and 3, has a similar effect to aluminium on the microstructure, 1.6% silicon being equivalent to 1% aluminium.

Some single-phase alloys may contain small amounts of iron, nickel, manganese or tin or combinations of these, iron and nickel being partially soluble in the alpha phase. The presence of small percentages of these elements does not basically affect the single-phase nature of the alloy although, if the nickel or iron content exceeds 1–2%, a finely dispersed precipitate forms within the alpha phase on cooling (see Chapter 12). Strictly speaking, such a precipitate, being visible in the microstructure, constitutes a separate phase but, since the alloy exhibits the working characteristics of a single-phase alloy, it is considered as such for practical purposes.

Iron refines the structure and increases the strength of the material without any adverse effect on ductility. Nickel improves resistance to erosion, corrosion and mechanical properties. The advantage of small addition of manganese in a single-phase alloy is open to debate. Some consider that the mechanical properties are improved by additions of up to 2% and that the proof strength or general toughness of the alloy is improved. As regards tin additions, research in the USA has shown that the susceptibility of alpha phase alloys to intergranular stress corrosion cracking in high pressure steam service can be eliminated by addition of 0.25% tin (see ASTM specification C61300 in Appendix 1). This tin addition does not adversely affect the hot working of the alloy.

Single-phase alloys can be much more extensively cold-worked than duplex and multi-phase alloys and it is possible to produce thinner sections in them. Thus only alloys in this group can be deep-drawn. Single phase alloys are also easier to extrude into tubes.

The most popular single-phase alloy is the above mentioned iron-containing **CuAl8Fe3** alloy. It offers the best combination of properties, thanks to the grain refining effect of iron and the fact that its aluminium content is at the top end of the range for single phase alloys. Although it has good cold working properties, it is normally hot worked within the range of 700 to 950° C. It is possible to produce strips as thin as 1.5 mm thick in this alloy.

Mechanical properties

As may be seen from Tables 5.3 and 5.4 and Appendix 1, single phase alloys are relatively weak in the annealed condition, but they have attractive mechanical properties when cold-worked. Mechanical properties well above those specified in standard specifications are achievable in practice depending on the degree of cold working. It is advisable therefore to consult manufacturers on the properties that can be achieved with any given alloy. Their strength and rate of work hardening is greater than that of alpha brasses and they are not, therefore, worked as readily and to the same degree.

Corrosion resistance

These alloys have excellent corrosion resisting properties since, being single phase, they are not susceptible to selective phase attack (see Chapter 8) or to transformation on cooling to a more corrodible phase.

Impact strength

Single phase alloys possess admirable resistance to shock loading and typical figures lie in the range of 70–95 Joules as measured by the Izod test. Cold working has comparatively little effect on impact strength and conversely, annealing of single phase alloys is relatively ineffective.

Fatigue strength and corrosion fatigue limits

Details of the fatigue strength of some single-phase alloys (mostly to ASTM compositions) for various forms and tempers are given in Table 5.5 and endurance limits in air and sea water in Table 5.6.

Applications

The following are typical applications for single-phase aluminium bronzes (mostly to ASTM compositions):

CuAl5: Condenser, evaporator and heat exchanger tubes, distiller tubes and ferrules.

CuAl7: Nuts and bolts, corrosion resistant vessels and tanks, components, machine parts, piping systems, heat exchanger tubes, marine equipment, explosive making and handling equipment.

CuAl8: Bolts, pump parts, shafts, tie rods, overlay on steel for wearing surface.

CuAl8Fe3: Nuts and bolts, corrosion resistant vessels and tanks, structural components, condenser tubes and piping systems, marine applications, explosive manufacture and handling. This alloy with a 0.25% tin addition, referred to above, is used increasingly in pressure vessels, condensers and heat exchangers for high pressure steam service.

CuMn12Al4.5Fe3Ni2 (equivalent to 6.5% nominal aluminium content): sheet, strip, tube and wire.

Table 5.5 Fatigue strength at room temperature of single-phase aluminium bronze alloys.[173]

Alloy	Form	Temper	Number of Cycles ×10⁶	Tensile Strength N mm⁻²	Fatigue Strength N mm⁻²
	Plate	Annealed	20	422	157[a]
CuAl5	25mm dia Rod	Rolled	100	495	132[b]
	Condenser tube	Annealed	100	392	108
CuAl7	12.5mm dia Rod	Hard	300	472	96.5[c]
	19mm dia Rod[e]	Extruded light drawn	52.52[h]	549	216[c]
	25mm dia Rod[f]	Rolled	100	598	167[b]
CuAl8	42mm dia Rod	Forged	50	515	181[b]
	Rod	11.5% Cold worked	300	672	152[c]
	Wire[g]	Annealed (0.16mm grain size)	10[h]	422	157[d]

(a) Rotating bending test (d) Push pull test (g) Alloy contains 7.9% Al
(b) Rotating cantilever test (e) Alloy contains 1.4% Zn (h) Unbroken specimen
(c) Rotating beam test (f) Alloy contains 9.1% Al

Table 5.6 Endurance limits of single-phase aluminium bronzes in air and sea water.[127]

Basic Composition %			Tensile Strength	Endurance limit in air N mm⁻²			Endurance limit in sea water or 3% NaCl N mm⁻²			Condition of material
(Cu rem.)			N mm⁻²	No of Cycles			No of Cycles			
Al	Fe	Ni		10⁷	5×10⁷	10⁸	10⁷	5×10⁷	10⁸	
5.5	–	–	495	–	–	131	–	–	–	Rod, half hard
7	2	–	–	–	–	144	–	–	–	Plate, 25 mm thick
7	2	–	526	–	–	167	–	–	104	–
7	2	–	618	–	–	226	–	–	165	–
8	–	–	464	155	–	114	–	–	–	Rod, cold-drawn 11%
8	–	–	572	–	–	204	–	–	–	Lightly worked

Duplex (twin-phase) alloys

Nature and working characteristics

Alloys with 8.0–8.4% aluminium are effectively on the upper limit of the ductile single-phase alloy range. As the aluminium content increases above 8.0–8.4%, a second high temperature solid solution begins to form which is known as the β (beta) phase. Under extremely slow cooling conditions, it is possible to retard the advent of this phase up to a 9.4% aluminium content, but commercial cooling rates in annealing and hot working operations are too rapid for this to be the case. Above 8.0–8.4% aluminium, the alloy therefore becomes a duplex alloy, that is a mixture of two solid solutions: the alpha and beta phases. The beta phase, which is harder and less ductile than the alpha phase, becomes increasingly present as the aluminium content is raised. Whilst the beta phase is hard and of limited ductility at room temperature, it becomes softer and more plastic than the alpha phase at temperatures above red heat. The duplex alpha/beta alloys containing more than 8.5–9% aluminium are therefore readily hot worked within the range of 700–800° C. Only a limited amount of cold working can be carried out on alloys with up to 10% aluminium.

The following alloys are duplex alloys:

- CEN alloys (Tables 5.1 and 5.2): CuAl6Si2Fe, CuAl7Si2, CuAl9Ni3Fe2, CuAl10Fe1 and CuAl10Fe3Mn2.
- ASTM alloys (Appendix 1): CuAl10Fe1 C 61800, CuAl9.5Fe4 (C 61900), CuAl10Fe3 (C 62300), CuAl11Fe3 (C 62400), CuAl13Fe4.3 (C 62500) and CuAl7Si1.8 (C 64200).

Some of these alloys contain 3–4% iron and/or smaller additions of nickel or manganese. As explained above in the case of single phase alloys, iron and nickel are partially soluble in the alpha phase. They are also partially soluble in the beta phase and will therefore form finely dispersed precipitate in both these phases on cooling (see Chapter 12). Although, strictly speaking, these precipitates constitute separate phases, the alloy retains the working characteristics of a duplex alloy and is therefore considered as such for practical purposes. The beneficial effects of iron and nickel are the same as for single-phase alloys mentioned above.

Gronostajski and Ziemba[81] carried out research on two wrought Cu–Al–Fe alloys of the following compositions and they conclude that it may be desirable to hot-work at higher temperatures than the 700–800° C mentioned above, if certain mechanical properties are to be achieved:

Alloy	Al	Fe	Mn	Impurities	Cu
A	9.90	3.64	1.63	0.5	bal
B	10.74	4.00	1.60	0.5	bal

Bearing in mind that the highest tensile properties are achieved if the wrought alloy has a banded fibrous structure, similar to the grain in wood, they report that, for this structure to be obtained and retained even after annealing, it is necessary to hot-work at a temperature at which the microstructure consists of the beta + kappa phases with little or no alpha phase. At a lower temperature, where there will be a significant proportion of the alpha phase present, the alloy recrystalises as it recovers from hot-working and the mechanical properties cease to be unidirectional (anisotropic). A fuller explanation is given in Chapter 12.

Although containing a maximum of only 7.6% aluminium, CEN alloys CuAl6Si2Fe and CuAl7Si2 and ASTM alloy CuAl7Si1.8 are in fact, as previously mentioned, duplex alloys because silicon is an aluminium substitute: 1% silicon being equivalent to 1.6% aluminium. These alloy, with their ~2% silicon, are therefore comparable to a CuAl10 aluminium bronze. They contain less than 0.7% iron, as a grain refiner, which remains in solution and does not therefore affect the duplex nature of the alloy. They are hot worked within a slightly lower range of temperatures: 600–800° C.

Mechanical properties

As in the case of single-phase alloys, mechanical properties of duplex alloys are very significantly improved by hot or cold working, especially by the latter, and this accounts for the wide variation of mechanical properties with section size shown in Appendix 1.

Impact strength

Duplex type alloys containing 9–10% aluminium and 2–3% other elements, have typical values varying from 27–54 Joules provided the gamma phase is avoided.

Fatigue strength

Details of the fatigue strength of some duplex alloys for various forms and tempers are given in Table 5.7 and endurance limits in air and sea water in Table 5.8.

Tests carried out under normal atmospheric conditions show the hot worked duplex-structured aluminium bronzes, containing additions of iron and nickel, to have endurance limits at 50 million stress reversals between 155 and 340 N mm^{-2}, depending on alloy composition. A progressive increase in these figures is obtained as the aluminium, iron, and nickel contents are raised (see multi-phase alloys), and it is probable that other elements exert a similar although less important effect.

The effect of heat treatment on the fatigue strength of a duplex alloy is dealt with in Chapter 6.

The threshold stress intensity range of silicon–aluminium bronze has been found to be considerably lower than those of copper-aluminium alloys of similar grain size, making it less suitable for fasteners.[141]

Applications and resistance to corrosion

The following are typical uses of some duplex aluminium bronzes:

CuAl7Si2: This alloy is used extensively in the United States because of its ease of machining and for its good wear and excellent corrosion resisting properties. It is suitable for valve bodies, stems and other components, gears, marine fittings, nuts and bolts. The corresponding CEN alloy is used primarily in naval applications for its low magnetic properties and good impact value in addition to its excellent corrosion resisting properties.

Table 5.7 Fatigue strength at room temperature of duplex aluminium bronze alloys.[173]

Alloy	Form	Temper	Number of Cycles ×10^6	Tensile Strength N mm^{-2}	Fatigue Strength N mm^{-2}
CuAl7Si2	12.5mm dia Rod	Hard	300	741	179[c]
		Annealed	300	649	207[c]
	Plate	Forged	40	520	>118[a]
CuAl9Mn2	25–50mm dia Rod	Cold worked	50	642	>235[a]
	50mm dia Rod	Forged	50	726	255[a]
	80mm dia Rod	Forged	20	540	186[a]
			20	490	177[a]
CuAl10Fe3	14mm dia Rod	10% drawn	100	641	196[c]
	25mm dia Rod	Rolled	100	683	241[b]

(a) Rotating bending test   (b) Rotating cantilever test   (c) Rotating beam test

Table 5.8 Endurance limits of duplex aluminium bronzes in air and sea water.[127]

Basic Composition % (Cu rem.)			Tensile Strength N mm^{-2}	Endurance limit in air N mm^{-2}			Endurance limit in sea water or 3% NaCl N mm^{-2}			Condition of material
				No of Cycles			No of Cycles			
Al	Fe	Ni		10^7	$5×10^7$	10^8	10^7	$5×10^7$	10^8	
9.3	–	–	–	–	153	–	153	121	–	Quenched and tempered
9.3	–	–	–	–	176	–	170	135	–	Quenched
9.3	2.1	–	–	291	–	255	258	–	162	Extruded and drawn rod

CuAl9Mn2: This a German alloy (similar to CEN CuAl10Fe3Mn2) with good corrosion resisting properties thanks to the manganese addition (seeChapter 12). It is suitable for marine applications and any application requiring medium strength.

CuAl10Fe3: This alloy has good wear resisting properties if the eutectoid gamma phase is allowed to form by slow cooling (see Chapters 10 and 12). As such it is used for bearings and bushing, valve guides and seats, worm gears and other gears, nuts and bolts, cams and pump rods. The presence of this gamma phase however makes it unsuitable for corrosive environments.

Multi-phase alloys

Nature and working characteristics

Compositions, forms and properties of multi-phase alloys are given in Tables 5.1 to 5.4 and in Appendix 1. The following are multi-phase alloys and are known as nickel-aluminium bronzes:

- CEN alloys (Tables 5.1 to 5.4): CuAl10Ni5Fe4 and CuAl11Fe6Ni6
- ASTM alloys (Appendix 1): CuAl10Fe2Ni5 (C63000) and CuAl9Fe4Ni5 (C 63200).

Nickel-aluminium bronzes

Nickel-aluminium bronzes contain substantial additions of both nickel and iron and their structure consists basically of a mixture of alpha and beta phases that contain dispersed precipitates of nickel, iron and aluminium compounds, known as κ (kappa) phases (see Chapter 13). These phases have a very marked effect on the hot-working characteristics of this group of alloys. Nickel-bearing alloys have the highest mechanical properties, as may be seen from Tables 5.2 to 5.4, but are more difficult to work. They have to be hot worked at a higher temperature (900–950° C) than the duplex alloys and may also be cold worked to a limited extent at the finishing stage. The excellent ductility (elongation) of nickel-aluminium bronzes at high temperature may be seen from Figure 5.5. Their correspondingly low tensile strength is an indication of their low general strength which, together with ductility, explains their workability at these temperatures.

Apart from the high manganese containing alloys, they are by far the most frequently specified aluminium bronzes due to their combination of mechanical strength and corrosion resisting properties.

Alloys with aluminium contents towards the upper end of the permitted range, such as alloy CuAl11Fe6Ni6, are hot worked at slightly lower temperature (880–920° C) and are easier to work, but their ductility and impact strength at room temperature are inferior. They have however excellent wear properties and are used as special bearing alloys. They are nevertheless likely to contain the gamma phase (see Chapter 13) which gives them their wear properties but also makes them unsuitable for corrosive environments.

Manganese-aluminium bronzes

Whilst essentially a casting alloy, the copper-manganese-aluminium alloy, CuMn13Al8Fe3Ni3, may also be hot-worked at temperatures in the range of 650°C-850° C in which range it is very soft and malleable. Its structure is similar to that of nickel–aluminium bronzes and consists basically of a mixture of alpha and beta phases that contain dispersed precipitates. In this case, the precipitates, also known as κ (kappa) phases, contain varying amounts of manganese, iron, aluminium and nickel (see Chapter 14)

The mechanical properties of the wrought manganese–aluminium bronzes depend on the degree and type of the forming operation and are comparable to those of nickel-aluminium bronzes.

Mechanical properties at elevated temperature

CEN specification EN 1653: 1997 specifies minimum mechanical properties over a range of elevated environmental temperatures for nickel aluminium bronze alloy CuAl10Ni5Fe4. Details are given in Table 5.9.

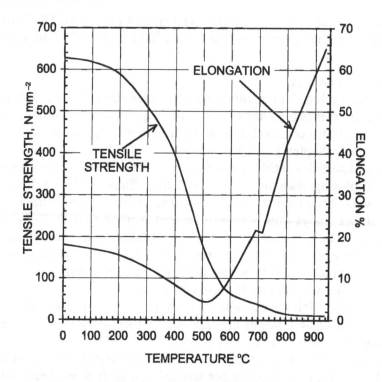

Fig. 5.6 Effect of temperature on the elongation and tensile strength of a nickel-aluminium bronze containing 9–10% Al and 5% each Ni and Fe.[184]

Table 5.9 Mechanical properties of CuAl10Ni5Fe4 alloy at elevated temperatures to CEN specification EN 1653:1997.

Material designation: Symbol: CuAl10Ni5Fe4 **Number:** CW307G						
Tensile Strength N mm⁻²	Thickness mm	Minimum 0.2% Proof Strength at temperature ° C $R_{p0.2}$ N mm⁻²				
min	up to and including	50° C	100° C	150° C	200° C	250° C
630	80	270	265	260	260	250

Impact strength

The Izod impact strength of multi-phase alloys, with the exception of alloy CuAl11Ni6Fe6, falls within the range of 14–27 Joules and, where severe shock loading may occur, care must be taken to avoid sharp notches and stress-raisers such as coarse machining marks.

Fatigue strength

Most fatigue tests have been carried out on the high strength CuAl10Ni5Fe5 type of alloy because of its more widespread use under highly stressed conditions. A summary of fatigue data on wrought aluminium bronzes is given in Table 5.10 and associated corrosion-fatigue results are given for comparison in Table 5.11. The slightly lower results obtained by McKeown[110] were attributed to the thin gauge to which the test specimens had been rolled. Additional data recorded from elevated-temperature tests and in manufacturers' trade literature confirm that the endurance limit in air at 50 million cycles may be taken as approximately 325 N mm^{-2}. There are strong indications, moreover, that the alloy has in fact reached a fatigue limit within this number of stress cycles. The above data is for reversed bending conditions. Williams[188] has reported that, under torsion, the endurance limit at a hundred million cycles is 15%–17% of tensile strength for the aluminium bronzes he had tested.

As with all high strength materials, great care must be taken to avoid notches and other stress-raisers as they may lower the fatigue life of a component. Murphy[136], however, has reported that, even in an extreme case, the reduction is only

Table 5.10   Fatigue strength at room temperature of multi-phase aluminium bronze alloys.[173]

Alloy	Form	Temper	Number of Cycles ×10^6	Tensile Strength N mm^{-2}	Fatigue Strength N mm^{-2}
	12.5mm dia Rod	11.5% Cold worked	300	860	221[c]
CuAl9Ni6Fe3	16mm dia Rod	Heat Treated	100	804	≈290[c]
	Not stated	Annealed	100	712	241[b]
	Not stated	Not stated	100	751	172[a]
					228
	6.4mm Flat	50% Hot rolled [e]	50	–	278[a]
		50% Hot rolled [f]	50	–	293[b]
	15mm dia Rod	15% Cold drawn	30	822	304[a]
	25–50mm dia Rod	Forged	20	883	324[b]
	35mm dia Rod	10% Cold drawn	30	711	265[a]
	45mm dia Rod	5% Cold drawn	30	738	289[a]
CuAl10Ni5Fe5	50mm dia Rod	Forged	30	723	285[a]
	Rod	Annealed	40	540	255[b]
	Rod	Annealed	35	711	275[b]
	Rod	Forged	20	736	294[a]
		20% Cold drawn	20	883	353 [a]
	Rod	Forged	56.8[d]	798	347[c]
	Rod	Forged	20	765	294[b]
			20	814	336[b]
	Not stated	Forged	57	507	221[c]
CuMn13Al8Fe3Ni3	Not stated	Forged	100	730	309

(a) Rotating bending test   (c) Rotating beam test   (e) Alloy contains 1.4% Zn
(b) Rotating cantilever test   (d) Push pull test   (f) Alloy contains 9.1% Al

Table 5.11   Endurance limits of multi-phase aluminium bronzes in air and sea water.[127]

Basic Composition %			Tensile Strength	Endurance limit in air N mm^{-2}			Endurance limit in sea water or 3% NaCl N mm^{-2}			Condition of material
(Cu rem.)			N mm^{-2}	No of Cycles			No of Cycles			
Al	Fe	Ni		10^7	5×10^7	10^8	10^7	5×10^7	10^8	
10	3	5	742	–	–	227	–	–	139	–
9.5	2.5	5	850	309	–	247	–	–	–	Rod, cold-drawn 11%
9	5	5	758	309	278	–	232	139	–	Hot rolled 6 mm strip
9.7	5.4	5	804	–	351	–	292	226	–	Forged
9.7	5.3	5.1	835	340	–	323	275	–	167	Rolled rod, 30 mm dia
10.5	5	5	943	309	294	–	209	155		Hot rolled 6 mm strip
10.6	4.7	4.6	866	356	–	–	255	–		As extruded
11	4	4	881	–	340	–	–	196		Quenched 890°C, tempered 620°C

of the order of 93 N mm^{-2} for the CuAl10Ni5Fe4 alloy which is significantly less than that experienced with steels of comparable strength.

It will be noticed that the reduction in the endurance limit, when the samples were tested in a corrosive environment (Table 5.11), is not as great as for the majority of other engineering materials. In comparing the fatigue properties of the CuAl10Ni5Fe4 type of nickel-iron-aluminium bronze with those of stainless iron and stainless steel, McKeown[110] and his co-workers found that although the ferrous materials had a higher tensile strength, their fatigue resistance under the corrosive conditions typical of sea water fell significantly below that of aluminium bronzes. At 50 million cycles the endurance limit for the stainless steels was 62–108 N mm^{-2} compared with 139–155 N mm^{-2} for the aluminium bronze.

Torsion

Little detailed information is available on torsional properties, but some useful data was obtained on large diameter shafting in a series of tests carried out for the British Ministry of Defence (Naval). The alloy used was of the 80/10/5/5 type and the tensile strength of the bars was between 711–742 N mm^{-2}. In every case, regardless of whether the bar was rolled, extruded or forged, the torsional limit of proportionality was found to be 178 ± 30 N mm^{-2}.[127]

As mentioned above, the endurance limit for several aluminium bronzes under torsion at a hundred million cycles is 15–17% of tensile strength. The torsional strength/tensile strength ratio of alloys in the same tests was in the region of 6/10 which is typical of most other engineering materials.

Creep strength

Creep is the slow plastic deformation that occurs under prolonged loading, usually at high temperature. It is of particular concern in the case of chemical plant and pressure vessels operating at high temperatures. The design of such plant and equipment is usually based on a service life of 100 000 hours (approximately 12 years). Tests of 30 000 hours duration, carried out by Drefahl *et al.*,[67] have shown that the creep strength of nickel–aluminium bronze compares favourably with that of copper and of other copper-based alloys, as may be seen from Table 5.12. Table 5.13 gives the CEN specified creep stress properties for alloy CuAl10Ni5Fe4.

Applications

The following are typical uses of multi-phase aluminium bronzes:

CuAl9Fe4Ni5 (C 63200): this alloy is used for naval applications because of its good impact values resulting from its lower aluminium content. It is used as nuts and bolts, valve seats, marine shafts, structural members, etc. in wide variety of equipment.

CuAl10Ni5Fe4 and CuAl10Fe2Ni5 (C 63000): These are very similar alloys used extensively in industrial, marine and naval applications.

CuAl11Ni6Fe6: This is an alloy that offers a good combination of hardness with tensile strength and elongation. It is suitable for bearings and gears and other applications requiring both good wear and strength properties but may be unsuitable for corrosive environments due to the presence of the $gamma_2$ phase.

CuMn13Al8Fe3Ni3: This alloy is available in a limited range of wrought forms, principally rod, bar and plate. At present, however, the tonnage of this wrought alloy is small compared with that of castings (see Chapter 3).

Temper

The 'temper' of a metal is its degree of softness/hardness or ductility/toughness resulting from hot or cold working or from heat treatment. Hot and, especially cold working, generally hardens and toughens the metal. Certain heat treatments involving quenching from a selected temperature also hardens or toughens the metal, whereas annealing softens it and makes it more ductile.

Temper falls under three broad categories:

● *Soft anneal*: this is achieved by a full or true anneal (see Chapter 6) and results in relatively low mechanical properties and hardness figure but generally improved ductility.

Table 5.12   Comparison of creep properties of nickel–aluminium bronze with copper and other copper-based alloys, by Drefahl *et al.*[67]

Alloy:	Phosphorus deoxidised Cu SF-Cu				Low leaded brass CuZn39Pb0.5				Aluminium brass CuZn20Al2			
	100° C	150° C	200° C	250° C	100° C	150° C	200° C	250° C	100° C	150° C	200° C	250° C
Minimum value of the 1% offset yield strength at room temperature												
at 1000 h	62	60	50	42	160	130	62	28	200	180	125	50
10,000 h	59	52	41	33	160	115	43	16	180	150	95	32
30,000 h	55	50	39	28	155	110	35	13	150	140	80	25
100,000 h	(52)	(44)	(30)	(21)	(250)	(90)	(28)	(7)	(130)	(120)	(60)	(20)
Minimum value of the tensile strength at room temperature												
at 1000 h	195	170	130	110	320	220	120	50	390	255	175	95
10,000 h	180	150	110	78	300	185	75	28	350	190	120	60
30,000 h	170	140	92	60	290	160	62	21	310	155	95	50
100,000 h	(160)	(120)	(71)	(38)	(270)	(130)	(40)	(13)	(270)	(125)	(70)	(40)
Stress for minimum creep rate of 1%												
at 10,000 h	180	125	90	55	280	170	70	25	350	180	125	60
30,000 h	165	110	80	45	265	150	60	20	330	170	110	50
100,000 h	115	60	40	15	210	110	30	6	260	140	80	18
Alloy:	Phosphorus deoxidised Cu SF-Cu				Low leaded brass CuZn39Pb0.5				Aluminium brass CuZn20Al2			
	100° C	150° C	200° C	250° C	100° C	150° C	200° C	250° C	100° C	150° C	200° C	250° C
Minimum value of the 1% offset yield strength at room temperature												
at 1000 h	165	150	120	50	200	160	100	45	340	310	280	230
10,000 h	145	130	85	23	185	130	69	28	320	280	260	180
30,000 h	130	120	65	16	180	120	60	22	310	270	250	160
100,000 h	(110)	(105)	(45)	(12)	(170)	(110)	(45)	(15)	(290)	(260)	(230)	(130)
Minimum value of the tensile strength at room temperature												
at 1000 h	390	320	240	120	340	290	150	61	610	520	400	280
10,000 h	380	290	150	65	310	215	110	38	590	470	330	220
30,000 h	360	255	115	50	300	180	90	30	580	430	310	190
100,000 h	(330)	(205)	(85)	(85)	(285)	(140)	(60)	(21)	(560)	(400)	(270)	(160)
Stress for minimum creep rate of 1%												
at 10,000 h	360	270	135	60	245	195	95	35	580	490	330	210
30,000 h	345	260	120	45	230	180	75	25	550	480	310	195
100,000 h	300	225	65	15	180	145	30	8	480	430	240	135

Values in brackets have been determined by extrapolation

Table 5.13   Creep Stress properties of CuAl10Ni5Fe4 alloy to CEN specification EN 1653:1997.

Temperature	1% Creep Stress for duration of N mm^{-2}			
°C	10 000 h	30 000 h	50 000 h	100 000 h
150	252	242	237	232
160	243	233	228	224
170	236	226	221	216
180	229	219	214	209
190	223	213	208	203
200	218	207	202	198
210	213	202	197	193
220	210	199	193	188
230	207	196	190	185
240	205	194	188	182
250	204	192	186	180

Material designation: Symbol: CuAl10Ni5Fe4 Number: CW307G

- *Half hard temper*: this is slightly cold worked and stress relieved below the recrystallisation temperature.
- *Hard temper*: this is more heavily cold worked to yield a high hardness figure.

The degree of temper is measured by tensile, elongation and hardness tests. Hardness is measured either by the Brinell, Vickers or Rockwell hardness test. The latter is used mostly in the United States and the other two mostly in Europe and Japan. The softer and more ductile the metal, the higher the elongation figure and the greater the difference between proof strength and tensile strength. Grain size is also a measure of softness since the larger the grain size resulting from an annealing treatment, the softer the metal.

American specifications are unique in stipulating the temper of a particular alloy and they divide into the following categories:

(a) Annealed Tempers: Grain size ranging from 0.100 mm to 0.015 mm, Soft Anneal and Light Anneal.
(b) Hot Finished Tempers: As Hot Rolled and As Extruded.
(c) Rolled or Drawn Tempers: ⅛, ¼, ½ and ¾ Hard and Extra Hard, Spring and Extra Spring, Light Drawn, Drawn and Hard Drawn (these apply to tubes).

Appendix 1 gives the specified tempers of American wrought aluminium bronzes. It shows the effects on properties of the tempers given to the metal by various degrees of working and by light and soft anneal. It also shows the effect of section size on properties. It is evident that the smaller the section, the more work has been done to the metal and the higher its proof and tensile strength. The effect of section size on elongation is not so evident. This information is given only to illustrate these effects and designers should consult the latest specification before choosing an alloy.

European and Japanese specifications do not specify the temper and the mechanical properties stipulated are minimum properties to be achieved by manufacturers using appropriate procedures. These properties are normally more than adequate to meet the needs of most engineering applications, although American specifications offer designers a wider choice of verifiable properties for special applications.

Manufacturers normally achieve the temper stipulated in American specifications by close control of the process without recourse to heat treatment other than for relieving internal stresses.

The shock resisting property of a component is also related to its temper and is measured by the Izod test which is usually required by European naval specifications.

There are two ways of using heat treatment to alter the temper of the metal:

(a) by annealing or tempering which softens the metal
(b) by quenching followed by tempering. This is normally done to strengthen and toughen the metal but, if carried out after cold working, may result in some degree of softening and a better balance of mechanical properties.

More details of these heat treatments are given in Chapter 6.

Factors affecting mechanical properties

The mechanical properties of wrought products are mainly a function of four factors:

(a) their chemical composition,
(b) their size and shape and the working process used in their manufacture (forging, rolling, extruding or drawing).
(c) whether hot or cold worked,
(d) the type of heat treatment applied.

The effects of these factors on mechanical properties are illustrated in Tables 5.2 to 5.4, 5.7 and 5.11.

Effects of composition

The effects of alloying elements on mechanical properties of aluminium bronzes has already been discussed in Chapter 1 and is basically the same for wrought as it is for cast alloys except that, in the case of wrought alloys, the effects of alloy composition may be obscured by the effects of hot and cold working and of heat treatment. Of all the elements, aluminium has the most pronounced effect, even within the limits of a given specification.

Effects of wrought process and of size and shape of the product

Both the size and shape of a product and the wrought process used in its manufacture (forging, rolling, extruding, drawing etc.) have an influence on the way its structure is distorted by hot or cold working and therefore on its mechanical properties. For this reason, standard specifications specify mechanical properties for given wrought forms (rod, bar, flats, sheets, forgings etc.) and range of sizes, as may be seen from Tables 5.2 to 5.4 and Appendix 1. In some cases the wrought process is also specified.

Effects of hot and cold working

Hot or cold working a metal distorts its crystalline structure and this has a very pronounced effect upon its mechanical properties. The more energy is used in altering the shape of a component, the greater the effect:

(a) cold working has a more marked effect on mechanical properties than hot working,
(b) hot working of a multi-phase alloy has a greater effect on properties than hot working a single phase or duplex alloy.
(c) the smaller the section, the more the alloy has been worked and, consequently, the higher the mechanical properties achieved in the as-cast and hot-worked conditions.[173]

Table 5.14 compares the mechanical properties of two alloys in the as-cast and hot-worked conditions. Hot working particularly increases the proof strength by comparison with the as-cast condition. It can be seen that the smaller the section – and therefore the more the metal has been hot worked – the higher the tensile and proof strength and the lower the elongation.

For special applications, manufacturers may be able to produce wrought sections or forged components with properties significantly different from those laid down in standard specifications and which may be better suited to the particular application. It is advisable therefore to consult manufacturers.

Heat treatment

A wrought product may be heat treated for any of the following reasons:

(a) to relieve internal stresses after cold working,
(b) to alter its 'temper', that is to say its degree of hardness/ductility,

Table 5.14 Comparison of the mechanical properties of two common alloys.

Nominal composition	Form	Size range mm	Condition	0.2% Proof Strength (N mm^{-2})	0.5% Yield Strength (N mm^{-2})	Tensile Strength (N mm^{-2})	Elongation %
CuAl10Fe3	Casting		as cast	170–200		500–590	18–40
	Rod	< 12	Drawn		345	620	12
		12–25	and		305	605	15
		25–50	stress		275	580	15
		50–80	relieved		255	525	20
		> 80	As rolled		205	515	20
CuAl10Ni5Fe4	Casting		as cast	250–300		640–700	13–21
	Plate	6–18	as rolled	400		700	10
		18–80	"	370		700	12
		> 80	"	320		650	12

(c) to change the phase composition of its microstructure to improve corrosion-resisting properties or other properties such as wear resistance.

Details of these forms of heat treatment and their effects on microstructure and mechanical properties are dealt with in Chapter 6.

6
HEAT TREATMENT OF ALUMINIUM BRONZES

Forms of Heat Treatment

Aluminium bronzes may be subjected to any of the following forms of heat treatment:

Annealing
Normalising
Quenching
Tempering or temper anneal

Annealing

Annealing consists in (a) heating the metal to a certain temperature, (b) 'soaking' at this temperature for sufficient time to allow the necessary changes to the micro-structure to occur and (c) cooling at a predetermined rate. The details of the annealing process and the reason for annealing vary with the type of alloy as will be seen below.

Annealing is normally done at a temperature at which the 'phase' composition of the microstructure will be altered and grain growth is encouraged in order to soften the metal. As explained in Chapter 5, a 'phase' is a constituent of an alloy which is either a solid solution of two or more elements in each other or a compound of two or more elements. It has a given characteristic appearance under the microscope and has certain specific properties which affect the properties of the alloy as a whole.

The purpose of annealing is essentially to soften the metal and improve its ductility at the expense of proof strength and tensile strength. Hence the term annealing is used even if the process involves rapid cooling by quenching, provided the end result of the overall treatment is to make the metal more ductile.

Other objects of annealing include:
(1) to relieve internal stresses induced either by cold working or by rapid cooling during a previous heat treatment,
(2) to improve corrosion resistance,
(3) to improve certain characteristics of the metal such as wear properties.

The same range of temperatures may be used for stress relief anneal as for an annealing designed to soften the metal, but the time of soaking will be different since the objective is different.

There are broadly three degrees of annealing:

a) *Full or true anneal:* this is designed to achieve maximum softening of the metal. It involves heating to a temperatures above the re-crystallisation temperature (typically above 650°C) and is used to soften the metal between stages of cold working. This is known as 'Process annealing'. Some finished products may also be required in this condition.

b) *Partial anneal:* this is to achieve a medium degree of softening and involves heating to a medium temperature range (typically 500–650°C).

c) *Temper anneal:* see below

The good resistance to oxidation of aluminium bronzes means that a very much lighter scale is formed during heat treatment than on other copper alloys. Hence protective furnace atmospheres are not so essential for annealing. The degree of oxidation varies little with different furnace atmospheres unless small amounts of sulphur dioxide are present, in which case a heavier scale is formed. For inter-stage anneals during cold-working operations, bright annealing may be advisable in order to maintain a very high standard of surface quality.

Normalising

Normalising is that form of annealing in which cooling is done in air in order that less grain growth occurs during the cooling period than with slower cooling. The end result is a compromise between ductility and tensile properties. For example, some nickel aluminium bronze castings may be normalised by soaking at 675°C for to 2–6 hours, depending on section thickness, followed by air cooling. This results in a change of microstructure which improves corrosion resistance and in restricted grain growth during cooling which ensures good mechanical properties. Normalising may also be done to relieve stresses with little or no change in grain size, and consists in heating the metal to a predetermined temperature, allowing it to soak at this temperature for a given period of time, during which the internal stresses will be relieved, and then allowing it to cool slowly in air. Some degree of grain growth and consequent softening will however occur, although this is not the object but an unavoidable consequence of the treatment. If this degree of grain growth is unacceptable, it may be necessary to reduce the soaking time and accept some degree of residual stresses.

Quenching

Quenching consists in (a) heating the metal above a certain critical temperature, (b) soaking at this temperature for sufficient time to allow the necessary changes to occur to the microstructure and (c) quenching (cooling rapidly in water or oil). The details of the quenching treatment depends on the type of alloy as will be seen below.

The purpose of quenching may be to increase the hardness of the metal in order, for example, to achieve better wear properties; or it may be done to improve strength at the expense of ductility.

Tempering or temper anneal

It was explained in Chapter 5 that the 'temper' of a metal is its degree of softness/hardness or ductility/toughness resulting from hot or cold working or from heat treatment. Hot and, especially, cold working hardens and toughens the metal. Similarly, as just explained, quenching from a selected temperature also hardens or toughens the metal. In both cases it is usually necessary to reduce the hardening effect in order to improve ductility/toughness and this is the purpose of tempering. The term tempering is also often used:

(a) when the main object is simply to relieve internal stresses after cold working or quenching and
(b) when the object is to obtain a more corrosion resistant microstructure.

Tempering is similar to normalising in that it involves air cooling. The temperature at which it is done depends on the reason for tempering:

● If it is a post-quenching treatment, designed to improve ductility at the expense of hardness, it is usually done at a relatively high temperature: between 500°C and 700°C (see Table 6.1).
● If it is a post hot or cold working treatment, designed to achieve a controlled reduction in hardness, it is typically done at about 400°C–540°C. Due to the careful planning and control of production processes, however, manufacturers normally achieve the required temper without recourse to heat treatment, other than stress relief anneal.
● If it is primarily to relieve stresses after cold working or quenching then it is usually done at about 350°C. This leaves the mechanical properties substantially unaffected but significantly improves shock resistance.

Reasons for Heat Treatment

The following are the principal reasons for heat treating aluminium bronzes:

● to relieve internal stresses,
● to increase ductility,
● to adjust hardness and tensile properties,
● to improve corrosion resistance
● to improve wear properties.
● to reduce magnetic permeability (Cu/Mn/Al/Fe/Ni. alloys only)

Relieving internal stresses

As aluminium bronzes can be cold worked to an extremely high tensile strength, certain manufacturing operations are liable to induce a high level of internal stresses in the material. Hot working is less liable to give rise to internal stresses. The finishing passes are often done, however, below the true hot working temperature and may therefore induce significant internal stresses. Stresses from cold working may be dangerously high in the case of drawn or bent components and may give rise to two effects:

(a) dimensional instability and distortion, particularly on machining or cutting, and

(b) stress-corrosion cracking in a corrosive atmosphere.

The latter subject is dealt with in more detail in Chapter 9, but a stress relief annealing or normalising will overcome both effects.

Increasing ductility

A cold working process, such as drawing, rapidly increases the hardness of the metal to the point that it cannot be further cold-worked. The object of annealing is to restore the ductility of the metal to allow further cold-working. The same heat treatment will also be applied if the final product is required in a ductile condition.

Increasing hardness and tensile properties

The hardness and tensile properties or 'temper' of a wrought product can normally be achieved by close control of the working process. There are cases however where these may need some adjustment and this is done by heat treatment.

Improving corrosion resistance

As explained in Chapters 11 to 14, a fast or slow rate of cooling from high temperatures can give rise to different corrodible structures in binary and complex alloys with relatively high aluminium contents. In practice, binary and tertiary alloys with duplex structures are chosen for their mechanical properties rather than for corrosion resistance which can nevertheless be improved in certain circumstances (see 'Duplex alloys' below). Cu/Al/Ni/Fe alloys significantly benefit from heat treatment (see 'Complex alloys' below).

Improving wear properties

The wear properties of duplex and complex alloys can be improved by a adjusting the structure so as to achieve a balance between too soft and too hard a structure (see Chapter 10). In some cases high wear properties can be obtained by encouraging the formation of the corrosion-prone gamma$_2$ phase which is very hard. This can be done by a combination of high aluminium content and slow cooling after heat treating at high temperature. Such alloys are of course unsuitable for corrosive environments.

Reducing magnetic permeability

This applies to the high manganese Cu/Mn/Al/Fe/Ni alloys which experience a considerable increase in magnetic permeability below 500°C. The heat treatment required to counter this effect is described below (see 'Improving magnetic properties – manganese–aluminium bronze).

Heat treating different types of alloys

See Chapter 5 for definition of single phase, duplex and complex alloys. For a fuller understanding of the effects of heat treatment on the structure of alloys and hence on their properties, it is necessary to consult Chapters 11 to 14.

Single phase alloys

(1) Softening the alloy

Appreciable softening of single phase alloys can be achieved at temperatures as low as 400°C, provided the soaking time is extended. For full annealing a temperature of 700°C–800°C must be maintained for about 1/2 hour per 25 mm thickness. In practice 500°C–650°C gives satisfactory results, although the final hardness is slightly above that obtained with higher temperature anneals.

Practical annealing times are rarely critical and depend primarily on individual furnace and charge considerations. Nor is the subsequent cooling rate important for alloys containing less than 6.5%–7% aluminium and air cooling (normalising) is commonly adopted. With 7.5%–9% aluminium, some beta phase can be formed in the structure at high annealing temperatures and controlled slow cooling through the range 600°C–400°C is necessary to avoid the retention of the harder beta phase at room temperature.

A typical softening curve for a CuAl7 binary alloy, which had been given a 50% cold reduction by rolling, is shown in Fig. 6.1. As this relates to the actual temperature of the metal, industrial furnace temperatures must necessarily be appreciably above the minimum value indicated by the graph. The presence of about 2% nickel or iron increases the annealing temperature by about 50°C–100°C compared with the straight binary alloys. For commercial manufacture of single phase alloys, it is important that inter-stage anneals should take the least possible time. Hence they tend to be done at a higher temperature (about 800°C) than the final anneal.

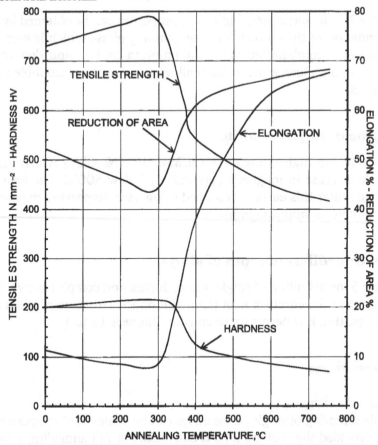

Fig. 6.1   Annealing curves for a CuAl7 copper-aluminium alloy previously
reduced 50% by cold rolling.[127]

A typical example of the use of full annealing is found in sheet production. In this
case, metal is rolled down to thin sections and the metal temperature gradually falls
from one pass to the next, to the point that, at the last pass, the process has become
one of warm rather than hot working. This may result in unacceptably low elongation and unnecessarily high proof strength. To restore the right combination of
properties, it is necessary to reheat to 860°C and quench in water. The metal is
then temper–annealed by soaking at 540°C for 3 hours and allowed to cool in air.
This relieves internal stresses and toughens the material, although strength and
hardness are reduced to some extent.

(2) Stress relieving
Stress relieving of single phase alloys containing up to 8% aluminium can be done
by annealing as above at temperatures from 400°C to 750°C, but the lower end of

this range should be used if appreciable softening is to be avoided. If little or no softening can be tolerated, considerable relief of internal stress can be achieved by temper annealing at a minimum temperature of around 300°C-350°C for 1/2 to 2 hours and cooling in air.

(3) Controlling hardness
Temper annealing at the same temperature is also used for achieving controlled reduction in hardness with appropriate soaking time. Temper annealing has the advantage over temper-rolling that the material is not left in a state of internal stress.

Duplex alloys

(1) Softening the alloy
The duplex-structured alloys rarely require full annealing as their properties in the hot-worked condition are suitable for most applications. However, after cold working, heating to 600°C–650°C followed by relatively rapid air cooling adequately softens them. With the binary or nickel-free alloys it is important to avoid the decomposition of the beta phase which accompanies very slow cooling, as it results in a less ductile and corrosion-prone structure containing the gamma$_2$ phase. Thus, water quenching is usefully employed particularly for heavier sections.

(2) Stress relieving
Among the duplex alloys, the silicon bearing alloy CuAl7Si2 is particularly susceptible to stress corrosion cracking. If the work piece has been subjected to significant amounts of cold working, it must be stress relief annealed at 350°C–450°C for at least half an hour followed by air cooling. With other duplex alloys, if the temperature is raised appreciably above 350°C, the alpha+beta phase may decompose on cooling and give rise to the corrosion prone gamma$_2$ phase. This is not a problem if the product is intended for a wear resisting application in a non-corrosive environment. If not, and if stress relieving at 300°C–350°C is not sufficient, a full annealing treatment may be necessary. This would consist in heating the alloy preferably to 600°C–650°C and quenching in water. If machining to very fine tolerance is to follow, the smaller quenching stresses may, if necessary, be removed by temper annealing at 300°C–350°C for 1/2 hour.

A duplex-structured alloy containing less than 2%–3% each of nickel and iron can be treated in a similar way by rapid air cooling or quenching from 600°C–650°C. For parts subject to extremely critical machining tolerances, the residual quenching stresses may be relieved by temper annealing at 300°C–350°C for 2 hours.

(3) Increasing hardness and mechanical properties
Table 6.1 shows the effect on the mechanical properties of rod, in a duplex alloy containing 9.4% aluminium of heating to 900°C, followed by quenching from

different temperatures. The as-drawn properties of a similar alloy containing
8.8%-10% Al and 4% max Ni+Fe, are given for comparison. It will be seen, by
comparison with the as-drawn properties,

- that the effect of quenching at 900°C is to increase the tensile strength but to
  reduce the proof strength,
- that subsequent tempering reduces the tensile strength but increases the proof
  strength which nevertheless remains less than in the as-drawn condition,
- that the effect of the tempering temperature on tensile strength is much more
  marked than on proof strength,
- that quenching at 900°C would seem to make little difference to elongation
  but subsequent tempering improves elongation significantly,
- that the higher the tempering temperature, the lower the hardness.

Table 6.1 Effect of different heat treatments on the mechanical properties
of rod in duplex alloy containing 9.4% aluminium.[127]

Heat Treatment			0.1% Proof Strength	Tensile Strength	Elongation	Hardness
Quenched	Tempered	For	N mm^{-2}	N mm^{-2}	%	HV
Duplex alloy containing 9.4% aluminium quenched only						
900°C	–	–	195	751	29	187
Duplex alloy containing 9.4% aluminium quenched and tempered						
900°C	400°C	1 h	212	750	29	185
900°C	600°C	1 h	238	699	34	168
900°C	650°C	1 h	223	646	48	150
As-drawn properties of similar alloy containing 8.8–10% Al and 4% max Ni+Fe						
As drawn			260–340	570–650	22–30	170–190

Table 6.2 Effect of tempering on the mechanical properties
of lightly drawn extruded rod in CuAl10Fe5Ni5 alloy.[127]

Form	Tempering conditions	Mechanical Properties			
		0.1% Proof Strength N mm^{-2}	Tensile Strength N mm^2	Elong. %	Hardness HB
Extruded Rod	None	455	798	23	248
lightly drawn	500°C for 1 h	510	832	18	262
2.5% reduction	600°C for 1 h	515	821	21	281
	700°C for 1 h	441	753	24	229
	800°C for 1 h	377	742	28	197

In conclusion, these figures show that it is possible to improve some of the properties of a duplex alloy by heat treatment and that various combinations of properties can be achieved by the appropriate choice of tempering temperature. Lowering the quenching temperature below 900°C would reduce properties generally except for elongation which would increase.

Although the above remarks refer to wrought duplex alloys, the properties of cast duplex alloys can be similarly altered to achieve certain desired properties.

(4) Improving fatigue strength

The tests by Musatti and Dainelli,[137] reproduced in Fig. 6.2, are of interest as they illustrate the influence of heat treatment on a duplex alloy containing 10.2% aluminium, 0.3% iron and 0.5% manganese. Their results clearly indicate the superiority of the quenched-and-tempered condition. Under corrosive conditions tempering at 400°C–500°C, however, would be undesirable due to the formation of the corrosion-prone eutectoid structure.

(5) Improving corrosion resistance

Aaltonen et al.[1] report that if a Cu/Al/Fe/Mn alloy, containing 10% Al, is heat treated between 600°C–700°C for three hours and air cooled, its corrosion resistance can be significantly improved (see Chapter 12).

Cu/Al/Ni/Fe type complex alloys

(1) Softening

Complex alloys of the Cu/Al/Ni/Fe type can be reduced to a condition of minimum hardness by annealing at 800°C–850°C and furnace cooling to 750°C. In this range most of the nickel-iron-aluminium kappa phase is precipitated in what is essentially an alpha matrix. Below 750°C air cooling is preferable but quenching does not give greatly inferior results and can be safely used. For most practical purposes soaking for 1/2–2 hours at 750°C followed by air cooling is adequate.

(2) Stress relieving

The complex CuAl10Fe5Ni5 type of alloys are generally stress relieved by soaking for about 30 minutes at a temperature within the range 400°C–600°C and air cooled. If this is insufficient to relieve all stresses, the higher temperature range of 650°C–750°C may be used, followed by air cooling. The softening induced by these high temperatures is rarely significant as the alloys are seldom work hardened, but if necessary, it can be minimised by limiting the time in the furnace to 1/2 hour at 600°–650°C.

In practice, some manufacturers obtain satisfactory stress relief by heating all nickel-aluminium bronzes to 450°C for one hour followed by cooling in air.

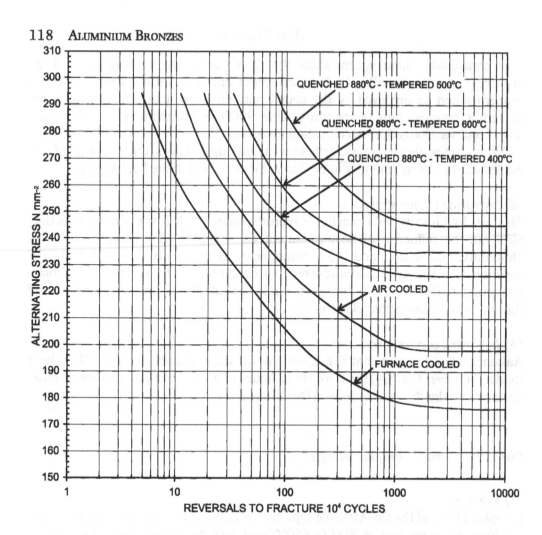

Fig. 6.2   Effect of heat treatment on the fatigue strength of a duplex aluminium bronze containing 10.2% Al, 0.3% Fe and 0.5% Mn.[127]

(3) Adjusting mechanical properties

(A) TEMPERING AFTER COLD WORKING

One of the most common forms of heat treatment consists in tempering after hot or cold working. This consists in re-heating to a certain temperature, typically 400°C–540°C, for 1–2 hours, then cooling in air. Figures given in Table 6.2, in the case of a CuAl10Fe5Ni5 alloy, show that tempering a lightly drawn rod at the moderately low temperature of 500°C can result in exceptionally high proof strength values with good elongation. Higher tempering temperatures do not give the same advantage.

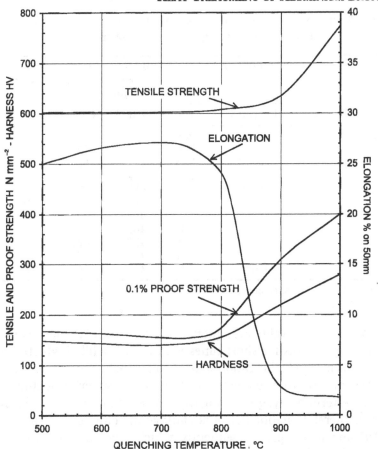

Fig. 6.3 Mechanical properties of a Cu110Fe5Ni5 alloy after slow cooling from 1000°C to various temperatures and quenching.[127]

(B) QUENCHING AFTER HOT WORKING

Fig. 6.3 shows the effect on the mechanical properties of a CuAl10Fe5Ni5 alloy of slow cooling from 1000°C to various temperatures and then quenching. It will be seen that quenching above 800°C dramatically reduces elongation but sharply increases hardness and proof strength. It also improves tensile strength.

(C) QUENCHING AND TEMPERING AFTER HOT WORKING.

The most common and readily controlled method of heat treatment involves quenching from a high temperature, followed by tempering at a lower temperature. This form of heat treatment is directly comparable to the tempering of steels.

Tempering quenched alloys at moderate temperatures increases both proof strength and hardness. This is illustrated by the properties quoted in Table 6.3 for samples tempered at various temperatures.

Table 6.3 Effects on the mechanical properties of a CuAl10Fe5Ni5 alloy of quenching followed by tempering at various temperatures[127].

Form	Heat Treatment			0.1% Proof Strength	Tensile Strength	Elonga-tion	Hard-ness
	Quenched	Tempered	For	N mm^{-2}	N mm^{-2}	%	HV
Rolled plate	1000°C	600°C	½ h	523	824	7	287
		800°C	½ h	303	773	9	221
		900°C	½ h	351	668	3	268
		500°C	2 h	464*	866*	2*	300*
		600°C	2 h	433*	819*	12*	260*
		700°C	2 h	340*	758*	14*	240*
Extruded rod	900°C	as quenched		362	932	8	235
		500°C	1 h	467	850	15	238
		600°C	1 h	470	827	18	244
		700°C	1 h	421	789	22	218
		800°C	1 h	371	767	26	192

* Values estimated from published curves

As may be seen from Table 6.3, quenching from 1000°C followed by tempering at different temperatures for short times shows that, while high strength may be obtained by this method, elongation figures can remain exceedingly low. However, by extending the period of tempering to two hours, superior ductility is obtained (see Fig. 6.4). Alternatively, quenching from lower temperatures will give improved ductility. It is therefore frequently of advantage in commercial practice to quench from 900°C and to temper subsequently which gives a much more favourable combination of proof strength and elongation values (see Table 6.3).

S. Lu et al.[123] investigated the effect on hardness of quenching and tempering some Cu/Al/Fe/Ni alloy castings at various combinations of quenching and tempering temperatures. Similarly they investigated the effect on hardness of hot-working and tempering forgings in the same alloy at various combinations of hot-working and tempering temperatures. The advantage of increasing hardness is principally to improve wear properties (see Chapter 10). The forging reduction was 40% in a single operation.

Their results are shown in Table 6.4. The highest hardness figures are obtained after tempering for 2–4 hours at temperatures in the range of 375°C-425°C, both in the case of quenched castings and of a hot-worked forgings. The higher the quenching temperature the higher the hardness figures. The effect of the hot-working temperature on the hardness of the forgings is similar and even more pronounced. Tensile strength and elongation are also shown but only for tempering at 400°C for 2 hours after quenching and hot-working. It will be seen that hardness is achieved at the expense of elongation figures which are very low. S. Lu et al.[123] point out, however, that by using the method of partial quenching, the excessive reduction in elongation can be avoided.

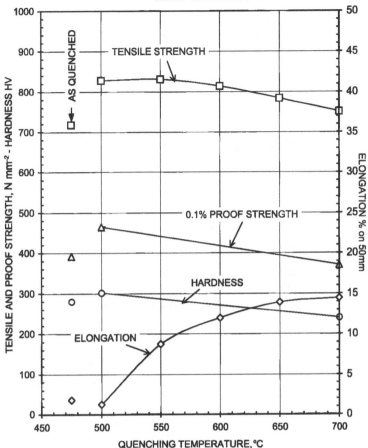

Fig. 6.4 Mechanical properties of **Cu10Fe5Ni5** alloy quenched at 1000°C and tempered for 2h at various temperatures.[127]

Since Rockwell Hardness C numbers, up to 40, are approximately 1/10th of corresponding Vickers numbers, it is possible to compare these results with those quoted in Table 6.3. This confirms that the best hardness achieved with a quenched and tempered casting or a hot-worked and tempered wrought product is obtained with at least 2 hours tempering time and at lower tempering temperatures than those given in Table 6.3. Tensile and elongation properties are however significantly less than those obtained at the tempering temperatures indicated in Table 6.3.

The effect on microstructure of tempering after quenching and hot-working is discussed in Chapter 13.

(4) Possible side-effect of heat treatment on creep strength and fatigue
Although various heat treatments may be employed to modify the mechanical properties, they are seldom required in commercial practice. This is due to close

Table 6.4   Effect of various heat treatments on the hardness of some Cu/Al/Fe/Ni alloy castings and forgings – by S. Lu et al.[123]

Tempering Temperature °C	Rockwell Hardness (HRC) Tempering time (hr)							Tensile Properties	
	1	2	3	4	6	8	12	Tensile Strength N mm⁻²	Elongation %
Casting quenched at 950°C and tempered									
Hardness as cast: 22.5									
Hardness as quenched: 31.3									
350	32	35	36.2	37.1	40	36.6	36.6		
375	39	39.5	39.2	39.1	38.2	37.5	37.3		
400	38	39	37.5	36	36.3	35.3	36	732*	1.84*
425	37.6	38.5	39.7	37.6	37.6	36.2	35.5		
450	37	36.8	34.7	34.5	34.2	34	28		
500	32.4	30.3	27.3	26.8	–	–	–		
550	29.6	28	27.7	27.4	–	–	–		
600	24.9	24	24.7	24	–	–	–		
Casting quenched at 850°C and tempered									
Hardness as quenched: 28.8									
400	35	33.6	34.2	34.5	34.5	32.5	28		
Forging hot-worked at 980°C and tempered									
Hardness as forged at 950°C: 27.7									
350	37	38	35.8	33.9	31.9	31.9	30.7		
375	35.9	37.5	37	37.3	35	33.3	–		
400	35.5	36.6	38.3	36.3	36.1	34.3	31.4	743*	1.96
425	34.8	36.3	35.5	35.6	36.5	35.8	–		
450	36.5	33.7	33.3	33.5	33.9	34.7	–		
500	31.5	29.5	30.5	28.5	–	–	–		
Forging hot-worked at 820°C and tempered									
400	32.5	37.7	32.6	31.5	30.3	–	31		
Forging hot-worked at 750°C and tempered									
400	27.6	–	28.5	28	–	26	26.5		

Above values are estimated from published curves and are based on average of 5–6 tests
* Tempered for 2 hours at 400°C

Composition of alloy (Wt %)			
Aluminium	**Iron**	**Nickel**	**Copper**
10.6	4.4	4.5	balance

control of production processes which generally give the required properties without recourse to heat treatment. In general, the properties of these alloys after working can usually be 'corrected' by heat treatment, but there are certain side effects which should be considered beforehand. McKeown et al[110] have found that the creep strength of heavily worked CuAl10Fe5Ni5 alloy is reduced by the heat treatment recommended earlier. Some care is therefore required in selecting, heat-treated material for creep resistant applications. R.E. Berry[127] has also suggested that fatigue properties can be similarly affected.

(5) Improving corrosion resistance

The corrosion resistance of Cu/Al/Ni/Fe alloys can be improved by annealing at 675°C for to 2–6 hours, depending on section thickness, followed by air cooling. Some claim that better results are obtained at 700°C[178–74]. This heat treatment is particularly advantageous after welding (see Chapter 13). B W Turnbull[178] reports that too slow a rate of cooling in air may reduce the elongation value below the specified value and he recommends cooling in an air blast immediately after withdrawing the casting from the furnace as quickly as possible. Restricting the aluminium content to 9% would probably achieve the same objective.

Cu/Mn/Al/Fe/Ni type complex alloys

(1) Stress relieving

Stress-relief annealing of the high manganese CuMn13Al8Fe3Ni3 alloy involves heating to 500°C-550°C for 2 hours though shorter times may be used at slightly higher temperatures.

(2) Adjusting mechanical properties

A) TEMPERING AFTER COLD WORKING

Whilst normally used in the as-cast or hot-worked condition, the high manganese CuMn13Al8Fe3Ni3 alloy may require tempering if cold-worked. This is best achieved by heating to 700°C-750°C for one hour followed by slow cooling to 500°C. The alloy with the higher aluminium content (8.75%) has a softening temperature approximately 50°C lower.

B) QUENCHING AND TEMPERING AFTER HOT WORKING.

Quenching from 825°C followed by tempering gives a useful increase in tensile strength at the expense of ductility. Although heat treatment is rarely required in practice, Table 6.5 indicates the properties attainable with a CuMn13Al8Fe3Ni3 cast or wrought alloy. For comparison, the properties of material before heat treatment are also given.

   Under all practical conditions, slow cooling of this alloy does not have an adverse influence on the structure or mechanical properties. Some change in hardness, however, occurs around 500°C; this is associated with atomic re-ordering of the beta phase and, as previously mentioned, it has a more pronounced effect on magnetic properties.

Table 6.5 Effects on the mechanical properties of a **CuMn13Al8Fe3Ni3** alloy of quenching from 850°C followed by tempering at various temperatures.[127]

Condition	0.1% Proof Strength N mm⁻²	Tensile Strength N mm⁻²	Elongation %	Hardness HB
As cast	309	696	25	180
Forged	464	773	25	200
Quenched at 825°C and re-heated at:				
350°C for 4 hrs	696	928	1	300
450°C for 4 hrs	572	850	5	270

Table 6.6 Effect of various heat treatments on properties of sections cut from a large propeller in Cu/Mn/Al/Fe/Ni alloy – by A Couture et al[56]

Annealing Treatment			Impact	Tensile Properties			Reduc. Area	Hardness Rockwell A		Magnetic Permeability μ*	
Temp °C	Time h	Cooling Rate °C/h	Sharpy test J	Tensile Str. N mm⁻²	0.5% YS N mm⁻²	Elong %	%	Before	After	Before	After
Alloy 1											
	As cast:		8	–	–	–	–	–	–	–	–
690	24	25	14	–	–	–	–	48	50	5.55	4.75
690	7	55	20	–	–	–	–	49	51	5.65	5.40
690	24	55	20	–	–	–	–	48	50	6.10	5.30
720	7	25	20	–	–	–	–	49	50	5.55	5.40
720	24	25	22	–	–	–	–	48	50	6.10	5.55
720	7	55	26	–	–	–	–	49	51	6.10	4.30
720	24	55	22	–	–	–	–	48	51	5.65	5.40
750	7	25	20	–	–	–	–	48	49	6.10	5.30
750	24	25	21	–	–	–	–	48	50	5.45	5.40
750	7	55	23	–	–	–	–	49	52	5.15	5.40
830	3	Quenched	20	–	–	–	–	–	56	5.40	0
Alloy 2											
	Test bar:			620	276	20	–	–	–	–	–
	Cast section:		10	522	241	20	20	–	–	–	–
690	24	55	19	586	270	20	18	48	50	2.95	3.40
720	5	55	19	596	277	20	20	48	50	3.02	3.50
720	24	25	18	567	257	23	24	48	49	3.12	3.50
720	24	55	18	585	266	22	24	49	50	3.40	3.50
750	5	25	18	578	258	24	26	48	49	3.35	3.50
750	24	25	18	517	254	15	20	48	49	3.35	3.60

* Converted from Magné-gage readings

Alloy composition (Wt %)

Alloy	Cu	Mn	Al	Fe	Ni	Pb	Al+Mn/6
1	72.1	14.68	7.59	3.55	2.14	0.022	10.04
2	73.2	13.50	7.44	3.22	2.51	0.046	9.69

(C) ANNEALING LARGE PROPELLER CASTINGS

Due to slow cooling following the casting operation, the properties of thick sections of large propellers may differ significantly from those of a standard test bar as was explained in Chapter 3. One property which is markedly affected is the impact strength. This is of particular concern in the case of icebreaker propellers which are sometimes used to plough up the ice to allow the ship to proceed. Couture et al.[56] carried out a series of heat treatment experiments on two high manganese aluminium bronze propellers to ascertain the heat treatment that would best improve impact properties. Their results are shown in Table 6.6. Sections cut out of a propeller to Alloy 1 were heat treated as indicated and tested for impact, hardness and magnetic permeability. Sections cut out from propeller to Alloy 2 were additionally subjected to tensile tests and the results can be compared with the tensile properties of the standard test bar of the same melt.

It will be seen that the impact values of the heat treated sections have increased by a factor of 2 to 3, although most are still significantly lower than those of separately cast test bars (34–48J – see Chapter 3). The most effective heat treatment, from the point of view of improving impact values, is annealing at 720°C for 7 hours and cooling at 55K h^{-1}. However, in order to avoid warping at high temperatures, Couture et al. recommend annealing at 720°C for 24 hours and thereafter cooling as fast as practicable.

The tensile properties of Alloy 2 are all higher than those of the cast sections although mostly lower than those of the separately cast test bar.

(3) Improving magnetic properties (manganese-aluminium bronze)

In the case of the high manganese Cu/Mn/Al/Fe/Ni alloys, some atomic re-ordering in the beta phase occurs at temperatures below 500°C. This results in a considerable increase in magnetic permeability when atomic re-ordering is allowed to go to completion at extremely slow cooling rates below 500°C. For example, the permeability may rise up to 15 from the more normal sand-cast values of between 2 and 5. Quenching from 550°C completely suppresses this reaction and the relative permeability then remains in the region of 1.03 which is appreciably lower than the magnetic permeability of Cu/Al/Ni/Fe alloys (1.40) and is comparable to that of the Cu/Al/Si alloy (1.04). This process has little effect on mechanical properties.

7
WELDING AND FABRICATION
(including metallic surfacing)

Welding Applications

One of the attractive features of aluminium bronzes is the relative ease with which they can be welded by trained welders. This weldability, combined with the fact that aluminium bronze can be hot or cold worked into a great range of shapes and sections, opens up the possibility of fabricating a variety of vessels and structures which, if necessary, can also incorporate aluminium bronze castings. Attachment of cast flanges or branches to pipes, vessels, etc. are common applications of this principle. Furthermore, aluminium bronzes can be welded to steel. It is therefore possible to reduce the overall cost of a structure by making in aluminium bronze only those parts exposed to a corrosive substance and the remainder in steel. Building-up and hard facing by welding layers of aluminium bronze on a metal surface are also possible as is metal spraying.

The weldability of aluminium bronze castings can be used to advantage in the case of a difficult complicated casting. In such a case, it may be more practical and more economical to weld certain separately cast or wrought parts to a basic casting. A further advantage of such a procedure is that, in cases where a basic body design (pump, valves etc.) is liable to be used in different installations, various mounting legs or lugs can be welded to the basic body as required. This can significantly reduce the cost of the finished component and reduce pattern making costs.

Welding is also used on occasions to repair casting defects, although this is no substitute for good foundry practice. Such repairs are not allowed on some categories of Naval castings used in high risk installations. One of the great advantages using aluminium bronzes for propellers is that they can be repaired by welding when damaged in service.

Correction of over-machining and reclamation of service-worn or damaged components are other advantageous applications of the weldability of aluminium bronze.

Aluminium bronzes can be joined by most welding processes, as well as by brazing and soldering, although the last method is seldom required for these alloys. Resistance (spot and seam) welding of sheet is possible but is seldom required. Friction welding of suitably shaped sections is practicable and is yet another means of joining aluminium bronze to other metals and alloys.

An understanding of the basic metallurgy of aluminium bronze is of assistance in selecting appropriate joining processes and reference to the chapters on metallurgy (11, 12, 13 and 14) is recommended.

Welding Characteristics

Aluminium-rich oxide film

The aluminium-rich oxide film that forms on the surface of aluminium bronzes gives them their outstanding corrosion resistance, but can impede welding. It has a high melting point and prevents the coalescence of molten drops of the alloy. The potential problem is the danger of entrapment of oxide in the weld. The right choice of welding process, correct welding practices and the experience of the welder should ensure that, as the oxide film on the base metal melts, it does not form inclusions in the weld and that the oxidation of the molten metal is prevented during welding. Pre-weld and inter-run cleaning of the metal is all important.

Thermal conductivity and expansion

Allowance must be made for the fact that aluminium bronze has a higher thermal conductivity and expansion than common low-alloy steels.

Higher thermal conductivity means that heat is dissipated more rapidly which accelerates the solidification of the weld metal. It also means that, since heat travels further, the zone in which the metallurgical structure of the base metal is altered by the heat of the weld, is wider. This is known as the heat-affected zone (see below: 'Effects of welding on properties'). The fact that the heat travels further, combined with the higher thermal expansion of the alloy, also means that the expansion and contraction caused by the weld is greater than for steel. It therefore causes potentially greater internal stresses in the metal. It is important therefore that no undue restraint be applied.

The effects of these differences can be catered for by correct joint design and jigging to avoid undue restraint and by restricting unnecessary heat spread by the correct choice of welding process and technique.

Ductility dip

Fig. 7.1 shows, for different alloys, the elongation that corresponds to the tensile strength (at which fracture would occur) at any given temperature, as each alloy cools from its solidification point to room temperature. This elongation goes beyond the elastic limit and represents the maximum permanent plastic deformation that the alloy can experience as it reaches the point of fracture. It could therefore be called the 'fracture' elongation and is therefore the limit of ductility of the alloy at a particular temperature. Ductility falls therefore from a high value at high temperatures, to a minimum at some intermediate temperature, before rising again as the temperature goes on falling to room temperature. The alloys are at their least ductile, and therefore most vulnerable to fracture, at temperatures where the

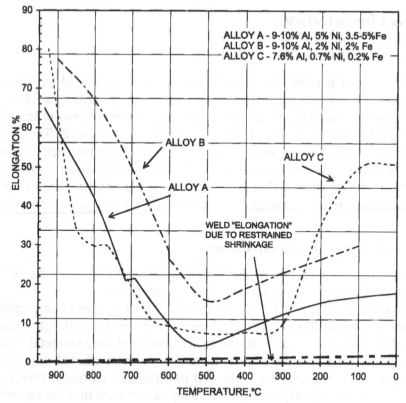

Fig 7.1 Variation of the maximum elongation with temperature of various aluminium bronze alloys in the as-cast condition showing the ductility dip, by Weill Couly.[184]

elongation curve reaches its minimum value. This phenomenon is known as the 'ductility dip'. In the case of nickel-aluminium bronzes with 9%–10% Al, the ductility dip occurs within the temperature range of 600°C–400°C and is most pronounced in the case of the alloys with the higher nickel content. In alloys with 7.6% Al and only small nickel and iron additions, the ductility dip extends over a wider range of temperatures from 650°C–300°C.

The maximum elongation shown on Fig. 7.1 is for alloys in the as-cast condition. The maximum elongation for weld metal of the same compositions is likely to be lower as indicated by the figures given in Table 7.4. Weld metal is therefore less ductile. This is because the rapid cooling of the weld metal results in a more brittle beta-phase microstructure (see Chapters 11–14).

Implications of ductility dip for welding
The implications of this phenomenon for welding are as follows:

1.   As the weld metal cools from the melting temperature to room temperature, it may be prevented from shrinking normally by the restraining effect of the

parent metal. It becomes, in effect, 'elongated' relative to what it would have been had its shrinkage not been restrained. The resultant 'elongation' is likely to be numerically similar to the percentage shrinkage that normally occurs on cooling and is shown at the bottom of Fig. 7.1. It will be seen that it lies below the 'fracture' elongation of the as-cast alloys at the lowest point on the curves, but the difference between the actual elongation of the weld metal and the 'fracture' elongation of the alloy is small, particularly in the case of the high nickel alloy 'A'. Furthermore, as mentioned above, the 'fracture' elongation of weld metal is likely to be appreciably less than that of the as-cast metal represented in Fig. 7.1. As long as there is a difference, however, between the actual elongation of the weld metal and its 'fracture elongation', weld cracking will not occur. But the fact that this difference is small over the range of temperature of the ductility dip means that there is a potential vulnerability to weld cracking over this temperature range for reasons that will now be given.

2. In practice, any impurities, inclusions or gas holes in the weld metal may lower the 'fracture' elongation of the weld metal below its restrained shrinkage 'elongation', thereby exceeding the tensile strength of the defective weld metal and resulting in weld fracture. In the case of alloy 'C', with 7.6% Al and only small nickel and iron additions, the lowest 'fracture' elongation is slightly higher than that of alloy 'A', but the ductility dip is very wide, making the alloy vulnerable to cracking over a wider temperature range. In fact these single-phase alloys are very difficult to weld under restraint.

3. The parent metal will have been heated, on average, to a lower temperature than the weld metal. The rate of cooling in air of the weld metal may therefore be higher than that of the parent metal which goes on receiving heat from the weld. Consequently the weld metal may tend to shrink faster than the parent metal resulting in a greater differential shrinkage between the weld metal and the parent metal. If this happens, it will cause an increase in the effective elongation of the weld metal and may lead to fracture. For this reason it is recommended not to preheat at all, if possible, and adjust the rate of welding to ensure that there is no significant build up of temperature. The inherent speed of MIG welding can easily lead to excess heat input in the case of short runs applied in quick succession. TIG welding is preferable to MIG welding in this respect. On critical weld repairs, Weill-Couly[184] recommends the use of a thermocouple whose terminals are welded to the work piece close to the weld to check that the temperature of the parent metal does not rise into the low ductility range.

4. Pre-heating is only advantageous if it reduces the temperature difference between the welded area and the remainder of the parent metal. At such high temperatures, it would make welding very uncomfortable.

5.   The problem of constraint leading to weld 'elongation' and corresponding internal stresses is unlikely to arise in a welded assembly provided the assembly does not apply a constraint on the seam being welded. Hence the importance of correct joint design and jigging.

Effect of alloying elements on ductility dip

ALUMINIUM

Aluminium has a very marked effect on elongation from high temperature to room temperature and therefore on the ductility dip. The elongation of a two-phase alloy, falls gradually to zero as the aluminium content increases from 8% to 12%. As a result, welding becomes increasingly difficult as the aluminium content is increased and is almost impossible beyond 11%. Weldability is good between 8.2% and 10.7% and optimum at 9.5%. (As shown in Fig. 1.1, Chapter 1, the best combination of mechanical properties in two-phase and multi-phase alloys is obtained at aluminium content of 9.4–9.5% and it will be seen from Chapters 12 and 13 that this level of aluminium content is desirable for corrosion resistance.)

In a single-phase alloy (i.e. less than 8% Al), the elongation is very high at room temperature but falls rapidly with rising temperature towards the critical zone (see Fig. 7.1). As mentioned above, single-phase alloys are very difficult to weld.[184]

IRON

According to Weill Couly, iron has no influence on weldability.[184] In the case of single-phase alloys, however, the strengthening effect of iron is considered by some to be beneficial in resisting weld cracking.

MANGANESE

A high percentage of manganese improves weldability but is kept low for reasons of corrosion resistance (except in the case of manganese-aluminium bronze).[184]

NICKEL

Nickel which improves strength and corrosion resistance makes welding more difficult as is evident from Fig. 7.1. With a 7% nickel content, the alloy is so crack-sensitive that welding is almost impossible. A 5.5% nickel content is a practical maximum for welding.[184]

FILLER METAL

A filler metal containing 8.5–9.2% Al, 2.7–3% Ni, 1–2% Fe, 1% Mn and bal. Cu, developed by Weill Couly,[184] has proved particularly suitable for crack-free welds and has been prescribed by the French navy for both repair and welded fabrications. It is, however, always preferable for corrosion resistance to use a matching filler metal.

IMPURITIES

For freedom from cracking, lead and tin and bismuth must be kept within close limits, preferably 0.01% maximum lead and 0.1% maximum tin.

Conclusions on effects of ductility dip on welding

1. The 'fracture' elongation of weld metal is likely to be less than that of the parent metal shown in Fig. 7.1.
2. Single-phase aluminium bronzes (approximately less than 8% Al) have the most extended ductility dip and are therefore at risk of cracking over a wider temperature range (650°C–300°C) on cooling.
3. Nickel aluminium bronzes have a narrow ductility dip (600°C–400°C) and, in this respect, are less vulnerable to weld cracking.
4. For best corrosion resistance, the nickel-aluminium bronze filler metal should be of matching composition to the parent metal and should preferably have a 9.5% aluminium content. If corrosion resistance is not critical, nickel should be as low as other considerations will allow. Manganese helps to reduce the risk of cracking but should not exceed 2% for corrosion resistance. Iron is thought to have little or no effect on cracking in the case of nickel aluminium bronze. In single-phase alloys small additions of iron are beneficial.
5. Lead should not exceed 0.01% and tin 0.1%. Other impurities should be as low as possible.
6. Avoidance, by good welding practice, of inclusions and gas holes in the weld metal is of critical importance to prevent weld cracking.
7. If shrinkage of the weld metal is likely to be restrained, as in most weld repairs of castings, heat input to the parent metal should be kept to a minimum by adjusting the rate of welding.
8. It is recommended not to preheat at all if the weld metal is likely to be restrained.
9. Pre-heating is only advantageous if it reduces the temperature difference between the welded area and the remainder of the parent metal.

Choice of Welding Process

There is now a wide choice of welding processes, of various degrees of sophistication, which are suitable for welding aluminium bronzes. Each has its particular advantages and limitations. Economic considerations and the welding characteristics of the particular alloy, together with the form of the components to be welded and their previous history, must all be taken into consideration when choosing the process and filler metal to be used.

Tungsten-arc inert gas-shielded (TIG) process

The TIG process is most suitable for the welding of thin plates, for localised weld fabrication or for casting repairs. In this latter application, it presents far less risk of slag inclusions than manual arc or gas-welding (see below) and is free of the problems of flux inclusions. It is also used to secure good and controllable root penetration for the root runs prior to MIG welding (see below).

In the TIG process a tungsten electrode is held manually to strike the arc which is shielded by inert gas. The welder holds in his other hand a filler rod of the appropriate alloy. The heat of the arc melts the base metal locally, forming a small pool of molten metal into which the filler rod is hand-fed. This manual control of the feed metal is slower than in the MIG process but is better for delicate operations.

Argon or helium gas is used as appropriate (see 'Welding practice' below)

The filler rod should be of the same quality as the filler wire employed for the MIG process, but of a bigger diameter, as the rate of metal feed is not dependent on the thickness and thin wire requires replacing too frequently.

There is a tendency with TIG welding for oxide inclusions to form at the root of the weld which may act later as crack initiators. To overcome this problem, Weill Couly[184] recommends coating the wire lightly by dipping it prior to use in a sodium silicate paste that contains cryolite. This effectively lowers the melting point of alumina from over 2000°C to 804°C and eliminates the defect without difficulty. Water is removed from the coating by short-circuiting the electrode to heat it.

There are two refinements to the basic TIG process: the Pulsed Current TIG process and the Plasma Arc process.

Pulsed Current TIG process
The object of this refinement to the TIG process is to obtain a better control of the heat input of the tungsten arc. It is of particular value in welding thin gauge material, in avoiding heat damage to insulation or to the surrounding structure and in the repair of castings.

In this technique, a low background current maintains the arc between the tungsten electrode and the work but is not large enough to cause melting. High current pulses are superimposed at regular intervals which provide the necessary increase in heat input for fusion to occur as well as the means of controlling the heat input of the conventional tungsten arc. Mechanisation of the process is essential for best results.

Plasma arc process
This refinement of the TIG process is used where close control of under-bead penetration is required, such as in high quality butt welds in thin gauge tube and sheet. Very accurate welding of thinner gauge material can be achieved by mechanising the process, although the above Pulsed TIG welding is a preferred option thanks to developments in its electronic control.

In the Plasma Arc process, a pilot arc is initiated between the nozzle and the tungsten electrode by a high frequency spark which ionises the inert gas stream (thereby creating a 'plasma') and causes the main arc to ignite between the electrode and the work. Since only a small amount of gas is needed to maintain the arc at a workable level, additional gas for shielding is provided to ensure complete protection of the weld area. This is supplied through an outer nozzle fitted with an arrangement designed to ensure a smooth laminar flow. Gas flow rates are in the range of $1.5–5.0$ min^{-1} for the plasma gas and $9–15$ l min^{-1} for the supplementary shielding gas. Both plasma and shielding gas are normally argon but, for certain special welding applications, argon/helium and argon/nitrogen mixtures are used for the plasma gas.

Power requirements are similar to those for the conventional TIG process, up to 400A being quite usual, although arc voltages are often considerably higher. A d.c. electrode-negative arc is generally used.

The nozzle, which is water-cooled, constricts the arc, thereby raising its temperature to a very high level. In effect, the arc becomes a uniform, high energy column of plasma which enables a hole to be melted right through the joint as welding proceeds. Molten metal flows in behind this hole by surface tension to give a smooth uniform weld and penetration bead. The welding current is adjusted so that the 'hole' action only just occurs. The circuits can be arranged so that the welding current is automatically tapered off to provide conditions suitable for starting and finishing the weld where run-on and run-off plates cannot be used.

A further refinement, known as the micro-plasma process, enables very thin material to be welded. A constriction is applied to the arc which has a stabilising effect that permits the use of very low welding currents (typically $0.6–15$A). Micro-plasma welding may be either manual or mechanised.

Pulse current techniques, with their attendant advantages, may also be applied to the plasma arc process.

Metal-arc inert gas-shielded (MIG) process

The fastest and most efficient way to weld aluminium bronze is by the MIG process. This process, which is widely used in industry, gives the most satisfactory results for heavy aluminium bronze fabrications. It lays down a deposit at a very high rate which minimises any tendency to weld cracking because of the low heat input in any given area. It is ideally suited for long weld runs. It is therefore less well suited to localised casting repairs than TIG welding although it is better for this purpose than manual metal arc welding or gas welding.

In the MIG process, the operator uses a hand welding-gun through which a filler wire electrode is automatically fed from a reel. A flow of inert gas also passes around the wire down the nozzle of the welding gun, shielding it from contact with air as in the TIG process. Argon is the most commonly used inert gas for the MIG process, but an argon–helium mixture may also be used according to the application (see 'Welding practice' below).

The filler wire is connected, as the positive electrode, to a direct current supply and the arc is struck between the end of the wire and the work-piece. The heat thus generated melts the base metal locally as well as the tip of the filler wire creating a spray of filler metal. The inert gas ensures that the weld metal is shielded from the oxidising action of air. The current, the arc voltage and the rate at which the wire is fed are pre-set by the welder to suit the wire diameter.

For thin gauge material the MIG process deposits metal at such a high rate that full control over the operation becomes difficult and the TIG process is more suitable.

As in the case of TIG welding, there is a refinement to the MIG process known as 'Pulsed-current MIG welding'.

Pulsed-current MIG welding

This refinement of the basic MIG process offers a means of controlling heat input in relation to both the quantity and quality of weld metal deposited and is particularly attractive for localised welding where good penetration with a small controllable weld pool is essential. Welds made by this process are uniform and free from spatter.

The process consists in superimposing pulses of high (spray transfer range) current on a low background current which is sufficient to pre-melt the electrode tip and maintain the arc. The high current pulses detach and transfer the metal as small droplets which are similar to those formed in free-flight spray transfer, and the process is normally adjusted so that each current pulse detaches one droplet. In this way, spray transfer conditions are achieved at very much lower average operating currents than in conventional MIG spray welding.

Square-wave pulse current sources are available which are more versatile than the conventional sine wave sources in that they offer control over the magnitude, duration and repeat frequency of the pulse, all of which has a profound influence on the weld bead characteristics.

Other electric arc processes

Some older processes are still used for maintenance and repair where MIG and TIG equipment is not available. They require the use of fluxes capable of dealing with the refractory alumina film and care must be taken to avoid entrapment of the residues in the weld metal to the detriment of physical properties. They are relatively slow processes.

Carbon-arc welding

In carbon-arc welding, a carbon electrode, connected to a d.c. supply, is used to strike the arc and weld metal is provided by a separately-held rod of filler metal. A powdered cryolite flux is used to protect the weld metal from oxide inclusions. It is generally considered that carbon-arc welding permits better control of the weld deposit than metal-arc welding and is therefore preferable to it. The value of this method for the repair of castings is that it is possible to make a relatively large deposit, to eliminate totally every trace of cryolite and to obtain self-preheating owing to the large size of the arc. [184] Carried out by a trained welder, it can give excellent results.

Manual metal-arc welding

In manual metal-arc welding, a flux-coated filler rod, connected to a d.c. supply, is used both to strike the arc and to provide the weld metal. Flux-coated filler rods are available commercially for the purpose. It is a less satisfactory process than the carbon-arc process as the electrodes do not run well. The electrodes also have a tendency to absorb humidity which changes into hydrogen, causing gas blowholes in the metal as it melts. Drying the electrodes at 120°C avoids this problem.

Electron-beam welding

Electron-beam welding consists in focusing a beam of high-velocity electrons on the weld area. The kinetic energy of the electrons is converted into heat energy which melts the metal. The process has to be carried out in a vacuum chamber since the presence of molecules from the air would interfere with the working of the electron-beam and would cause oxidation. No filler wire is needed.

The advantage of this process is that the rapid melting restricts the heat-affected zone and hence the possibility of distortion. The disadvantage is that the speed of the process results in a corrosion-prone beta structure (see effect on corrosion resistance below).

A work table within the vacuum chamber allows three-axis linear and rotational movement, permitting high speed accurate welding under clean conditions. Large chambers can be made to accommodate large structures.

Since the weld is very narrow and penetration deep, plain butt joints are normal, but edge finish and fit up must be good. There is usually no joint gap but small gaps, up to one millimetre may be required for thicker sections.

Friction Welding

Friction welding has been used for joining similar and dissimilar metals for several decades and can be used to make joints between aluminium bronzes and other metals.

Originally the process was only applied to those joints where at least one component was circular and could be rotated against the other to provide the frictional heat for welding before a forging end-force was applied. Later developments have consisted of orbital friction welding where one component is moved, like an orbital sander, in a small orbit against another surface. In this way a variety of sections can be joined. A more recent development is 'friction stir-welding' in which a welding wire or rod is rotated or 'stirred' between the surfaces to be joined.

In all these processes, the relative movement of the interfaces causes the asperities to heat up, soften and plastically deform. When the relative motion is quickly stopped and a load is applied normal to the interfaces, surface oxides are squeezed out into a 'flash' which can be cropped off. This leaves a metallurgically clean and sound weld, with the metal in a forged rather than cast condition.

Oxy-acetylene gas welding

Oxy-acetylene gas welding, which uses a gas torch to melt the base metal and the separately held filler rod, is barely practicable on any but very thin sections. It has effectively been superseded by the various forms of electric arc welding. The principal difficulty in gas welding is caused by the oxide film, particularly that of the liquid metal. Disturbance of the surface of the weld pool produces more oxide and, to prevent oxidation, relatively large quantities of highly reactive fluxes are required. These must be maintained at a high temperature to dissolve the alumina and the oxy-acetylene flame does not generally provide sufficient heat. Furthermore, the high thermal conductivity of the base metal removes heat from the weld area and it is almost impossible to avoid a series of defects ranging from oxide inclusions to lack of fusion with weld cracking often resulting from stress concentrations at defects.

Welding Practice: General

The aim of good welding practice[184] is to achieve:

- a compact weld zone free of oxide inclusions, blow holes etc.
- a stress-free weld metal and heat-affected zone,
- a weld metal free from cracks, microcracks and susceptibility to cracking,
- a weld metal and heat-affected zone with the required mechanical properties,
- a weld metal and heat-affected zone with no significant tendency to corrosion.

Weld procedure and welder approval

In the case of companies approved to naval or to certain commercial quality standards, it is necessary to have an approved welding procedure and to submit welders to qualifying tests. This includes the preparation and completion of a specified weld specimen which is micro-etched with a reference number and subjected to dye-penetrant testing, tensile tests and, possibly, to radiographic examination. Welding should not be attempted without approval to a recognised welding procedure or standard *applicable to the alloy.*

Cleanliness and freedom from grease

For high quality work and consistent results, the area within and around the weld to a minimum distance of 50 mm, and in particular the edges of the material, must be free from grease, dirt, non-volatile marker, dye-penetrant testing compounds, visible oxides and all other form of surface contamination. The weld area should be brushed with a stainless steel wire scratch brush to expose clean metal. The wire brush should not have been previously used to clean other materials. This is followed by de-greasing with petroleum, ether or alcohol, which is normally sufficient. Even though the surface appears bright, there is a risk of superficial oxide films entrapping moisture. Degreasing should be carried out if no other surface preparation is necessary, otherwise oil films will cause porosity. It is generally preferable to err on the side of over-cleaning.

The use of the wire brush between runs to remove as much as possible of the oxide film formed during welding (and flux residues if applicable) is essential to avoid the risk of weld porosity and lack of fusion.

Selection of filler metal for TIG and MIG welding

Whilst it is possible to weld aluminium bronzes without additional filler metal by careful application of inert gas shielding to the top and underside of the weld area to prevent atmospheric contamination, or by placing reliance on the 'built-in' deoxidising characteristics of the alloy, it is generally advisable to use the filler metal specially developed for gas-shielded arc welding processes. Where possible, filler metals should match as closely as possible the chemical composition of the parent metal to avoid corrosion problems. They are commercially available in a standard range of diameters in straight rod form for TIG welding and in reels of wire for MIG welding.

Filler metals must be scrupulously clean to avoid introducing contamination, and, in the case of MIG welding, to provide good current contact with the contactor tube of the welding gun.

Table 7.1 gives details of some standard aluminium bronze filler metals. As explained above, harmful impurities in the base and filler metals result in low strength and ductility at intermediate temperatures. It is therefore essential that elements such as lead, tin and bismuth are kept to a minimum if cracking is to be avoided.

(a) Single-phase alloys

Single-phase alloys are alloys containing less than 8% aluminium. As previously mentioned, they are most vulnerable to cracking due to the wide ductility dip. Additions of iron reduce the cracking tendencies and the rapid solidification that occurs with MIG welding is a help in eliminating cracking.

Although they contains less than 8% aluminium, copper-aluminium-silicon alloys are not single-phase alloys, but in fact duplex alloys due to the effect of the silicon which is an aluminium equivalent.

The following are the single-phase alloys and their matching filler metals:

- Cu/Al alloys with less than 8% aluminium, filler metal: CuAl8
- Cu/Al/Fe alloys with less than 8% aluminium, filler metal: CuAl8

Thin gauge material in single-phase alloys, requiring a single run of weld, are unlikely to suffer from the ductility dip and may be welded using a matching CuAl8 filler in order to obtain the same corrosion resistance in the weld as in the base metal.

Table 7.1 Recommended Filler Metals to CEN standards

CEN Symbol	CEN Number	Nearest former BS 2901	Al	Fe	Mn	Ni	Si	Others specified max
CuAl8	CF309G	C12	7.0–9.0	0.5 max	0.5 max	0.5 max	0.2 max	Pb 0.02 Sn 0.1 Zn 0.2
CuAl10Fe1	CF305G	C13	9.0–10.0	0.5–1.5	0.5 max	1.0 max	0.2 max	Pb 0.02 Zn 0.5
CuAl6Si2Fe	CF301G	C23	6.0–6.4	0.5–0.7	0.1 max	0.1 max	2.0–2.4	Pb 0.05 Sn 0.1 Zn 0.4
CuAl9Ni4Fe2Mn2	CF310G	C26	8.5–9.5	2.5–4.0	1.0–2.0	3.5–5.5	0.1 max	Pb 0.02 Zn 0.2
CuMn11Al8Fe3Ni3	–	C22	7.0–8.5	2.0–4.0	11.0–14.0	1.5–3.0	–	–

Cu: remainder — Total of other elements not specified: 0.2 max

With thicker gauge materials, requiring more than one run, it is recommended that a filler with a higher aluminium content be used for root runs, such as CuAl10Fe1 which is more ductile and therefore less vulnerable to the ductility dip. To obtain the same corrosion resistance however, a capping run of CuAl8 is advisable. If weld beads are subsequently machined flush, care should be taken not to expose the CuAl10Fe1 material which would be vulnerable to corrosion.

(b) Duplex alloys

Duplex alloys are two-phase alloys which contain more than 8% aluminium (including aluminium equivalent). They have good ductility and are the least prone to cracking problems. With the exception of the Cu/Al/Si alloys, they are however vulnerable to corrosion due to the presence of the beta phase. The following are the main categories of duplex alloys with their matching filler metals:

- Cu/Al and Cu/Al/Fe alloys with more than 8% aluminium, filler metal: CuAl10Fe1
- Cu/Al/Si alloys, filler metal: CuAl6Si2Fe.

These alloys are normally satisfactorily welded using their matching filler metal.

Cast Cu/Al/Fe alloys are mostly die-cast and not suitable for corrosive environments. Cast Cu/Al/Si alloys have good machining and corrosion resisting properties and low magnetic permeability. These properties make them popular for certain Naval and commercial applications. The use of matching filler is essential.

(c) Complex alloys
Complex alloys have four or more alloying elements. They are the strongest alloys and have good corrosion resistance.

The following are the main categories of complex alloys and their matching filler metals:

- Cu/Al/Ni/Fe alloys, filler metal: **CuAl9Ni4Fe2Mn**2.
- Cu/Mn/Fe/Ni alloys, filler metal: former BS2901 C22

Most aluminium bronze castings are made in the Cu/Al/Ni/Fe alloys. As previously mentioned, a filler metal containing 8.5–9.2% Al, 2.7–3% Ni, 1–2% Fe, 1% Mn and bal. Cu, developed by Weill Couly[184], has proved particularly suitable for crack-free welds and has been prescribed by the French navy for both repair and welded fabrications.

Wrought and cast Cu/Al/Ni/Fe alloys are stronger but less ductile than duplex alloys during cooling after solidification and therefore more prone to cracking during welding under restraining conditions. Although the permitted aluminium content for this type of alloy is 8.5–10.5%, the following is recommended:

(a) that the relationship of the aluminium to nickel content should be according to the relationship: $Al \leq 8.2 + Ni/2$ in order to prevent the formation of undesirable and embrittling metallurgical phases (see also 'Effects of welding on properties' below), and

(b) that the aluminium content should be within 9.4%–9.8% for optimum combination of tensile strength and of ductility for welding purposes.

If cracking is likely to be experienced, it may be advisable to use a CuAl10Fe1 filler for the root and filling runs, followed by a capping of CuAl9Ni4Fe2Mn2 as appropriate to provide matching corrosion resistance. If the capping metal is subsequently machined flush, care should be taken not to expose the underlying CuAl10Fe1 material which would be vulnerable to corrosion. For this reason this practice is not normally acceptable to Naval authorities.

The manganese–aluminium bronzes are in general easier to weld than the nickel-aluminium bronzes.

Selection of shielding gas

(a) In TIG process
Argon is the standard shielding gas. It has the lowest arc voltage and thus the lowest heat input for a given welding current. The arc voltage for helium is higher. It gives therefore a greater heat input than argon. It can be substituted for argon completely, or used in a mixture with argon to increase the heat input in order, either to reduce the pre-heat temperature, or to increase penetration or welding speed.

Table 7.2   Typical operating data for TIG and MIG butt welds
(Maximum pre-heat: 150°C).[38]

TIG Butt Welds (Alternating current, 3.2 mm dia electrode, Argon shielding)

Thickness (mm)	Filler Rod Diameter (mm)	Gas Nozzle Diameter (mm)	Gas Flow Rate (1 min)	Welding Current (A)
1.5	1.6	9.5–12	5.8	100–130
3	2.4	9.5–12	5.8	180–220
6	3.2	12–18	8–10	280–320
9	3.2–4.8	12–18	8–10	320–400
12	3.2–4.8	12–18	8–10	360–420

MIG Butt Welds (Direct current, 1.6 mm dia filler wire, Argon shielding)

Thickness (mm)	Welding Current (A)	Arc Voltage (V)	Gas Flow Rate (1 min)	Wire Feed Rate (m/min)
6	280–320	26–28	9–12	4.5–5.5
9	300–330	26–28	9–12	5.0–6.0
12	320–350	26–28	12–17	5.8–6.2
18	320–350	26–28	12–17	5.8–6.2
24	340–400	26–28	12–17	5.8–6.2
>24	360–420	26–28	12–17	6.0–6.5

Table 7.3   Recommended current settings and voltage for Metal-arc and Carbon-arc
welding processes.[127]

Process	Electrode dia (mm)	Current (A).	Arc voltage (V)
Metal Arc	2.5	60–100	22–28
	3	90–160	24–30
	4	130–190	24–32
	5	160–250	26–34
	6	225–350	28–36
	8	275–390	28–36
	10	325–450	30–38
	13	450–600	34–44
Carbon Arc	5	60–80	25–30
	6	80–130	30–35
	10	130–250	35–40
	13	230–350	35–45
	19	350–500	40–50
	25	500–700	45–60

With argon, d.c. welding can be used but an a.c. arc gives better results as its cyclical reversing polarity disperses the oxide film on the weld pool. This is satisfactory for routine work on uncomplicated thin gauge material (< 5 mm), but it is necessary to use high-frequency re-ignition injection circuitry to keep the arc established.

Refractory oxide films prove less troublesome in helium and for this reason, d.c. working in helium is a suitable alternative to normal a.c. working in argon for welding aluminium bronzes.

Helium shielding is recommended for more complex structures of any gauge material and especially in the welding of thick to thin sections, wrought to cast materials and where it is difficult to avoid restraint being applied. Although helium is a more expensive gas, its use is justified because, using d.c. electrode-negative working, it gives very clean welding conditions, a hotter arc and faster welding. The overall heat input is therefore less and the weld likely to be cleaner and free of porosity and other defects.

(b) In MIG process

Even at very high current densities, argon is the only gas which, on its own, can ensure a normal spray transfer with the MIG process. It is associated with d.c. electrode-positive working. The addition of up to 50% helium to argon improves the arcing behaviour and increases the heat input without destroying spray transfer. The higher cost of helium is in many cases offset by the significant improvement in welding speed and is the main reason for using this mixture.

Current settings, voltage and other operating data

Typical operating data is given in Table 7.2 for TIG and MIG processes and in Table 7.3 for manual arc processes.

Fluxes

Fluxes for the coating of metal-arc electrodes, for carbon-arc welding and for oxy-acetylene welding are hygroscopic and must, therefore, be dried before the metal is melted to avoid hydrogen porosity.

Welding technique

TIG

Although inert gas shielding prevents the formation of oxides during welding, it does not remove them from the surface being welded. This can result in small oxide inclusion in the weld which can act as crack initiators. This problem can be avoided by coating the wire lightly with a sodium silicate paste containing cryolite. This is done simply by dipping the wire in the paste prior to use and the water in the coating is removed by short-circuiting the electrode to heat it. This procedure reduces the melting point of alumina from 2000°C to 804°C, that is to say, well below the melting point of aluminium bronze, and results in the avoidance of oxide inclusions.[184]

Tungsten inclusions, stemming from the TIG-welding electrode, are caused by too high a current and may also result from lack of skill on the part of the welder.[184]

MIG

In MIG welding, it is desirable to lay down the deposit as quickly as possible. For heavy sections it is therefore advantageous to build up the deposit from a large number of narrow runs, each less than about 3 mm deep. The deposit is applied without weaving, the rate of deposition being controlled to give good 'melting in' or 'wetting' of sides of the weld pool. The leftward method has definite advantages and, with satisfactory conditions of deposition, the weld is smooth and uniform. In view of the rapid rate of welding, an experienced operator can weld from all angles which is not possible with other processes. For best results in multi-run welds, the thin oxide film, which forms on the surface of the deposit, should be lightly dressed by scratch brushing before the next run is made.

Metal-arc welding

Generally speaking, satisfactory metal-arc welding with coated electrodes is difficult and best avoided. It is necessary to use high quality coated electrodes to obtain satisfactory results. As previously mentioned, the electrode coating has, however, a tendency to absorb moisture which changes during fusion to hydrogen for which copper alloys have a great affinity and which gives rise to gas holes on cooling. Electrodes should therefore be dried by preheating at 120°C before use. The work should be arranged so that welding can be carried out in the downhand or flat position.

The current should be close to the maximum permissible for a given thickness of base metal. It is essential in applying weld metal to keep base metal penetration to a minimum (especially when welding other metals with aluminium bronze).

During welding the electrode should be subjected to a weaving action to give a width of weld approximately three to five times the diameter of the electrode. The weld pool is kept just molten by directing the arc into the pool and by weaving sufficiently rapidly to thoroughly 'wet' each side of the joint. This may be achieved by a slight pause during weaving as each side of the weld pool is reached. If large drops of molten metal form on the end of the electrode, they should be dabbed into the weld pool to prevent excessive oxidation, and if trouble should be encountered due to weld cracking, the 'stringer-bead' technique is a recommended alternative.

Following each run, all flux residues must be removed before further metal is applied as flux inclusions cause serious loss of strength and ductility of the joints. A certain amount of dressing may also be desirable to remove protuberances which can act as slag traps and prevent penetration of superimposed runs.

Oxy-acetylene welding

As explained above, oxy-acetylene welding is hardly practicable for any but very thin sections and gas-shielded or carbon-arc welding are far preferable.

Welding Practice: Joining Wrought Sections

General

Preheating is not normally required for weld assemblies unless aluminium bronze components are welded to copper, in view of the high thermal conductivity of the latter.

In joining wrought sections, peening after each run, as recommended below in the case of weld repairs, need only be used where welding strains may be high as in the case of restrained joints or of welding aluminium bronze to steel.[184]

With MIG welding, too large deposits can induce stresses, due to the high cooling rate of aluminium bronze, which can in time give rise to cracks.

Welding sheets together, which have been extensively cold drawn without prior annealing, should be avoided. This is because of the high level of stresses in cold worked sheets.

Design of joints and weld preparation

The edge preparation chosen for a particular welding operation depends on the following factors:

(1) the alloy,
(2) the thickness of the joint,
(3) the welding process,
(4) the welding position and accessibility of the joint area,
(5) the type of joint: i.e. whether butt or fillet,
(6) whether there is a likelihood of distortion and control is required,
(7) the control required on the profile of the penetration bead,
(8) economic aspects of weld metal consumption and wastage of metal in edge preparation.

It is therefore not generally possible to specify precise joint configurations for any given set of circumstances. Fig. 7.2 gives recommendations only of a general nature for routine butt welding.

Joints must be designed to allow free thermal expansion and contraction so that the material is not placed under excessive restraint during the weld cycle.

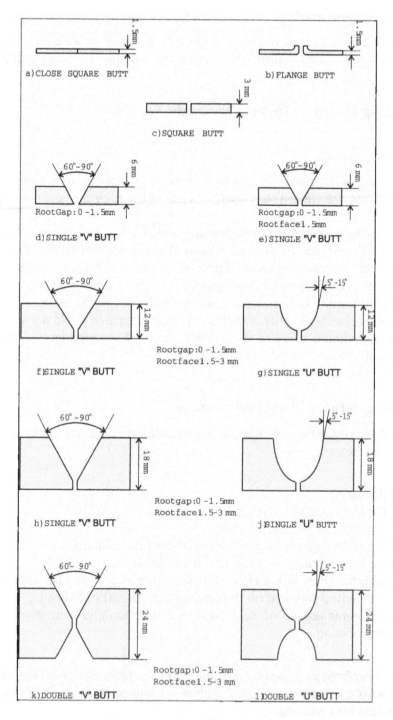

Fig. 7.2 Recommended edge preparation for MIG and TIG butt welds.[38]

It will be seen from Fig. 7.2 that for normal thicknesses, a 1.5 mm root gap is recommended to ensure satisfactory weld metal penetration, although the greater penetration obtainable with the MIG process may make a root gap unnecessary.

Edge preparation of butt joints must receive particularly careful consideration when the metal is thicker than about 10 mm. Wide joints facilitate ready fusion but an increase in the amount of metal deposited raises the total heat input and thus tends to increase the stress across the weld.

For sheets or plate from 6 to 18 mm thick the edges should be chamfered to form a 60° V for MIG and metal-arc welding, but a wider V up to 80° or 90° is favoured for carbon-arc welding because the arc is less concentrated. The V joint should always be formed through four-fifths to two-thirds of the thickness of plate, and it is preferable for the bottom of the V to be rounded as in the case of a U joint. A double-U joint is necessary for material over 24 mm thick and this should be extended for two-fifths of the thickness from each side, though some welders prefer an unequal double-U, approximately two-thirds and one-third of the thickness respectively in order to minimise distortion.

When a gap is required between the two sheets or plates it should not be greater than 1.5 mm at the point of welding. If this is exceeded, there is risk of break-through of molten metal. Tack welds should be made from the back of the joint, whenever possible, to prevent the parts from moving during welding and to assist regular penetration of weld metal. For sections of intermediate thickness involving single V preparation, the need for a backing-plate may be avoided by means of a single run of TIG weld from the reverse side before commencing the main weld. Particular care should he taken to obtain full penetration at the root when this technique is employed.

Jigging and backing techniques

Jigging and backing are arranged to ensure that the parts to be joined are accurately positioned to prevent excessive distortion during welding and to control and support the weld penetration bead. The design of jigs and supports will depend upon the likely heat input, the section thickness and the type of joint.

If the underside of the joint is accessible, it is possible to control penetration by means of a suitably grooved copper, mild steel or stainless backing bar coated with colloidal graphite or a proprietary anti-spatter compound to prevent it fusing with the weld bead. Ceramic coated strip used in conjunction with the backing bar will allow a smooth flush penetration and will also prevent heat dissipating too rapidly from the joint area. Where accessibility is limited, backing bars of matching composition may be used which are intended to fuse into the weld and become an integral part of the joint.

Tack welds, which are most conveniently done by TIG welding, ensure the correct alignment of the joint and root gap but must be made with the same filler. They must be cleaned to ensure full fusion with the first main weld run.

Movable clamps are used both circumferentially and longitudinally to position the root gap accurately, particularly on long seams. They are moved along the joint as welding proceeds.

Welding Practice: Joining and Repairing Castings

Weld preparation

See note above on 'cleanliness and freedom from grease'.

The first requirement for the successful welding of castings is a sound metal base from which to work. Gas and shrinkage porosity in the casting can seriously impair the achievement of a sound weld deposit. A thorough removal of all traces of defective metal is therefore of the utmost importance. Defects are removed by machining or with tungsten carbide burrs or by pneumatic chipping and/or grinding using rubber/resin bonded alumina or silicon carbide grinding wheels. The excavation must be finished smooth and should taper off at a minimum slope of 1/3 to permit welding access to the excavated base.

Some form of non-destructive testing technique must be applied to check that all defects have been removed. Using initially dye-penetrant testing to check that all uncovered defects have been removed, followed by radiography to ensure that there are no remaining sub-surface defects, is the best combination. Similarly, part-weld tests are strongly recommended to ensure that no weld defects have arisen in the weld so far. The use of ultrasonic testing on castings, other than for thickness checks, is not satisfactory because of the varying grain size of the cast structure.

It is important not to remove more of the parent metal prior to welding than is strictly necessary. It is also a classical error to fill the cavity partially by melting too much of the parent metal. Both these practices result in excessive heat input during the welding operation with the attendant dangers of distortion, cracking (see 'Ductility dip' above) and undesirable effects to the microstructure.

The weld preparation must be smooth, clean and of a profile that enables access of the welding electrode to the root, since complete fusion at this point is essential for satisfactory results. All traces of metal grindings, dirt, grease and dye-penetrant fluid must be removed from the weld area before welding commences.

Preheat and inter-run temperature control

If preheat is required, it must not make the weld more susceptible to solidification cracking due to the ductility dip mentioned above. It should therefore be either well below 400°C or well above 800°C. In addition to the discomfort of welding at a high preheat temperature, the resultant coarsening of the grain of the parent metal weakens it and makes it difficult to deposit weld metal satisfactorily. Excessive preheat will also lead to an unacceptably wide heat-affected zone, and may give rise to distortion and corrosion problems. This is particularly important in the case of casting repair since, because of their shape and bulk, castings are likely to present more restraint than wrought components and are therefore less likely to 'give' under thermal stress.

It is seldom necessary in practice to apply preheat of more than 150–200°C, and it is sufficient in most cases to heat the work just sufficiently to drive off dampness and prevent further condensation. A 50–100°C preheat, checked with a contact pyrometer, is recommended. Inter-run temperatures should be limited preferably to 150°C (and be no more than 200°C) by allowing the work to cool between runs and checking with a contact pyrometer. The temptation to 'puddle in' large quantities of weld metal, when repairing a casting, should be resisted. With TIG welding, preheating is essential for ease of striking and maintaining the arc.

Weld deposit

The TIG and carbon-arc processes provide better control of the weld deposit. If the repair is done by the MIG process, a gentle entry and exit gradient is essential and this significantly increases the repaired area. Any defective metal must be completely removed and the clearance of the defects monitored by NDT methods. Similarly the soundness of the weld should be checked by NDT after welding and fettling. The welded area should be over-filled to allow for machining back to fully sound metal below the uneven weld metal surface.

When weld repairing a casting, it is recommended to peen the weld for 10 seconds after each run.[184] It has been shown that the stresses developed by a weld run, without peening, can be as high as 200 N mm^{-2}, whereas, with peening, they are of the order of 20 N mm^{-2}.

If the excavation of a defective area has gone through the wall thickness, it will be necessary to use a colloidal graphite backing plate.

Joining one casting to another or to a wrought part

The above recommendations to ensure that the part of the casting being welded is sound, applies to joining as it does to repair. In other respects the welding practice for joining is broadly as outlined above for wrought fabrications.

Inspection and Testing

Depending on the severity of the operating conditions in which a fabrication or repaired casting will be used, certain inspection and testing requirements for welds are laid down by the relevant specifications. The British Standard Specification for 'Visual inspection of fusion welded joints' is BS 5289:1976. In addition to visual inspection, dye penetrant testing, radiography and/or ultrasonic inspection may be called for.

Aluminium bronzes absorb X-rays and gamma-rays to a greater extent than steel. Test conditions are therefore different. Typically, 300kV X-rays can be used up to

50mm thickness and ^{60}Co gamma-rays up to 160 mm thickness depending on the degree of detection required to reveal unacceptable levels of porosity and other defects.

Ultrasonic testing of aluminium bronze castings is unsatisfactory due to the variation in grain size and damping capacity, but it can be used for the thinner wrought sections of wrought materials.

Effects of Welding on Properties

As previously mentioned, the heat generated by welding and the subsequent rate of cooling cause changes in the microstructure of the zone adjoining the weld, known as the heat-affected zone. They also affect the structure of the weld metal itself. This may have a potentially deleterious effect on physical properties and on corrosion resistance in the affected area. For this reason, heat treatment (see below) to restore mechanical properties and/or corrosion resistance may be advisable and is a requirement of certain specifications.

Effects on metallurgical structure and on corrosion resistance

The corrosion resistance of a welded assembly or of a weld repair depends first of all on the choice of a corrosion-resisting base alloy and on a proper match between the filler and base metal. If non-matching metal is used to part-fill the weld groove or cavity, great care must be taken to ensure that an adequate protective layer of matching weld metal remains after machining or dressing. It is always preferable, however, to weld entirely in a matching filler.

The following is a summary of the effects of welding on the different types of aluminium bronzes.

(a) Single phase alloys
Cu/Al alloys with an aluminium content of less than 8% and with, possibly, a small addition of iron, solidify into a single alpha-phase. They have excellent corrosion resistance but unfortunately are difficult to weld due, as mentioned above, to a tendency to crack caused by the wide ductility dip. Special attention must be paid to the purity of the filler and the conditions of deposition must be controlled to give thorough wetting of the side of the prepared plate. Because of the sensitivity of the alpha alloys to cracking when under restraint, there is a possibility that cracking of the base metal might occur in spite of a specially selected filler metal. The condition and chemical composition of these alpha materials are critical, and it is therefore advisable to state clearly when ordering this material that it is required for fabrication by welding.

The temperature of the heat-affected zone and the subsequent rate of cooling will affect grain size and hence the mechanical properties, but will not affect the resistance to ordinary corrosion of a single-phase alloy. Welding is liable however to leave internal stresses which, if not relieved by heat treatment, may lead to intergranular stress corrosion cracking. As mentioned in Chapter 9, single phase alloys used in high pressure steam service are particularly prone to this problem but experience has shown that susceptibility to this type of attack can be eliminated by the addition of 0.25% tin to the alloy (e.g. American specification UNS 61300). Unfortunately, although the presence of tin is advantageous to prevent stress corrosion cracking, it is liable to make the weld metal more liable to cracking under restraint as previously mentioned. The weld metal should therefore not contain more than 0.1% tin.

(b) Duplex alloys

Cu/Al alloys with an aluminium content greater than 8% and with, normally, a small addition of iron and sometimes of manganese, solidify into a mixture of two phases: the alpha and beta phases. These alloys are inherently less corrosion resistant in sea water than single-phase alloys because of the difference in electro potential of the two phases which may give rise to selective phase attack (see Chapter 8). Furthermore, if the aluminium content is above 9.5%, they are liable, if cooled too slowly, to change to an even more corrodible combination of alpha and $gamma_2$ phases. In such a case, welding of relatively thick sections is likely to result in slow cooling of the weld metal and of the heat-affected zone and therefore to give rise to this more corrodible phase combination. The presence of the $gamma_2$ phase also renders the alloy more brittle.

Heat treatment will restore the alpha plus beta structure, but even that structure is not ideal for critical components in a corrosive situation. On the other hand if the aluminium content is less than 9.5%, and although rapid cooling in welding of thin sections will retain this same alpha plus beta structure, subsequent heat treatment, involving slow cooling, will transform this structure into the more corrosion resistant single alpha-phase structure.

Even a small nickel addition will improve the corrosion resistance of a duplex alloy provided the aluminium content is less than 8.2+Ni/2[184] and that post-weld heat treatment is carried out to counter the adverse effect of rapid cooling after welding.

Because of the difficulties, just described, of obtaining corrosion resistant structures with the Cu/Al/Fe type of alloys, the standard cast or wrought Cu/Al/Si alloys offer an attractive alternative. They have very good weldability and good corrosion resisting properties which are not likely to be adversely affected by welding, provided the rate of cooling of the weld metal and heat-affected zone is not so fast as to cause a significant retention of the beta phase. They do not, therefore, normally require post-weld heat treatment except perhaps for stress relieving in cases of rapid cooling of the weld metal and heat-affected zone.

(d) Complex alloys

As explained in Chapter 13, the main effect of welding on the corrosion resistance of Cu/Al/Ni/Fe alloys is that the heat generated by the weld raises the temperature of the adjoining heat-affected zone to the point where the alpha phase and the various kappa precipitates reconstitute the high temperature beta-phase. The subsequent rapid cooling converts the high temperature beta-phase to large areas of martensitic beta-phase in the area adjacent to the weld. It is this martensitic beta-phase which is particularly vulnerable to corrosion. The effect of cooling rate on microstructure and hence on corrosion resistance is discussed in Chapter 13. The greater the distance from the weld area, the less these transformations occur.

What has just been said shows the need for the heat treatment (annealing at 675°C for 2 to 6 hours followed by cooling in still air) described below to be carried out after welding to restore the heat-affected zone to its pre-welded structure and to relieve thermal stresses which profoundly affect the corrosion behaviour of the alloy.[62]

The rapidity of MIG-welding and of electron-beam welding to join wrought or cast sections together can be such as to result in a predominantly beta microstructure. Some post-fabrication heat treatment may therefore be advisable.

The high manganese Cu/Mn/Al/Ni/Fe alloys have good weldability and do not suffer from intermediate temperature brittleness to the same extent as the normal aluminium bronzes. They do, nevertheless, require heat treatment after welding to restore corrosion resistance and mechanical properties (see 'Heat treatment' below).

Effects on Mechanical Properties

The strength of a sound weld normally compares fairly closely with that of the base metal. The high strength of aluminium bronzes is only slightly reduced by welding, provided a satisfactory deposit is obtained. The welds have a lower ductility, however, which is closely related to weld quality.

Effect of rate of cooling from high temperature

Belyaev and al.[24] carried out experiments to simulate the effect of cooling from high temperatures and of subsequent annealing on the mechanical properties of the heat-affected zone of a welded component. They experimented with two alloys: a nickel aluminium bronze of nominal composition CuAl9Ni4Fe4 and a high manganese aluminium bronze of nominal composition CuMn14Al7Fe3Ni2. The conditions of the experiments and their results are shown on Table 7.4. It will be seen that, for both alloys, the tensile strength, proof strength and hardness of the heat-affected zone are higher than those of the parent metal as cast, whereas elongation and impact strength are lower. Elongation is significantly reduced by cooling from high temperature but markedly improved by subsequent annealing, although still falling short of the as-cast elongation. Annealing after cooling at 3.8°K s^{-1} reduces the impact strength of the CuAl9Ni4Fe4 alloy whereas it increases it after cooling at 60°K s^{-1}. In the case of the CuMn14Al7Fe3Ni2 alloy, on the other hand, annealing results in a lower impact strength after cooling at either cooling rate.

Effect of welding on fatigue strength

Belyaev and al.[24] also investigated the effect of welding the above mentioned alloys on their fatigue strength in sea water. Their tests were carried out on specimens of 12

Table 7.4   Effect of cooling rate and annealing on mechanical properties of heat-affected zone of two alloys, by Belyaev and al.[24]

Heated to °C	Condition	Tensile Strength N mm⁻²	0.2% Proof Strength N mm⁻²	Elongation %	Hardness HV	Impact kJ m⁻²
		$N\ mm^{-2}$	$N\ mm^{-2}$	%	HV	$kJ\ m^{-2}$
	CuAl9Ni4Fe4					
	Parent metal as cast	634	270	25.8	165	500
~950	Cooled at 3.8°K s⁻¹	654	327	13.2	191	402
	annealed*	675	357	14.1	192	361
~950	Cooled at 60°K s⁻¹	720	420	4.6	231	151
	annealed*	776	496	16.3	218.5	270
~1000	Cooled at 3.8°K s⁻¹	638	336	11.3	211	372
	annealed*	614	355	7.0	195	351
~1000	Cooled at 60°K s⁻¹	575	437	2.2	268	195
	annealed*	664	315	20.7	234	275
	CuMn14Al7Fe3Ni2					
	Parent metal as cast	612	293	27.0	167	590
~850	Cooled at 3.8°K s⁻¹	740	392	19.0	190	570
	annealed*	662	382	19.3	188	440
~850	Cooled at 60°K s⁻¹	765	664	4.0	247	355
	annealed*	738	542	11.0	201	285
~900	Cooled at 3.8°K s⁻¹	580	400	3.8	192	470
	annealed*	682	406	20.3	172	440
~900	Cooled at 60°K s⁻¹	792	688	4.3	245	285
	annealed*	589	571	3.1	211	245

*Annealed at 550°C for 4 hrs

mm dia. and of 75 × 50 mm section. They concluded that welding these alloys reduces their fatigue strength in sea water but that, in the case of a small section such as the 12 mm diameter, annealing at 550°C for 4 hours increases the fatigue resistance of the weld and heat-affected zone almost to that of the parent metal. In the case of a thicker section, such as the 75 × 50 mm section, this annealing increases the fatigue resistance of the welded joint to 70–80% of that of the parent metal.

Post-weld heat treatment and its effects

Stress relief anneal

A simple stress-relief anneal may be carried out at temperatures as low as 300°C – 350°C for a time dependent on section thickness but normally between 30 minutes to one hour.

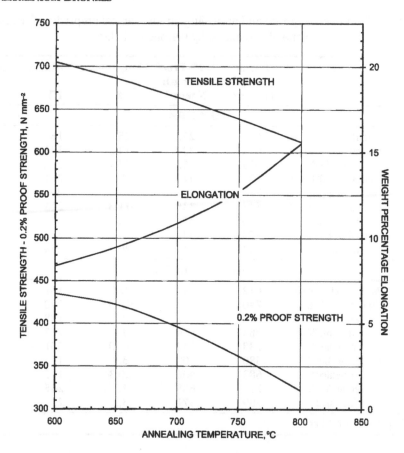

Fig. 7.3   Effect of different annealing temperatures on the mechanical properties of CuAl9Fe5Ni5 alloy – by Soubrier and Richard.[165]

Full anneal

Full anneal consists in heating the metal above the re-crystallisation temperature (usually in the range of 650°C to 750°C) and cooling at a predetermined rate. It is normally done to soften the metal after hot working but, as a post-weld heat treatment, it is done to make the structure of the weld metal and of the heat-affected zone conform to that of the parent metal. It also improves the structure of the parent metal and its corrosion resistance. The temperature needed to obtain the desired combination of corrosion resistance and mechanical properties depends on the type of alloy and section thickness. Expert advice should be sought. The British Naval specification for post-weld heat treatment of castings in an alloy of the type **CuAl9Ni5Fe4Mn** calls for soaking at 675°C for six hours and cooling in air. It is claimed by some that, in the case of thick sections, better results are obtained at

700°C.[178–74] An alternative treatment recommended by some is to soak for two hours at 850°C, slow cool to 750°C and then cool in air. The effect of different annealing temperatures on the mechanical properties of a CuAl9Fe5Ni5 alloy is shown on Fig. 7.3, from which it can be seen that soaking at around 675°C to 700°C results in a good combination of tensile and elongation properties.

Certain Naval specifications call for dye-penetrant testing and radiography after post-weld heat treatment.

Some specifications for the post-weld heat treatment of propellers call for a rate of cooling not exceeding 50°K h⁻¹ (slower than cooling in warm sand) down to 100°C. Heat treatment of propellers needs very special care because of the risk of distortion. For this reason, annealing at the lower temperature of 550°C for 4 hours, as shown in Table 7.4, will bring about improvements in elongation. Although it will not bring about any significant reduction in the corrosion-prone aluminium-rich beta phase, this phase is likely to be more uniformly distributed and less likely to form a continuous path for selective corrosion. One treatment specified for continuous cast tubes, with cast flanges welded on, specifies soaking at 700–730°C for six hours, followed by a cooling rate not exceeding 250°K h⁻¹ (air cooled).

In some cases, where improving mechanical properties is the main consideration rather than restoring corrosion resistance, one or other of the heat treatments described in Chapter 6 may be applied as appropriate.

The Cu/Mn/Al/Fe/Ni alloys are straightforward in their welding behaviour, but post-weld heat treatment at 600–650°C, followed by air cooling is recommended to restore fully the corrosion resistance of the weld and heat-affected zone structure.

Arc Cutting of Aluminium Bronze

Oxy-gas cutting of aluminium bronze is not satisfactory, but the tungsten arc method of cutting described by Cresswell[57] gives outstandingly good results. Cutting is also possible by carbon-arc, oxy-arc, iron powder cutting or even ordinary mild steel cutting electrodes, although the cut edge is inferior in all cases to that obtained by the tungsten arc.

Use of Aluminium Bronze in Joining Dissimilar Metals

The ease with which some of the aluminium bronzes are welded, plus the high weld strength obtained, lead to their use for joining a wide variety of different metals.

Aluminium bronze fillers are ideally suited for joining aluminium bronzes to steel or cast iron. No difficulty is normally experienced. It is desirable to 'butter' the ferrous surfaces with a fairly thick coating of filler metal before joining to avoid ferrous contamination of the bulk of the weld. This also prevents the arc from being deflected towards the steel or iron surface. For the same reason it is good practice to continue the 'buttering' about 12 mm. beyond the joint faces to avoid instability of the arc near the top of the joint.

When joining dissimilar metals, fusion of the 'foreign' metal must be minimised to avoid the formation of brittle inter-metallic layers. At high temperatures, molten copper alloys can sometimes penetrate steels to cause cracking of the ferrous material. It is recommended that a long arc technique should be used when joining steel to any metal by the use of an aluminium bronze filler. The arc should be directed into the weld pool, thus allowing the filler metal to 'braze weld' to the steel. Aluminium bronze may be successfully used for joining copper and aluminium bronze to both mild steel and stainless steel, and for welding and repairing high tensile brasses, silicon bronze and cupro-nickel. It is necessary, however, to ensure that a ferrous or nickel-base material is in a stress-free condition before welding.

Surfacing with Aluminium Bronze

Surfacing by weld deposit of aluminium bronze

Provided the considerations described in the previous section are taken into account, aluminium bronzes are particularly valuable for surfacing steel and other metals. As the weld deposit has a hardness and strength comparable with steel, it may be used for repairing worn cast iron or steel surfaces, fractured gear teeth, valve seats, pump impellers, etc. It has the added advantage of not dissolving carbon from cast iron to form hard deposits which are difficult to machine. It may also be used for coating new components to improve corrosion resistance or abrasion and wear resistance. Because of the different contraction rates on cooling, particularly between aluminium bronze and ferrous materials, it is necessary to lay the deposit as evenly as possible to minimize distortion of thin sections.

When multi-layers of aluminium bronze are applied some cracking may occur. This can be reduced by applying the first layer as thinly as possible and by melting only the minimum quantity of the previously deposited material in subsequent runs.

Surfacing by Spraying Aluminium Bronze

Aluminium bronze is the most satisfactory copper alloy for metal spraying because of its ease of application and the properties of the deposit. The majority of applications involve facing of steel or cast iron components to give increased wear resistance and freedom from galling in sliding contact with other ferrous surfaces. Aluminium bronze sprayed deposits are used also to build up worn shafts, bearings and similar faces subjected to mechanical wear.

Surface preparation is extremely critical to ensure maximum bond strength. Thorough degreasing should be followed by removal of surface metal by machining, grinding or by abrasion with clean emery paper. Bonding to steel may be achieved by spraying a thin bonding coat of molybdenum before commencing to spray the aluminium bronze. For bonding to copper alloys, including aluminium bronze, the surface must be grooved and the edges burred over to provide a mechanical key for the deposit.

The sprayed deposit has a moderate degree of porosity (7% – 10%) which makes it unsuitable for corrosion prevention, except in the form of thick deposits over small areas. The porosity renders the deposit ideal for machining, and it may be given a high degree of finish without difficulty. The pores in the sprayed deposit also enable a lubricant film to be retained more readily.

Typical mechanical properties of a sprayed deposit of CuAl10Fe3: tensile strength 200 N mm^{-2}, elongation 0.5%, hardness 145 HV.

Other Joining Processes

The joining of aluminium bronze by methods other than welding are generally not as satisfactory and result in weaker bonds. There may however be applications where brazing or soft soldering is the only practicable option.

Capillary brazing using silver-based brazing alloys

There are inherent problems in brazing aluminium bronzes due to the presence of the aluminium oxide (alumina) film and to the formation of this film during brazing. The time of contact between the molten brazing filler and the parent metal must be kept to a minimum to prevent an excessive transfer of aluminium to the filler metal. The oxidising of aluminium can cause non-wetting and poor bond formation. Special attention needs to be given to pre-cleaning and fluxing, using special fluxes for aluminium containing alloys (there is a special aluminium bronze grade of flux which can give good results with care). Advice on the most suitable type of filler metal and fluxes should be sought from suppliers of wrought aluminium bronze products.

Brazing aluminium bronze to a ferrous material may result in the diffusion of the aluminium through the molten brazing filler metal. It may then combine with oxygen dissolved in iron oxide on the surface of the steel. This would form a brittle layer of alumina at the steel/brazing filler interface. Copper or nickel plating of the aluminium bronze may be effective in preventing this diffusion.

The heat produced by brazing is likely to have similar effects to those discussed above in the case of welding. It may therefore be necessary to apply similar heat treatment to restore physical and corrosion resisting properties after brazing.

Soft soldering

Due to the tenacious alumina film which forms so rapidly, aluminium bronzes can only be soldered if the surfaces are first copper-plated or by using special techniques for soldering aluminium alloys.

8

MECHANISM OF CORROSION

Resistance to corrosion

The resistance to corrosion of aluminium bronze alloys depends on a combination of the following factors:

- the protective oxide film,
- the avoidance, if possible, of corrodible 'phases' (see below),
- if corrodible phases are unavoidable, ensuring that they are not continuous.

The protective oxide film

The most common form of corrosion is one which is in fact beneficial to aluminium bronzes: it is the reaction of oxygen with a metal to form an oxide film on its surface. In the case of aluminium bronze, this oxide film adheres very firmly to the alloy, it is only slightly permeable to liquid corrodant and it readily reforms when damaged. It consequently provides a high measure of protection against further corrosive attack. It is thought to reduce the corrosion rate by a factor of 20–30.[161] By contrast, the oxide film of some metals is loose or easily removed, allowing corrosion to proceed.

This tenacious oxide film is therefore the 'first line of defence' against corrosion of aluminium bronze alloys and explains their resistance to corrosion in a wide variety of corrosive environments.

It may, however, be damaged in the following ways:

- by physical means such as impact of a hard object, wear, abrasive substances (mostly sand) in fluid flow, excessive flow velocity and turbulence, cavitation etc.; provided these effects are not sustained, the protective film has the ability to reform itself;
- by internal stresses which may result from too rapid a rate of cooling in a mould or during heat treatment, from various wrought processes, from welding etc., or due to fatigue; experience has shown that internal stresses are liable to facilitate the ingress of corrodants. Annealing will relieve such stresses;
- by chemical attack such as that of sulphides and caustic alkaline solutions; if the time of exposure to such attack is limited, the protective film will reform.

Avoidance of corrodible 'phases'

Certain constituents of the microstructure of an alloy known as 'phases' are vulnerable to corrosion if the oxide film defence is breached. As explained in Chapter 5, a 'phase' is a constituent of an alloy which is either a solid solution of two or more elements in each other or a compound of two or more elements. It has a given characteristic appearance under the microscope and has certain specific properties which affect the properties of the alloy as a whole.

The mechanism of corrosion of these phases is explained below. We shall see in Chapters 11 to 14 which are these phases and how they can be eliminated or reduced by alloy composition, cooling rate and heat treatment.

Avoidance of continuous corrodible phases

It is not always practicable to eliminate completely certain corrodible phases but by making sure that they do not create a continuous path for corrosion through the microstructure, corrosion can be limited to an acceptable shallow surface attack.

Nature of protective oxide film

A. Schüssler and H. E. Exner[161] have established and Ateya et al.[16] have confirmed that the oxide film of nickel–aluminium bronzes contains both aluminium oxide (Al_2O_3) and copper oxide (Cu_2O). It is aluminium-rich adjacent to the parent metal and richer in copper in the outer layer. This is because, if the surface is initially oxide-free, aluminium will oxidise preferentially, but aluminium oxide then acts as a barrier to the diffusion of aluminium from the parent metal leaving copper to oxidise preferentially on the surface of the corrosion product. Copper hydroxide, $Cu(OH)_2$, is the major surface compound. There are also oxides of iron, nickel and magnesium and traces of copper salts and copper hydro-chlorides: $Cu_2(OH)_3Cl$ and $Cu(OH)Cl$ which form after longer exposure to the corrosive medium. R. Francis and C. R. Maselkowski[73] found that the protective film contained less copper and aluminium than the base metal and more nickel and iron. They also found that the more nickel and iron the film contained, the more protective it was.

The protection against corrosion provided by the oxide film is due to its following properties:

- It adheres firmly to the base metal.
- It is relatively thick (although it is under 1/1000 mm thick[161]), and is only slightly permeable to liquids.
- It is resistant to most chemical attack.
- It has good resistance to flow velocity: weight loss corrosion studies show that, provided the flow velocity does not exceed a certain limit, the protective film goes on improving until a long-term steady rate of ~0.05 mm per year is reached. The flow velocity limit for nickel–aluminium bronze is 22.9 m s^{-1} and 15.2 m s^{-1} for duplex aluminium bronze.[179]
- It has good resistance to abrasive attack (erosion). Alumina (Al_2O_3), which is the main constituent of the inner layers of the oxide film, is a very hard abrasive substance.

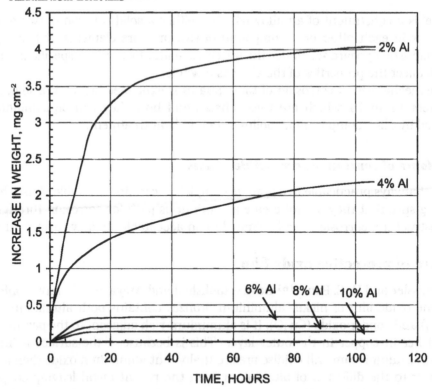

Fig. 8.1   The influence of aluminium content on the rate of oxidation of Cu/Al alloys at 650°C.[127]

- Should the oxide film be damaged, it reforms rapidly in the presence of oxygen.
- The presence of copper oxide (Cu_2O) in the outer layer of the oxide film is a deterrent to deposits of marine organisms on the surface of aluminium bronze components. Copper is poisonous to marine organisms as it dissolves into sea water. Due to their lower solution rate, however, aluminium bronzes are more susceptible to biofouling than copper and the less corrosion-resistant copper alloys.[39] The significance of this will be understood later (see 'Crevice corrosion' below).

Ateya et al.[16] have established that, in the case of an alloy containing 7% Al, subjected to 3.4% sodium chloride solution, there is an initial loss of surface weight of 7.5mg cm^{-2} as the oxide film forms over a period of approximately 16 days and virtually no weight loss thereafter.

Oxidation resistance at elevated temperatures

The protective oxide film, which forms on the surface of aluminium bronzes, endows the alloys with excellent resistance to further oxidation at elevated temperatures.

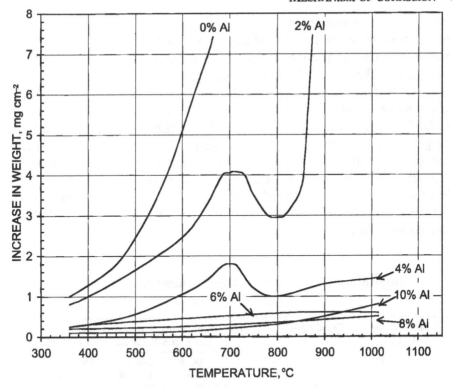

Fig. 8.2   The influence of temperature on the rate of oxidation after 24 hours of various percentage aluminium additions to copper.[127]

The higher the aluminium content of the alloy the greater is the protective nature of the oxide, although, as shown in Fig. 8.1 taken from the work of Dennison and Preece,[127] only minor improvements occur with aluminium contents in excess of 6%.

At temperatures below about 500°C the oxidation is extremely slight and often scarcely darkens the surface of the alloy, but at higher temperatures a grey scale will form after an extended period, although even this is protective to further heavy oxidation. The influence of temperature on oxidation rate is shown in Fig. 8.2.

Oxidation at 400°C

Hallowes and Voce[85] included several aluminium bronzes in a series of intermittent oxidation tests at 400°C. Test atmospheres included dry and moist air, and air contaminated with acid gases. Weighed specimens were heated for 5-hour periods in these atmospheres and re-weighed after cooling and brushing to remove all non-adherent scale. This cycle of operations was repeated until a constant loss of weight per cycle was attained. The results, expressed in terms of thickness of metal removed during ten 5-hour cycles, are given in Table 8.1. All the aluminium bronzes were resistant to dry and moist air at 400°C though slight adherent scales were formed, especially on the 2% and 5% aluminium alloys.

Selective oxidation of aluminium

As mentioned above, the reaction of oxygen with aluminium bronzes results in the formation of an oxide film which is composed predominantly of aluminium oxide (Al_2O_3) next to the base metal with a higher proportion of copper oxide (Cu_2O) in its outer layer. Selective oxidation of aluminium to give a surface oxide film completely free from copper further increases the oxidation resistance of the alloys to a remarkable extent. Price and Thomas[146] found that the alumina film formed on heating a 95/5 alloy to 800°C for 15 minutes in a slightly moist hydrogen atmosphere was sufficiently protective to maintain a bright scale-free appearance during subsequent exposure to oxygen for 4 hours at 800°C. Additional research on the subject has been sponsored by the International Copper Research Association[99] but a number of practical difficulties remain and the process has not been applied commercially.

Table 8.1   Oxidation and scaling of aluminium bronzes at 400°C compared with copper.[127]

% Composition (Rem: Cu)		Condition	Thickness (mm) of metal removed at 400°C per ton 5-hour heating cycles				
Al	**Fe**		**Dry Air**	**Air + 10% H_2O**	**Dry Air + 0.1% SO_2**	**Dry Air + 5% SO_2**	**Moist Air + 0.1% HCl**
2.06	0.01	50% Cold drawn	Nil	Nil	0.008	0.056	0.135
5.66	0.008	50% Cold drawn	Nil	Nil	Nil	0.290	0.038
9.76	0.039	Extruded	Nil	Nil	Nil	0.020	0.053
10.13	2.80	Extruded	Nil	Nil	Nil	Nil	0.028
11.10	0.006	Extruded	Nil	Nil	Nil	0.018	0.018
12.06	0.02	Extruded	Nil	Nil	Nil	Nil	0.023
Copper for comparison (Impurities: 0.46% As – 0.07% P – 0.06% Ni – 0.002% Fe – 0.01% Pb)							
Copper		50% Cold drawn	0.015	0.013	0.020	0.038	0.686

Mechanisms of Corrosion

Electro-chemical action

A metal corrodes when it discharges tiny positively charged particles, known as ions, into a corrosive liquid or moist atmosphere. The rate of discharge of ions – i.e. the corrosion rate – depends on the difference of electrical potential between the metal object and the corrosive medium, known as its electrical potential in that medium. Different substances have different inherent electrode potential values relative to a particular medium.

Fig. 8.3, known as an *electro-chemical or galvanic series*, shows the range of electrode potential values of a number of metals and alloys in natural sea water at 10°C (and also at 40°C in some cases). The electrode potential value is expressed in volts or millivolts relative to a Standard Calomel Electrode (SCE). It will be seen from Fig. 8.3 that most alloys experience a wide range of potentials in sea water, depending on conditions: water temperature, degree of aeration, turbulence of the water, pH value, biofouling, presence of chlorine etc. The potentially more severe corrosive condition of having two or more different metals immersed in the same electrolyte will be discussed below (see 'Dissimilar metals – Galvanic coupling').

The presence of an oxide film on the surface of a metal object prevents, or at least greatly reduces, the discharge of ions and is said to render the metal object 'passive'. In the case of aluminium bronze it is the layer of aluminium oxide which acts as an ion barrier and causes passivation.[161] In the case of stainless steels and nickels alloys, the oxide film is more 'noble' than the parent metal and consequently more cathodic and less vulnerable to corrosion: it renders the alloy 'passive'. If the film is eroded or physically damaged, the damaged area becomes anodic to the remainder of the metal surface and therefore corrodes; it renders the alloy 'active' (see Appendix 3: 'Pitting Corrosion'). The oxide film may also be chemically attacked, allowing freer discharge of ions, as in the case of copper alloys under prolonged exposure to stagnant sea water when the protective film may break down and a porous sulphide film may form, as will be seen later.

Table 8.2   Effect of chlorine addition on the electrochemical potential of certain materials in sea water at room temperature, by R. Francis.[74]

Alloy	Electrochemical potential at room temperature mV (SCE)	
	in natural sea water	in chlorinated sea water
Nickel aluminium bronze	−260 to −100	−250 to −50
Cupro–nickel	−250 to −100	−100 to 0
Stainless steel (active)	−300 to 0	−100 to +150
Stainless steel (passive)	+250 to +350	+500 to +700
Alloy 400 (65/35 Ni–Cu)	−150 to +200	−150 to 0
Alloy 625 (high strength nickel alloy)*	+160 to +250	+290 to +500

Note: the above figures were estimated from a graph
* Alloy 625: Ni68, Cr20, Mo9, Nb 3

The presence of chlorine in sea water has a marked effect on electrode potentials as may be seen from Table 8.2. In the case of nickel aluminium bronze, the presence of chlorine slightly narrows the range of variation in potential without significantly altering the mean potential. In the case of cupro-nickel and stainless steels, on the other hand, the mean potential is raised significantly but the effect on the range of variation in potential is narrower for cupro-nickel but wider for passive stainless steel.

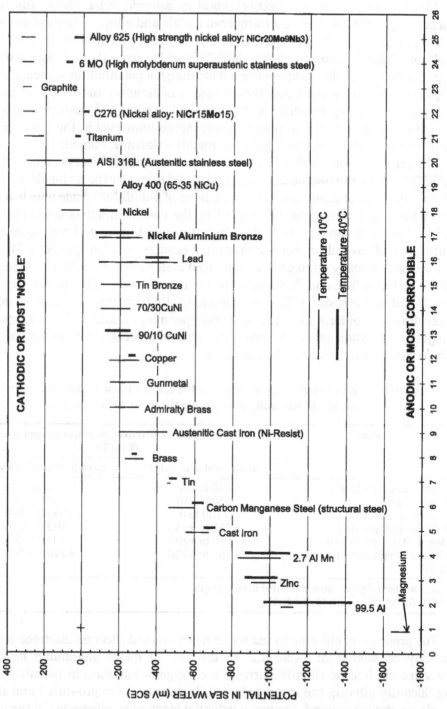

Fig. 8.3 Galvanic series in natural sea water at 10°C and some at 40°C.[74]

Corrosive effect of acids
If a metal object is immersed in an acid, the corrosive effect of the acid will depend on the difference in electrode potential between hydrogen in the acid and the metal object immersed in the acid. Hydrogen lies between tin and copper in the galvanic series. If the metal is cathodic to hydrogen (as in the case of copper), it will not discharge ions into the acid because of the adverse potential difference, unless the acid is an oxidising acid (see 'b' below). If the metal is anodic to hydrogen (as in the case of aluminium), it will release ions into the acid which will displace the hydrogen in the acid to form a salt. The released hydrogen ions will collect on the metal object and create an opposing hydrogen electrode which, if undisturbed, will eventually stop the corrosion of the metal object unless the following other factors come into play:

a) any *movement* will cause the hydrogen to escape as bubbles,
b) the presence of *oxygen* or of *an oxidising agent* such as nitric acid.

This will react with the hydrogen ions to form water, thus exposing the metal object to continued attack by the acid. In the case of aluminium bronze, this overrides the tendency of the oxygen or oxidising agent to restore its protective oxide film. This is why aluminium bronze is an unsuitable alloy for use in processes that involve contact with nitric acid or other oxidising agents.

Summing up therefore, the severity of attack by an acid depends on its strength (i.e. its hydrogen ion concentration), on whether the acid is an oxidising agent such as nitric acid, and on the presence of oxygen or of some other oxidising agents, unless these produce an inert film on the surface of the metal.

Corrosive effect of salt solutions
If an aluminium object is immersed in a solution of a salt of a metal, such as iron, copper or mercury, which is cathodic to aluminium, the aluminium ions will displace these metals from their salts to form aluminium salts. This would cause corrosion of an aluminium bronze object immersed in a solution of these salts. In the case of a sodium chloride solution (sea water), however, sodium is anodic to all common metals and the sodium chloride is therefore unaffected by the presence of other metallic ions. There is nevertheless a difference in electro-potential between the metal object and the hydrogen ions in the water, and, if the metal is anodic to hydrogen (e.g. aluminium), it will go slowly into solution and its ions will displace the hydrogen ions from the water to form an oxide or hydroxide of the metal. The displaced hydrogen ions will collect on the metal object as in the case of an acid (see above). Provided there is no oxidising agent or no dissimilar metal object present (see below), the corrosive effect in the case of aluminium bronze would be negligible. Aluminium bronzes are therefore not significantly liable to corrosion in sea water unless other factors come into play, as will be explained below (galvanic couples, differential aeration etc.).

Corrosive effect of caustic alkaline solutions
The corrosive effect of caustic alkaline solutions is of a different kind to that of acids and salts and is dealt with below under 'chemicals which attack the oxide film'.

Dissimilar metals (galvanic coupling)
The tendency for a single metal object to corrode in a corrosive medium is usually relatively small but, if two different metals are electrically connected to each other and are immersed in a solution which has good electrical conductivity, a 'galvanic couple' is created and the tendency to corrode is greatly increased. Metals that are more electropositive than a given metal in the electro-chemical series (see Fig. 8.3) are said to be *cathodic* to it (or more '*noble*') and any that are more electronegative are said to be *anodic* to it. The more 'noble' metal therefore becomes the 'cathode', the other metal becomes the 'anode' and the solution the 'electrolyte'. The two metals are then said to be 'polarised'. An electrolytic cell is thereby created and a flow of (positively charged) ions flows through the solution (the electrolyte) from the anode to the cathode, thus corroding the anode, and a corresponding stream of (negatively charged) electrons passes directly from the anode to the cathode via the metal to metal connection (although the flow of electrons is, as just explained, from the anode to the cathode, the 'electric current' is said by convention to flow in the opposite direction).

It will be noted that, due to the range of potentials experienced by alloys in sea water (see Fig. 8.3), the relative electropotential of one alloy to another may be anodic or cathodic depending on prevailing conditions (water temperature, degree of aeration, turbulence of the water, pH value, presence of chlorine, biofouling etc). The state of turbulence of the water is particularly significant since it may remove the corrosion products.[183] According to Soubrier and Richard,[165] the potential of aluminium bronze in non-aerated quiet water can be as much as 200 mV lower than in turbulent water, making the alloy more anodic and therefore more corrodible. Furthermore, the undisturbed deposit of corrosion products is liable to give rise to crevice corrosion (see below).[155] On the other hand, in the case of chemical attack, the undisturbed corrosion products can reduce the rate of attack (see 'Sulphides' below).

It was widely, but wrongly thought at one time, that the difference in electrode potential of uncoupled metals in an electrolyte, such as sea water, was an indication of the rate of corrosion that would occur if these metals were coupled together. In fact, it is the current resulting from the area ratio of the cathodic metal to the anodic metal, the electroconductivity of the electrolyte (e.g. sea water is a better conductor than fresh water) and the resistance of the metal to metal connection which determine the rate of corrosion of the anodic metal. Although the galvanic series (Fig. 8.3) is a general indication of which metal of a galvanic couple is likely to be anodic and which cathodic, the resulting rate of corrosion is often not related to the difference in uncoupled electrode potentials. Modern development in microelectronics is now making it possible to measure galvanic current and to predict likely corrosion rates.[15]

The more rapidly the anodic metal is attacked, the more the nobler metal is protected by the deposit of ions. This is the reason for the use of a 'sacrificial anode' to protect an expensive material from corrosion. A sacrificial anode is a lump of inexpensive metal which, as the term indicates, is anodic to the metal to be protected. It is connected to the nobler metal and immersed in the same medium where it corrodes and protects the nobler metal from corrosion as explained above. It is standard practice on offshore oil rigs to fit sacrificial anodes that are designed to be replaced during major maintenance and, in some cases, to last the life of a rig.

The ions from the anode will not in every case go direct from the anode to the cathode but may displace ions from the electrolyte which in turn will be deposited at the cathode. Thus if the electrolyte is a salt solution and the anode is anodic to the metallic constituent of the salt, as in the case of a ferrous anode in a copper sulphate solution, the ions from the anode will displace the metallic constituent of the salt to form a new salt and the ions released from the salt will be deposited on the cathode. Similarly, an aluminium anode, being anodic to iron, copper and mercury will displace these metals from solutions of their salts and will therefore corrode in the process. This is why solutions of these salts are a potential corrosive environment for aluminium bronzes.

If the anode is anodic to hydrogen, as in the case of aluminium, the following will occur:

(a) If the electrolyte is an acid solution, the ions from the anode will displace the hydrogen ions from the acid, forming a salt, and the released hydrogen ions will collect on the surface of the cathode.
(b) If the electrolyte is a solution of a salt, such as sodium chloride (sea water), whose metallic constituent is anodic to the metal anode, the salt will remain unaffected and the ions from the anode will displace hydrogen from the water and the released hydrogen will collect on the surface of the cathode.

As explained above, any movement or the presence of oxygen or of an oxidising agent will increase the severity of attack in cases (a) and (b).

The implications of electo-chemical action in the case of aluminium bronze, will now be discussed.

Selective phase attack – de-alloying – de-aluminification
Alloys solidify as a mass of crystals that have grown simultaneously and which are strongly 'glued' together by the last film of metal to solidify around them. A crystal is composed of a solid solution of one or more constituent metals in one another. It may contain some particles of intermetallic compounds that have precipitated within it. Duplex (twin-phase) and complex (multi-phase) alloys, solidify as an agglomerate of crystals composed of different solid solutions which may contain intermetallic precipitates within the crystals and/or around them. The different compositions of these crystals make them distinguishable as different 'phases',

when seen in cross-section on a photomicrograph. Intermetallic precipitates also constitute distinct phases visible in the microstructure. Phases, being of different compositions, have different electrochemical potentials and there is consequently always a tendency for the most anodic phase to be corroded preferentially. This difference in electrochemical potential between phases can be very significant: e.g. in excess of 100 mV between the α-phase and the γ_2-phase.[11-12] The resultant corrosion is known as 'selective phase attack' which may occur in two ways:

- between phases in the one component, due to the different compositions of adjoining phases,
- when one metal object forms a galvanic couple with a component of a more 'noble' metal. The less noble component is then vulnerable to corrosion and is preferentially attacked in its most anodic phase.

In the case of aluminium bronze alloys, as the anodic phase, which is richer in aluminium than other phases, goes into solution in the electrolyte, the (anodic) aluminium ions are attracted to the cathode whereas the (cathodic) copper ions re-deposit at the anodic corroded phase. This re-deposited copper has a honeycomb structure which is weak, porous and occupies the space previously occupied by the corroded phase. The external appearance of the component is thus basically un-changed, except for a slight discoloration and the depth of corrosion attack may often not be detected other than by destructive methods such as the preparation of metallographic sections.[63] Other alloying elements than aluminium are also re-duced by selective phase attack. The term 'de-alloying' is therefore more strictly correct than the more frequently used expression 'de-aluminification'.

If all the crystals of an aluminium bronze alloy consist of the same solid solution with no intermetallic precipitates, the alloy is known as a 'single phase' alloy. The last metal to solidify, and which forms a boundary around the crystals or 'grains', is richer in aluminium because of its lower melting point. The grain boundaries of single phase alloys are consequently anodic to the adjoining crystal and are there-fore liable to corrode preferentially as in the case of selective phase attack.

The extent to which corrosion occurs in aluminium bronze alloys depends upon how great the potential difference is between the anode and the cathode and upon their respective exposed surface areas. If the cathode is large relative to the anode, the latter will corrode more severely. The rate of corrosion also depends upon the intrinsic corrosion resistance of the anodic phase and its distribution in the struc-ture. If it is fragmented, the effect of corrosion may be negligible whereas if it is continuous, corrosion may significantly weaken the structure. The anodic phase of wrought alloys is more likely to be fragmented due to the effect of the hot or cold working process.

As will be seen in Chapters 11 to 14, certain alloy compositions may give rise under certain conditions to phases that are significantly more anodic than other phases and are therefore particularly vulnerable to selective phase attack. But the corrosive effect may be negligible unless an aluminium-rich anodic phase is present in a continuous form. The most corrosion prone phase is the aluminium-rich 'γ_2 phase'. Less rich in aluminium but still significantly corrodible is the 'martensitic β phase'. If good corrosion resistance is a design requirement, the formation of these phases is avoided by suitable control of composition and/or cooling rate or is corrected by heat treatment. By controlling the composition, the γ_2 phase can normally be avoided and the β phase considerably reduced or made discontinuous. There is however another phase combination in nickel–aluminium bronze, known as the '$\alpha + \kappa_3$ eutectoid', which is less corrodible than the two phases just mentioned, but liable nevertheless to give rise to selective phase attack, specially in the heat affected zone of welds and under 'crevice corrosion' conditions (see below and Chapter 13).

Under free exposure conditions in fresh waters or sea water, nickel–aluminium bronze alloys with the correct balance of aluminium to nickel content (see Chapter 13) do not show signs of significant selective phase corrosion thanks to their protective oxide film which is only slightly permeable to liquids.

The high manganese containing alloy, **CuMn13Al8Fe3Ni3**, contains essentially two phases, known as alpha and beta, where the beta phase is more susceptible to selective phase corrosion. This does not occur, however, to any significant extent under free exposure and rapidly flowing water conditions such as exist on marine propellers. In static sea water service, severe selective phase corrosion of the beta phase can occur under some conditions and, if the beta phase is continuous, it can cause serious deterioration. The susceptibility of this alloy to selective phase de-alloying corrosion is greater than that of the low manganese aluminium bronzes which should always be used in preference to it for applications involving static or shielded area conditions in sea water and for acidic environments.

The most commonly encountered examples of selective phase corrosion in other alloys are the duplex brasses such as free machining brass, diecasting and hot stamping brasses, Muntz metal, naval brass and the high tensile brasses commonly called manganese bronzes. The beta phase in all these alloys is anodic to the α phase and forms a continuous network providing a continuous path of low corrosion resistance by which attack can penetrate deeply into the alloy. This selective phase attack on the beta phase takes the form of dezincification with effective removal of zinc and formation of a weak copper residue. This dealloying susceptibility is greater than that of high manganese–aluminium bronze and much greater than that of low manganese multi-phase aluminium bronzes.

Galvanic coupling of aluminium bronze with other metals
   a) Coupling with other copper alloys
      Tests have shown that most aluminium bronzes are slightly more noble than other copper alloys with the exception of 70/30 copper–nickel. The differences are, however, small and the corrosion of aluminium bronzes due to galvanic coupling to copper-nickel or the corrosion of other copper alloys due to galvanic coupling to aluminium bronzes is usually insignificant.

(b) Coupling with iron, aluminium and zinc

The galvanic potential developed between aluminium bronze and non-copper-base metals is similar to that developed by other copper alloys. Aluminium bronze is more noble than iron, aluminium and zinc based alloys which therefore tend to corrode when coupled galvanically with aluminium bronze. The resultant deposit of ions on aluminium bronze protects it from other corrosive attack and this is the reason for the use of 'sacrificial iron anodes' as explained above.

(c) Coupling with nickel alloys

In contact with the more cathodic nickel alloys, aluminium bronze is not so adversely affected as other copper alloys. This is due presumably to the high resistivity of the surface oxide film which partly prevents the flow of galvanic currents.

(d) Coupling with titanium

Titanium is more noble than aluminium bronze and is aggressive to most metals, but the degree of acceleration of attack produced by coupling to this material is normally only slight.[55] Tube plates in nickel–aluminium bronze are commonly used for heat exchangers with titanium tubes and experience has confirmed that galvanic attack on the tube plate is negligible. The absence of significant galvanic effects under these conditions depends partly upon the effective exposed area of the titanium tube ends being not greatly in excess of that of the aluminium bronze.

(e) Coupling with stainless steels.

Stainless steels with approximately 13% chromium are anodic to aluminium bronze while the position of the 18/8 austenitic stainless steels depends on service conditions. Normally, aluminium bronze and austenitic stainless steels have little galvanic effect on each other, but when oxygen is limited the steel corrodes at a slightly accelerated rate, while in highly aerated solutions the copper alloy suffers an increased rate of corrosion under adverse surface area ratios.[181] In natural (non-chlorinated) sea water, the biofilm on stainless steel renders the galvanic reaction between the cathodic stainless steel and the anodic nickel–aluminium bronze so active that deep pitting occurs in the latter even with an area ratio of one to one. Why the biofilm should have this effect is a matter of conjecture. In chlorinated seawater systems, however, it is possible to couple nickel–aluminium bronze valves with superaustenitic stainless steel piping, provided crevice areas, such as valve-flange joints, are protected with high integrity coatings. Such a combination can be very significantly cheaper than an all-superaustenitic stainless steel system.[74]

In situations where aluminium bronze is used in contact with and in close proximity to much larger areas of more noble materials such as titanium, stainless steel or nickel–copper alloys, accelerated attack is likely to be experienced. In this connection, R. Francis[74] has found that, in the case of nickel–aluminium bronze, the iron and nickel content has a significant effect: the standard alloy with nickel content higher than iron content, such as CuAl10Ni5Fe4, gives satisfactory result in chlorinated sea water, provided the area ratio is not too unfavourable; whereas the alloy with low nickel and iron contents, such as CuAl9Ni3Fe2, is liable to suffer localised corrosion under the same conditions.

Effect of differential aeration

For various reasons, the degree of aeration in fresh or sea water is not always uniform. Dr U. R. Evans of Cambridge University has shown that, if there is a difference in oxygen concentration at two different points on the surface of a metal object, or on the surfaces of two different components of the same metal in contact with each other and immersed in the same electrolyte, the less aerated surface becomes anodic to the better aerated surface. It has been found that, if the alloy contains some corrosion-prone phases, this differential aeration is significant in causing corrosion. It also aggravates the conditions of localised corrosion discussed below (crevice corrosion and pitting). Internal defects such as porosity or oxide inclusions, exposed by machining or fettling, are liable to be subject to differential aeration effect and consequent corrosion.[165]

Effect of Electrical Leakage

Situations are sometimes met in service where aluminium bronze components are inadvertently exposed to electrical leakage currents either due to electrical faults resulting in current passing to earth, via a submerged pump for example, or due to incorrect positioning of impressed current cathodic protection equipment resulting in current passing from the water on to the metal equipment at one point and leaving it again at another. These conditions will accelerate electrochemical attack of practically all metallic materials whether the current concerned is DC or AC. Aluminium bronze under these conditions may show local corrosion in the region affected by the current leakage. The avoidance of this type of attack is obviously a matter of correct design and maintenance of the electrical equipment concerned.

Chemicals that attack the oxide film

Sulphides

The principal threat to all copper-based alloys and especially aluminium bronze, is that presented by concentration of sulphides in certain fresh and sea water environments. Wherever organic matter containing sulphur undergoes putrefaction, hydrogen sulphide gas is given off which partly dissolves in sea water. Sulphides are powerful reducing agents and react with the cuprous/aluminium oxide film to form copper sulphide. Whereas copper oxide is virtually impermeable to liquid and adheres firmly to the base metal, copper sulphide is porous and therefore permeable to liquid. It does not adhere to the base metal and it consequently erodes away when subjected to high flows. The damage done by this type of corrosion is therefore strongly dependent on flow velocity. In quiet water conditions the undisturbed copper sulphide corrosion products significantly reduces the rate of attack. The permeable nature of copper sulphide, however, allows corrosion to take place by electrohemical action which is accelerated by the presence of dissolved oxygen. The combined action of sulphide and oxygen is even more severe if the protective oxide film has previously been eroded away. In common with other copper alloys, if a component that has been subjected to sulphide attack is transferred to non-polluted sea water, the protective copper/aluminium oxide film does not readily reform as it does when it is physically damaged in non-polluted sea water. The component therefore remains vulnerable to electrochemical corrosive action unless and until high flow removes the copper sulphide layer. This may take a long time.

More information on the corrosion of nickel–aluminium bronze in the presence of sulphide pollution may be obtained from the result of an investigation by A. Schüssler and H. E. Exner.[161]

Caustic alkaline solutions

A peculiarity of aluminium and of silicon is that, like sulphur and carbon, they form acidic oxides: Alumina AL_2O_3 and Silica SiO_2 respectively. These oxides react with caustic alkaline solutions (sodium, potassium and calcium hydroxides) to form aluminate and silicate salts. Thus in the case of sodium hydroxide (caustic soda), the reactions for aluminium and silicon oxides are as follows:

$$2\ NaOH + AL_2O_3 = 2NaAlO_2\ (\text{Sodium aluminate}) + H_2O$$
$$2NaOH + SiO_2 = Na_2SiO_3\ (\text{Sodium silicate}) + H_2O$$

Since aluminium oxide is part of the protective oxide film of all aluminium bronzes and silicon oxide is part of the protective oxide film of silicon-aluminium bronze, concentrated caustic alkaline solutions will attack the protective oxide films of these alloys. This means that aluminium bronzes are not suitable for use in processes which handle concentrated caustic alkaline solutions.

Types of corrosive and erosive attack

Corrosion attacks can be divided into two main types:

- Uniform or general corrosion
- Localised corrosion

Uniform or General Corrosion

Uniform or general corrosion is that which succeeds in permeating to some extent the protective oxide film under electrochemical action or which directly attacks this film chemically. In seawater, significant chemical attack occurs only in polluted

Table 8.3   Comparison of resistance of various copper alloys to general corrosion and erosion/corrosion in sea water.[41]

Alloy	% Composition (bal Cu)					Depth of Impingement Attack mm		General Corrosion Weight Loss mg cm^{-2} per day	
	Al	Fe	Ni	Mn	Zn	28-day Jet Impinge-ment 20°C	14-day Brownsdon & Bannister 20°C	Water in slow motion	At 10 m s^{-1} Water Speed
Copper–aluminium	8.2	1.7	–	–	–	0.04	0.19	0.15	0.17
Nickel–copper–aluminium	8.2	2.9	4.3	2.4	–	0.00	0.32	0.04	0.10
Nickel–copper–aluminium	8.8	3.8	4.5	1.3	–	0.00	0.28	0.04	0.16
Mang–copper–aluminium	7.6	2.8	3.1	10.0	–	0.01	0.24	0.04	0.11
High tensile brass	0.8	0.8	0.2	0.5	37.0	0.03	0.08	0.09	0.73
	Sn	Zn	Pb						
Gunmetal	9.7	1.4	0.6			0.02	0.32	0.14	0.74
Gunmetal	5.1	5.0	4.3			0.23	0.39	0.22	1.66

waters containing hydrogen sulphide and is aggravated by the presence of dissolved oxygen. As previously mentioned, concentrations of sulphides in water directly attack the copper oxide in the oxide film and are consequently detrimental to copper alloys and particularly to aluminium bronzes. If this condition is sustained and severe, it is considered to be a special type of corrosion and not general corrosion.

As long as the oxide protection is not undermined, aluminium bronzes are very little affected by electrochemical action in sea water and fresh water, except in the special cases which will be discussed below. In the case of other non-ferrous metals or alloys in normal commercial use, the amount of metal removed by general corrosion in sea water or fresh water is insufficient to cause significant damage to components. In some aggressive waters, pure copper and some of the high copper content alloys such as bronzes can, however, introduce sufficient copper into the water to cause increased corrosion of galvanised steel or of aluminium alloys downstream of the copper alloy components, but this problem is not experienced in connection with aluminium bronze components. A comparison of general corrosion rate of aluminium bronzes with other copper and ferrous alloys is given in Tables 8.3 and 8.4. The figures for Table 8.4 were determined using freely exposed specimens fully immersed for one year beneath rafts in Langstone Harbour, Great Britain.

Aluminium bronzes are very little affected by non-oxidising acids and are widely used, for example, for handling sulphuric acid, whereas acidic solutions cause relatively rapid dissolution of many other copper alloys.

The corrosive effects, if any, of a variety of chemical substances on aluminium bronze are discussed in Chapter 9.

Table 8.4   Comparison of resistance of various copper and ferrous alloys to general corrosion, crevice corrosion and erosion in flowing seawater.[41]

Alloys	General Corrosion Rate mm/year	Crevice Corrosion mm/year	Erosion/ Corrosion Resistance m s⁻¹
			$m\ s^{-1}$
Wrought Alloys:			
Phosphorus deoxidised copper C106–7	0.04	< 0.025	1.8
Admiralty brass CZ111	0.05	< 0.05	3.0
Aluminium brass CZ110	0.05	0.05	4.0
Naval brass CZ112	0.05	0.15	3.0
HT brass CZ115	0.18	0.75	3.0
90/10 copper–nickel	0.04	< 0.04	3.7
70/30 copper–nickel	0 025	< 0.025	4.6
CuAl5 copper–aluminium	0.06	< 0.06	4.3
CuAl7 copper–aluminium	0.05	< 0.05	4.3
CuAl10Fe3 copper–aluminium	0.06	0.075	4.6
CuAl10Ni5Fe4 nickel–copper–aluminium	0.075*	< 0.5 for 3 to 15 months, then 1.0	4.3
CuAl7Si2 silicon–copper–aluminium	0.06	< 0.075	2.4
17% Cr stainless steel 430	< 0.025	5.0	> 9.1
Austenitic stainless steel 304	< 0.025	0.25	> 9.1
Austenitic stainless steel 316	0.025	0.13	> 9.1
Monel	0.025	0.5	> 9.1
Cast Alloys:			
Gunmetal LG2	0.04	< 0 04	3.7
Gunmetal G1	0.025	< 0.025	6.1
High tensile brass HTB1	0.18	0.25	2.4
CuAl10Fe3 copper–aluminium	0.06	< 0.06	4.6
CuAl10Ni5Fe5 nickel–copper–aluminium	0.06*	< 0.5 for 3 to 15 months, then 1.0	4.3
CuMn13Al8Fe3Ni3 manganese–copper– aluminium	0.04	3.8	4.3
Austenitic cast iron (AUS 202)	0.075	0	> 6.1
Austenitic stainless steel 304	< 0.025	0.25	> 9.1
Austenitic stainless steel 316	< 0.025	0.125	> 9.1
3% or 4% Si Monel	0.025	0.5	> 9.1

* Under ideal conditions a black film slowly forms on nickel–copper–aluminium in sea water which reduces the corrosion rate in accordance with an equation of the form:
Corrosion rate ~ (time)$^{-0.2}$.

Localised corrosion

In most circumstances the external oxide film on aluminium bronze components protects them from corrosive attacks. There are however circumstances in which this protection is undermined and the following are the forms of local corrosion which may occur as a result:

Pitting
Crevice Corrosion
Impingement Erosion/Corrosion

Cavitation Erosion/Corrosion
Stress Corrosion Cracking
Corrosion Fatigue

For the most severe conditions of service it can be beneficial to heat-treat nickel–aluminium bronze castings and hot rolled plates for six hours at 675°C followed by cooling in still air. For thicker sections, annealing at 700°C may be preferable. This improves both the resistance to corrosion and the mechanical properties.

Pitting
Pitting is an example of the effect of differential aeration mentioned above. Due to localised damage to the protective oxide film, or due to internal defects uncovered by machining or fettling, a recess or 'pit' is created at a given point on a metal component which is inaccessible to oxygen. It may be caused also by a non metallic inclusion in the metal component going into solution in the liquid medium or by sulphide attack mentioned above. The 'pit' may initially be little more than a scratch on the surface of the component but it increases in size as its surface corrodes. Once pitting corrosion has started it becomes self sustaining. This is because, according to V. Lucey[124], a layer of cuprous oxide forms a bi-polar membrane across the mouth of the pit. Behind this membrane, the surface of the pit is anodic and ions go into solution which then migrate through the cuprous oxide membrane and deposit around the mouth of the pit where they form a mound of corrosion products which is cathodic. The greater the potential difference between the inside surface of the pit and that of the aerated outside surface, the faster the rate of corrosion. Corrosion is also accelerated by the fact that the aerated area is considerably greater than the inside surface of the pit. Furthermore, the accumulation of corrosion products at the mouth of the pit and the above mentioned membrane of cuprous oxide across the mouth of the pit, further restricts the access of oxygen and prevents re-oxidation of its inside surface.

Pitting corrosion is important because of its localised character which can result in perforation of the wall of a valve, pump casting, water tube or other vessel in a relatively short time. All common metals and alloys are subject to pitting corrosion to a greater or lesser extent under certain conditions of service, but aluminium bronzes and copper alloys in general are not normally vulnerable to significant pitting in sea water service. Cathodic protection will reduce the risk of pitting occurring.

Crevice (Shielded area) Corrosion
A crevice is a 'shielded area' where two components or parts of the same component are in close contact with one another, although a thin film of water can penetrate between them: for example between flanges, within fasteners and at 'O' ring joints. A crevice can also be created by marine growth (biofouling) or other deposits on the surface of the component.

The crevice is starved of oxygen and therefore becomes lower in oxygen than its surrounding. In the case of stainless steels, this low oxygen area becomes the anode of an electrolytic cell and the higher oxygen concentration outside the crevice becomes the cathode. Consequently, corrosion may occur within the crevice. It is another example of the effect of differential aeration.

Nickel-aluminium bronze, which is not cathodically protected in its vicinity by steel structures or by a 'sacrificial anode', is susceptible to crevice corrosion. There is therefore a significant advantage in providing cathodic protection. J. C. Rowlands[155] carried out experiments on the crevice corrosion of nickel–aluminium bronze in sea water. He observed that the copper-rich α phase was initially anodic to the aluminium-iron-nickel rich precipitate known as the κ_3 phase (see Chapter 13) and corroded preferentially for a time but at a low rate. Meanwhile the hydrogen concentration in the crevice gradually increased, transforming the sea water in the crevice, over a period of five months, from very slightly alkaline (pH 8.2) to markedly acidic (pH 3). At this point, the κ_3 phase had become anodic to the α phase and was corroding at the rate of 0.7–1.1 mm/year. This was accompanied by the deposition of metallic copper in the corrosion zone which masked the corrosion damage. The continuous nature of the κ_3 phase means that its corrosion can significantly reduce mechanical properties. It was observed that the crevice corrosion effect was independent of whether the sea water was aerated or non-aerated.

Copper ions from the corrosion film of copper-rich alloys normally dissolve into sea water and, being poisonous to marine organisms prevent biofouling. Due to their lower solution rate, however, aluminium bronzes are more susceptible to biofouling than other copper-rich alloys. Cathodic protection, however, prevents the discharge of copper ions and therefore makes copper alloys more vulnerable to biofouling, but protects them from crevice corrosion by galvanic action. Calcium salts or oxide deposits may also prevent copper going into solution and therefore encourage biofouling. The best protection is provided by a combination of cathodic protection (by sacrificial anode if necessary) and chlorination to deter biofouling as is the practice on offshore oil rigs.

J. C. Rowlands also report that seam-welded nickel–aluminium bronze tubes, subject to low continuous or intermittent flow, were liable to corrode slightly in the weld area and that the corrosion products, deposited on the heat affected zone, created crevices which then led to severe crevice corrosion. This did not occur at high flows since the corrosion products were swept away.

Culpan and Rose[62] report, on the other hand, that in crevice corrosion tests which they carried out on nickel–aluminium bronze castings, corrosion occurred *around* the crevice and was very similar to that seen at the heat affected zone of a welded specimen.

Practically all metals and alloys suffer accelerated local corrosion either within or just outside a crevice but it is very rare in the case of cathodically protected aluminium bronze. The risk of crevice corrosion can often be avoided by sealing the joint between two components and using sealing washers under bolt heads and nuts (this practice is not recommended, however, for stainless steel).

A comparison of the resistance to crevice corrosion of various copper and ferrous alloys is given in Table 8.4. These figures were determined using samples fully immersed for one year beneath rafts in Langstone Harbour, Great Britain. The specimens held in Perspex jigs, providing crevice conditions between the metal sample and the Perspex.

Impingement Erosion /Corrosion

All common metals and alloys depend for their corrosion resistance on the formation of a superficial layer or film of oxide or other corrosion product which protects the metal beneath from further attack. Under conditions of service involving exposure to liquids flowing at high speed the flow generates a shear stress at the metal surface which may damage this protective film, locally exposing unprotected bare metal. This is a form of wear (see Chapter 10) which becomes more severe with a high degree of local turbulence or if the flow contains abrasive particles such as sand. The continued effect of erosion, preventing permanent formation of a protective film, and the corrosion of the bare metal consequently exposed, can lead to rapid local attack causing substantial metal loss and often penetration. This type of attack is known as erosion/corrosion or impingement attack. Nickel–aluminium bronze is the most resistant to erosion/corrosion of the copper-based alloys.[113]

Because of the configuration of pumps and valves and of the resultant turbulence, flow velocities at certain points are much higher than mean velocity. The successful use of aluminium bronzes in these items, despite the turbulence, demonstrates their excellent resistance to erosion/corrosion.[74]

Erosion/corrosion can be avoided,

(a) by choosing an alloy that can withstand the flow velocities of the particular equipment. The allowable design impingement velocity of clean water with aluminium bronzes is around 4.3 m s^{-1} (14 ft/sec). J. P. Ault[17] found that the annual erosion/corrosion rate of nickel aluminium bronze in fresh unfiltered sea water varied logarithmically with velocity. Thus at 7.6 m s^{-1} (25ft/sec) it was 0.5mm/year and at 30.5 m s^{-1} (100 ft/sec) it was 0.76mm/year. Even at 7.6 m s^{-1} it could rise locally to 2mm/year. He found, however, that 'cathodic protection to −0.60 Volt versus silver-silver chloride essentially stopped corrosion of the coupon exposed to low and high flow rates (7.6 m s^{-1} and 30.5 m s^{-1} respectively)'. It is nevertheless advisable not to exceed the design limit. He also observed that turbulence intensity affected corrosion rates and that significant pitting occurred under highly turbulent flow conditions.

(b) by fitting filters or strainers at the inlet of pumps and turbines which are likely to handle water contaminated with sand or other abrasive substances.

A comparison of the resistance of aluminium bronze to erosion/corrosion with that of other alloys is shown in Tables 8.3 and 8.4. The erosion/corrosion resistance tests results given in Table 8.4 were carried out using the Brownsdon and Bannister test. The specimens were fully immersed in natural sea water and supported at 60° to a submerged jet, 0.4 mm diameter placed 1 – 2 mm away, through which air was forced at high velocity. From the minimum air jet velocity required to produce erosion/corrosion in a fourteen-day test, the minimum sea water velocity required to produce erosion/corrosion under service conditions was estimated on the basis of known service behaviour of some of the materials.

The highest resistance to erosion/corrosion is shown by alloys that have a protective film resistant to erosion and which reforms very rapidly if it should suffer mechanical damage. Stainless steels are particularly resistant to this type of attack. Unalloyed copper is relatively poor but all copper alloys are substantially more resistant than copper itself and nickel aluminium bronze is among the most resistant of all the copper alloys. The British Defence Standard Data Sheets suggest slightly higher erosion/corrosion resistance for CuAl10Fe3 than for CuAl10Ni5Fe4 and much lower resistance for CuAl7Si2. Practical experience indicates, however, that the nickel–aluminium bronzes are superior and silicon-aluminium bronze only marginally inferior to other aluminium bronzes in this respect. It is perhaps significant that the Defence Standard Data Sheet figures for erosion/corrosion resistance were derived from Brownsdon and Bannister test results. Table 8.3 compares other Brownsdon and Bannister test results with those of jet impingement tests which are considered to be more representative of service behaviour.

Cavitation Erosion/Corrosion
Under certain water flow conditions the phenomenon of cavitation may arise. Rapid changes of pressure in a water system, as may occur with rotating components such as propellers and pump impellers, cause small vapour bubbles to form when the pressure is lowest. These bubbles then tend to migrate along the pressure gradient until the pressure suddenly increases causing them to collapse violently on the surface of the metal. Hence cavitation damage tends to occur at a point some distance from the low pressure point which caused the bubble to form. For example, in the case of a propeller, the bubbles form near the hub and then migrate along the blades and usually implode about a third of the way from the centre of the propeller.[74] The effect is most severe when the lowest pressure in the system is below atmospheric pressure. The stresses generated by the collapse of bubbles (cavitation) can be quite severe and may locally remove the protective oxide film of certain alloys. It may tear out small fragments of metal from the surface – usually by fatigue. The soundness of the alloy is of critical importance in resisting cavitation erosion since any sub-surface porosity may collapse under the hammering effect of cavitation.

The metal freshly exposed as a result of cavitation will of course be subject to corrosion and the resultant damage is due to a combination of corrosion and the mechanical forces associated with the bubble collapse. In view of the magnitude of the mechanical forces associated with cavitation damage, the contribution made by the associated corrosion is, however, relatively small.

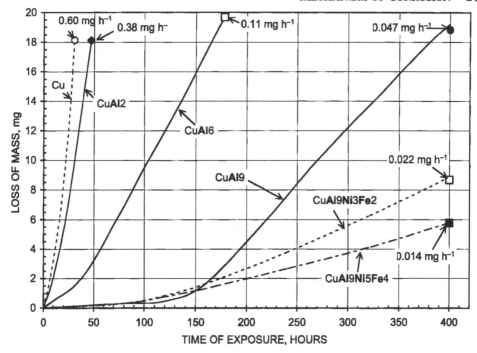

Fig. 8.4 Effect of composition on metal loss resulting from cavitation erosion by J. L. Heuze et al.[91]

The effect of alloy composition on resistance to cavitation erosion has been investigated by J. L. Heuze et al.[91], using a cavitation vortex generator. The results are shown in Fig. 8.4. They show that, in the case of binary alloys, the higher the aluminium content the greater the resistance to cavitation erosion. The best resistance to cavitation erosion is obtained with alloys containing nickel and iron. The effect of microstructure on resistance to cavitation erosion is explained in Chapter 13.

Shalaby et al. and Al-Hashem[162] have carried laboratory experiments on the cavitation erosion/corrosion of a standard nickel aluminium bronze. They report that cavitation made the surface of the material very rough, with large cavities and some ductile tearing. The rate of mass loss under cavitation was 186 times that of quiescent conditions. With cathodic protection the mass loss was reduced to 53% of the non-protected rate of loss. It is likely therefore that corrosion at the grain boundaries, which occurred in the absence of cathodic protection, facilitated the dislodging of grains by cavitation erosion, resulting in a much greater rate of mass loss. They gave no indication of how artificially created cavitation is likely to compare with cavitation encountered in service.

Nickel–aluminium bronze has extremely good resistance to cavitation damage and is consequently the principal, high performance alloy for small or large marine propellers. It is also extensively used in water turbines and high duty pump

Table 8.5   Cavitation Erosion in 3% NaCl Solution.[41]

Material	Depth of Attack
CuAl10Fe5Ni5 aluminium bronze	< 0.025 mm in 7 hours
Austenitic stainless steel 321	0.305 mm in 7 hours
High tensile brass	0.280 mm in 6 hours

Table 8.6   Cavitation Erosion Rates in Fresh Water.[41]

Material	Cavitation Erosion Rate mm^{3}h^{-1}
CuAl10Fe5Ni5 aluminium bronze	0.6
CuAl10Fe3 aluminium bronze	0.8
Cu**Mn**13Al8Fe3Ni3 aluminium bronze	1.5
High tensile brass	4.7
Gunmetal G1	4.9
Monel K500 – cold drawn	2.8
Monel K500 (aged)	1.2
Austenitic stainless steel 321	1.7
Austenitic stainless steel 316	1.7
Cast martensitic stainless steel 420	1.7
Cast austenitic stainless steel 347	1.0
Spheroidalgraphite cast iron	1.3
Ni-resist cast iron	4.4

impellers. Although less resistant to cavitation erosion than cobalt-based hard-facing alloys, titanium, series 300 austenitic and precipitation hardened stainless steels, nickel–chrome and nickel–chrome–molybdenum alloys, nickel–aluminium bronzes closely approach the cavitation resistance of these alloys.[179]

Tables 8.5 and 8.6 give comparisons between the cavitation/erosion performance of aluminium bronzes and that of some other alloys.

Stress Corrosion Cracking

Stress corrosion is a highly localised attack occurring under the simultaneous action of tensile stresses in a component and a particular type of corrosive environment. Thus low alloy austenitic stainless steels, such as types 304 and 316, are vulnerable to stress corrosion in warm chloride solutions (sea water) and so are single-phase aluminium bronzes[151] whereas duplex and complex aluminium bronzes are not affected. All copper alloys, however, are susceptible to stress corrosion cracking in the presence of moist sulphur dioxide, nitrites, ammonia, and ammonium compounds.

Traces of sulphur dioxide are found in the atmosphere of industrial areas as a result of the burning of coal and oil.

Nitrites, which are used as inhibitors to prevent steel corrosion either as an addition in solution or added to a polymeric coating, react with copper at the

Table 8.7   Effect of pH on time to failure of two CuAl alloys, strained at a rate of 0.33/sec in solution of (NH$_4$OH), (NH$_4$)$_2$SO$_4$ and (Cu SO$_4$) by H Leidheiser.[116]

pH	Time to failure min	
	CuAl4	CuAl8
4.0	NF	NF
5.8	NF	NF
6.8	NF	NF
7.3	NF	NF
8.3	2250	1260
10.2	2250	1260
11.2	930	230
12.1	130	95

NF = No failure after 5000 min exposure

surface of all copper alloys to produce a small amount of ammonia. The combination of nitrite and ammonia is particularly aggressive even at very low stresses.[74] The use of copper alloys in these applications is not advisable.

Ammonia and ammonium compounds are formed by the action of bacteria on organic matter and are given off by urine. They are soluble in water. Industrially manufactured ammonia is used in the production of fertilisers and explosives and as a refrigerant. The pH value has a critical effect on stress corrosion failure, as is illustrated in Table 8.7 in the case of two binary copper aluminium alloys immersed in a solution containing ammonium hydroxide (NH$_4$OH), ammonium sulphate (NH$_4$)$_2$SO$_4$ and copper sulphate (Cu SO$_4$).[116] The effect becomes very pronounced as soon as the solution changes from acid to alkaline but decreases as the pH values increases. It will also be noted that the 4% Al alloy is less vulnerable than the 8% Al alloy. This does not agree, however, with the findings of A. W. Blackwood et al.[27] who found that, of three binary alloys containing 1.5%, 4% and 7% Al, the 4% alloy was the most vulnerable. They also report that a transition from intergranular to transgranular cracking occurred at 4% Al. The copper content of the solution is related to the pH value and liability to failure decreases as the copper content increases.[27]

Aluminium bronzes have better resistance to stress corrosion cracking than brasses, though not as good as copper–nickel. Nickel–aluminium bronze is preferable to the high–manganese–aluminium bronze in sea water applications and, for this reason, is tending to supersede it for ships' propellers.

The total amount of corrosion is very small but the local weakness it creates leads to cracking under stress which occurs in a direction perpendicular to that of the applied stress and may cause rapid failure. It is not clear why one corrosive environment is more effective than others in this respect, but it could be that the stored energy in the component may be a contributing factor in the mechanism of corrosion as well as in the consequent cracking. For this reason, components that have been hot or cold worked or subjected to welding should be stress-relief heat-treated to minimise the risk. This is particularly important in the case of single-phase aluminium bronzes.[151] It is also advisable to keep assembly stresses in fabricated equipment as low as possible by accurate cutting and fitting of the component parts.

Table 8.8   Atmospheric Stress Corrosion Tests on Copper Alloys.[41]

Alloy	Temper % Cold Rolled	Time to Failure	
		New Haven	Brooklyn
70/30 brass	50	35–47 days	0–23 days
Leaded alpha – beta brass	50	51–136 days	70–104 days
Admiralty brass	40	51–95 days	41–70 days
Aluminium brass	40	221–495 days	311–362 days
Aluminium bronze (9.7% Al, 3.86% Fe)	40	> 8.5 years	> 8.5 years

Table 8.9   Comparison of stress corrosion resistance of brasses, copper–aluminium and copper–nickel alloys.[41]

Alloy	Time to 50% Relaxation (hours)
Arsenical admiralty brass	0.30
Muntz metal	0.35
Naval brass	0.50
70/30 brass	0.51
Aluminium brass	0.60
5% aluminium bronze	4.08
8% aluminium bronze	5.94
90–10 copper–nickel	234
PDO copper	312
70–30 copper–nickel	> 2000

Service stresses are, however, frequently unavoidable and, where these are likely to be high, the low susceptibility of duplex and complex aluminium bronzes, and especially of the nickel aluminium bronze, to stress corrosion is an important consideration.

Stress corrosion cracking may follow a transgranular or intergranular path depending upon the alloy and the environment. In the presence of ammonia, stress corrosion cracking of aluminium bronze follows a transgranular path. Intergranular stress corrosion cracking can occur, however, in single phase alloys in high pressure steam service or in hot brine. Research in the USA has shown that susceptibility to this type of attack can be eliminated by the addition of 0.25% tin to the alloy (American specification UNS 61300). Such a tin addition is liable however to cause cracking in welding (see Chapter 7).

Table 8.8 gives the results of atmospheric tests of U-bend specimens exposed to two different industrial environments.

Table 8.9 shows the results of tests carried out under very severe conditions, i.e., a high ammonia content in the atmosphere and very high stress levels (including plastic deformation) in the samples and would not be representative therefore of the performance of the alloys tested under normal service conditions. They are nevertheless of interest as a comparison of the resistance to stress corrosion of these alloys. The very significant difference in resistance to stress corrosion of 90–10 copper–nickel as compared to that of the single-phase 5% and 8% copper-aluminium alloys should be noted.

These tests were carried out using loop specimens of sheet material exposed to moist ammoniacal atmosphere. The ends of the loops were unfastened once every 24 hours and the extent of relaxation from the original configuration was measured. This is a measure of the progress of stress corrosion cracking on the outside surface of the loop. Table 8.9 gives the time to 50% relaxation for various alloys tested.

Corrosion Fatigue

Corrosion fatigue strength is an important consideration in the choice of cast and wrought alloys used in propellers and in pumps, piping and heat exchangers used in deep diving submersibles, undersea equipment and certain oil production activities. Much of the latter equipment is subject to low-cycle fatigue that can occur with repeated operation at great depths, or to high-cycle fatigue occurring in rotating machinery or to both.[179]

Metals can fail by fatigue as a result of the repeated imposition of cyclic stresses well below those that would cause failure under constant load. In many corrosive environments the cyclic stress level to produce failure is further reduced, the failure mechanism then being termed corrosion fatigue. The relative contributions to the failure made by the corrosion factor and the fatigue factor depend upon the level of the cyclic stress and upon its frequency, as well as upon the nature of the corrosive environment. Under high frequency loading conditions such as may arise from vibration or rapid pressure pulsing due to the operation of pumps, etc., the corrosion resistance of the alloy is of less importance than its mechanical strength but under slow cycle high strain conditions both these properties become important.

Because of their combination of high strength with high resistance to normal corrosive environments, aluminium bronzes, and particularly the nickel-aluminium bronzes (which are the best in both these respects), show excellent corrosion fatigue properties under both high frequency and low frequency loading conditions. The corrosion resistance of nickel aluminium bronze is the primary factor that affects its corrosion fatigue resistance.[45] It is not surprising, therefore, that heat treating nickel aluminium bronze components at 700°C for 6 hours

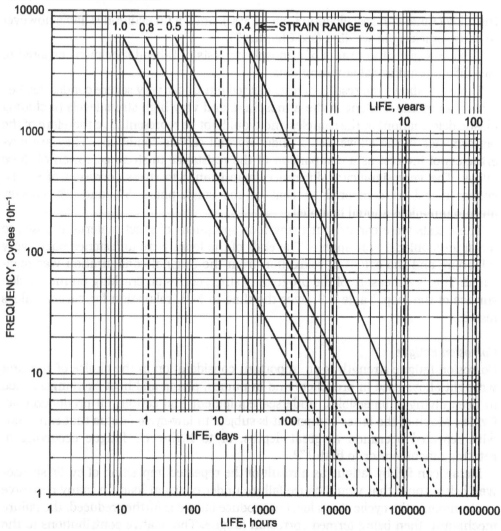

Fig. 8.5 High strain/low cycle corrosion fatigue results for heat-treated cast nickel–aluminium bronze to DGS 348 (CuAl9Ni5FeMn).[41]

followed by air cooling has been shown to improve its fatigue strength in both air and 3% sodium chloride solution.[133]

Figure 8.5 shows results of corrosion fatigue tests carried out in sea water at 32°C on nickel aluminium bronze specimens which were strained by bending about a zero strain mean position.

Corrosion Associated with Welds
Welding can adversely affect the corrosion resistance of many alloys in different ways. Galvanic coupling can result from differences in composition or of structure

between the filler and the parent metal. The metallurgical structure of the heat-affected zone adjoining the weld may be changed for the worse, giving rise to a more anodic phase, especially in multipass welding in which the time at elevated temperature is relatively long. Welding under conditions of restraint can also introduce stresses in the weld metal and in the heat-affected zones of the parent metal which may lead to stress corrosion cracking.

The aluminium bronzes most commonly used under conditions where welding is required are the single phase alloy CuAl8Fe3 ('Alloy D'), and the nickel–aluminium bronzes CuAl9Ni6Fe3 (wrought) and CuAl9Ni5Fe4Mn (cast). The welding of aluminium bronzes is dealt with in chapter 7 and only those aspects directly concerned with corrosion resistance will be discussed here.

Since problems of weld cracking can arise in welding the CuAl8Fe3 alloy with a matching filler, unless the impurity levels in both the filler and parent metal are closely controlled, it is common practice to use a duplex alloy filler containing ~10% Al. To avoid selective phase corrosion of the beta phase in the filler on subsequent service in sea water or in acid solutions, it is recommended that an overlay with a composition matching the parent metal should be applied on top of the duplex filler. If a matching filler is not available an overlay of nickel–aluminium bronze is used.

The possibility of tensile stresses and consequent increased susceptibility to stress corrosion cracking arising as a result of welding under conditions of restraint has already been mentioned. A further factor to be watched in welding the CuAl8Fe3 alloy is the formation of micro-fissures in the heat-affected zone during welding which can act as stress raisers and so further increase the danger of stress corrosion cracking in subsequent service.

No serious corrosion problems are introduced in welding nickel–aluminium bronze CuAl9Ni6Fe3. The use of an approximately matching filler ensures that galvanic effects between the filler and parent metal are reduced to a minimum, although the aluminium content of the weld bead will usually be higher than that of the parent metal. The good high-temperature ductility of the CuAl9Ni6Fe3 alloy also means that there is little likelihood of micro fissuring occurring and the level of stress in the heat-affected zone, arising from welding under restraint, is also likely to be less than in the CuAl8Fe3 alloy welded under similar conditions.

Nickel–aluminium bronze castings may be welded to repair small areas of casting porosity, etc., or in the manufacture of large components or water circulating systems. The welding is usually carried out using a filler with approximately the same composition as the parent metal but, under conditions of severe restraint, care must be taken to avoid weld cracking.

As explained in Chapter 13, changes in the microstructure of nickel–aluminium bronze in the heat affected zone of a weld can make a welded component more vulnerable to corrosion in sea water service. This can be aggravated by the presence of internal stresses in a casting or wrought component caused by welding, which could lead to stress corrosion cracking. The likelihood of this can be eliminated by heat-treatment, although it must be said that welded aluminium bronze components, which have had no post-weld heat treatment, are widely used in sea water and other environments without difficulty. This is particularly so in the case of nickel–aluminium bronze propellers that are routinely repaired in service without giving rise later to stress cracking or de-aluminification.[179] Under severe service conditions, however, a post-weld heat treatment consisting of six hours at 700°C ± 15°C followed by cooling in still air may be advisable.[178–74]

9
ALUMINIUM BRONZES IN CORROSIVE ENVIRONMENTS

Introduction

Few metals or alloys are totally immune to corrosion. Most will corrode under some conditions and some are very much more resistant than others. Apart from the physical properties required, the choice of an alloy for a particular application depends therefore on the environmental conditions in which the metal component is to be used. The choice will also be influenced by cost in relation to the required life span of the equipment and, in some cases, by the relative weldability of the various alloys under consideration.

Most aluminium bronze alloys have excellent resistance to corrosion, but not all. It is therefore important to choose an alloy that is appropriate to the corrosive environment in which it is to be used. For corrosive environments in which certain ferrous parts are not suitable, some aluminium bronze alloys offer a corrosion-resistant alternative with a strength equal to that of low alloy steels. Hence many ferrous components, such as machine-tool parts, hydraulic valves and bearing surfaces, can be directly replaced by aluminium bronze without the necessity of complete redesign.

Marine fittings are required to withstand aggressive attack from sea water and spray without significant deterioration over long periods of time. Under these conditions the appropriate aluminium bronze alloy has been found to be an ideal material, even where relatively high-velocity water is encountered, and its reliability may be gauged from the numerous pumps, valves, stern-tubes, nuts, bolts and other deck and underwater fittings in service today. Propellers provide the largest single tonnage with some weighing over 70 tonnes as-cast.

Most dilute acid, alkaline and salt solutions are safely handled and some aluminium bronze alloys show an outstanding resistance to sulphuric acid at concentrations up to 95%. At moderate strengths this acid has an economically low rate of attack, even at temperatures up to the boiling point. Good results have also been reported with pumps handling hot concentrated acetic acid. C. P. Dillon[65] confirms that aluminium bronze can be used in alkaline chemical processes if the conditions are properly understood and controlled.

Since by far the greatest tonnage of aluminium bronze used is in sea water applications for which the high strength nickel–aluminium bronze is generally specified, a comparison between this alloy and competing ferrous alloys is of special interest. A comparison is therefore given in Appendix 4 between the mechanical, physical and corrosion resisting properties of these alloys.

Table 9.1   Summary of environments for which aluminium bronze is suitable.[127-41]

Corrosive environments for which aluminium bronze is suitable	Exceptions
Industrial, rural and marine atmospheres	Atmospheres containing concentrations of ammonia, ammonium compounds and sulphur dioxide
Sea water and hot sea water	Sea water containing concentrations of sulphides
Steam	Steam containing concentrations of sulphur dioxide and chlorine
Acids Some concentrated acids: Sulphuric Acid up to 95% concentration Acetic acid Most dilute acids including: Hydrochloric acid up to 5% concentration and at ambient temperature (unless given cathodic protection) Phosphoric acid Hydrofluoric acid	Oxidising acids such as nitric acid. Aerated acids or acids containing oxidising agents such as ferric salts and dichromates
Most alkalis	Concentrated caustic alkaline solutions and alkalis containing concentrations of ammonia or its derivatives.
Most salts	Salts of iron, copper and mercury. Oxidising salts such as permanganates and dichromates.

Suitability of Aluminium Bronzes for Corrosive Environments

The resistance to general corrosion of aluminium bronze in various corrosive environments will now be considered. A summary of environments for which aluminium bronze is suitable is given in Table 9.1

Atmospheres

Atmospheric exposure-tests of up to twenty years' duration have proved the good resistance of aluminium bronze to industrial, rural and marine atmospheres.[54-176-7]

Table 9.2 gives a comparison of the corrosion rates of various copper alloys after 15 to 20 years exposure to marine, industrial and rural atmospheres in the US. Unfortunately, only one aluminium bronze (a silicon–aluminium bronze shown in bold) was included in the test. It will be seen that this alloy had the lowest corrosion rate. There was some intergranular corrosion to a depth of 0.05 mm and the tensile strength of the alloy was reduced by 5.85% whereas that of other copper alloys was reduced in most cases by less than 2%. As one would expect, the table shows that, after 20 years, an industrial atmosphere is the most corrosive.

Table 9.2   Corrosion rates of various copper alloys after 15–20 years in marine, industrial and rural atmospheres by L. P. Costas.[54]

Alloy	Composition									Corrosion rate in various atmospheres μm/yr			
										20 Yrs	15 Yrs	20 Yrs	20 Yrs
	Cu	Zn	Sn	Ni	Mn	Fe	Al	Si	Other	Mar.	Mar.	Ind.	Rur.
65	98.71	<0.10	1.18						P: 0.11	1.1	0.38	1.4	0.65
66	88.23	0.10		10.12	0.32	1.23			Pb: <0.02	1.4	0.97	2.3	1.1
67	95.72	0.04	4.08			0.005			P: 0.16 Pb: 0.001	1.7	0.75	2.0	0.70
68	88.37		1.83			0.015		0.01	P: 0.10 Pb: 0.003	0.53	0.52	1.7	0.80
69	**91.65**	**0.02**				**0.02**	**6.40**	**1.85**	**Pb: 0.001 As: 0.06**	**0.46**	**0.22**	**1.2**	**0.54**
70	84.71	15.25	<0.01			0.02			Pb: <0.05	0.61	0.44	1.5	0.68
71	70.60	29.38	<0.01	<0.01		0.014			Pb: <0.05	0.54	0.39	1.8	0.91
74	99.94								O: 0.042 S: 0.003	1.1	0.43	1.3	0.70
75	97.99			1.85	0.02	0.03				0.78	0.60	1.4	0.83
76	93.73			6.34	0.01	0.01				0.67	0.84	1.3	0.80
77	90.35			9.16	0.27	0.18				0.74	1.4	0.04	0.69
78	77.18			22.76	0.01	0.04				0.77	1.0	1.5	0.70
79	55.38			42.75	1.67	0.30				0.55	0.53	1.8	0.53
80	82.62			4.13	12.83	0.20				0.72	1.1	1.7	0.69

Tests by Tracy,[176–7] showed that a 92/8 aluminium bronze alloy tested showed no initial advantage over copper in industrial atmospheres but corroded at an average rate of only 17 μm/year after ten years. In maritime atmosphere aluminium bronze showed a significant advantage over copper and corroded at only one-fifth of its rate in industrial atmospheres in spite of the salt spray.

After prolonged atmospheric exposure, aluminium bronze usually has a grey or black protective film, although sometimes a green film is formed which is not as attractive as the patina on copper. The golden colour of the alloy can, however, be retained by wax polishing or lacquering.

Castings or wrought material of almost any composition give good service, but cold-worked products should be annealed or stress-relieved to prevent any possibility of stress- corrosion in polluted or ammonia-contaminated atmospheres. Where conditions are particularly corrosive, such as in railway tunnels or near factory chimneys, it is particularly important that the alloy should be free from the γ_2 phase mentioned in Chapter 8. Several aluminium bronzes were included in an extensive series of tests undertaken by the Association of American Railroads to assess materials for overhead electrification systems where acid condensate corrosion is a serious problem.[37] Similar tests were carried out in the UK by Britton,[35] who subjected various materials to an extremely corrosive railway tunnel

Table 9.3   Oxidation and scaling of aluminium bronzes.[127]

| % Composition | | | Condition | Thickness of metal removed at 400°C per ten 5-hour heating cycles (mm) | | | | |
Cu	Al	Fe		Dry Air	Air + 10% water	Dry Air + 0.1% SO$_2$	Dry Air + 5% SO$_2$	Moist Air + 0.1% HCl
Rem	2.06	0.01	50% Cold drawn	Nil	Nil	0.3	2.2	5.3
Rem	5.66	0.008	50% Cold drawn	Nil	Nil	Nil	11.4	1.5
Rem	9.76	0.039	Extruded	Nil	Nil	Nil	0.8	2.1
Rem	10.13	2.80	Extruded	Nil	Nil	Nil	Nil	1.1
Rem	11.10	0.006	Extruded	Nil	Nil	Nil	0.7	0.7
Rem	12.06	0.02	Extruded	Nil	Nil	Nil	Nil	0.9

Copper (for comparison)

Cu	Fe	As	P	Ni	Pb						
Rem	0.002	0.46	0.07	0.06	0.01	50% Cold drawn	0.6	0.5	0.8	1.5	27.0

atmosphere for three years. During this period, a copper–aluminium alloy containing 8.6% Al, 0.3% Fe with only traces of the γ_2 phase gave excellent results, but an alloy with higher aluminium content and a continuous γ_2 phase suffered more severe attack.

Atmosphere heavily polluted with ammonia and ammonium compounds can be detrimental to aluminium bronze particularly in the case of stressed components. This can give rise to stress corrosion cracking unless the component is stressed relieved.

Table 9.3 shows that, although an adherent scale forms in dry and moist air at 400°C, the presence of 0%-1% hydrochloric acid gas is deleterious. Sulphur dioxide at a concentration of 0.1% caused no attack but significant deterioration occurred when the concentration was raised to 5%. Aluminium bronzes, however, were found to be much more resistant than any other copper alloy to these corrosive atmospheres.

In the same tests, selective oxidation by the Price and Thomas technique (see Chapter 8) was found to protect a 95/5 copper–aluminium alloy from atmospheric oxidation at temperatures up to 800°C but was not effective against the acidic atmospheres at 400°C.

Sea Water

The principal constituents of seawater that affect the corrosion performance of metal alloys are:

- 3% solution of sodium chloride (common salt) which has good electrical conductivity and therefore acts as an electrolyte in electrochemical corrosion,
- dissolved oxygen which restores the protective oxide film when damaged; if unevenly distributed, it can give rise to electrochemical reaction and its presence aggravates the effect of sulphides,
- nutrients and bacteria which give out sulphide and ammonia emissions that are very corrosive in concentration,
- biofouling organisms, sediment, waste and debris which give rise to crevice corrosion,
- residual chlorine from chlorination which narrows the electro-potential range in the case of aluminium bronze.[179]

Aluminium bronzes have excellent resistance to seawater corrosion and have been widely used at normal and elevated temperatures for low and high water velocity conditions. Provided the corrosion-prone β and γ_2 phases are avoided, most commercial aluminium bronzes have good general corrosion resistance, but some have an exceptionally high resistance to cavitation and impingement attack. No cases of stress-corrosion in sea water are known and pitting attack is uncommon. In moderately polluted waters, however, pitting of heat-exchanger tubes has been encountered particularly under deposits. Due to their lower solution rate, aluminium bronzes are slightly more susceptible to biofouling than the less corrosion-resistant copper alloys (see Chapter 8). A comparison is given in Appendix 4 between the corrosion resistance of nickel aluminium bronze and that of competing ferrous alloys.

The remarkable results obtainable from aluminium bronzes are well illustrated by tests made at the US Naval Civil Engineering Research and Evaluation Laboratory, Port Hueneme, California.[36] A 5.5% Al copper–aluminium alloy suffered an average corrosion rate of only 0.013 mg mm^{-2} per day without any noticeable pitting after two years continuous immersion. These figures were less than those for any other copper alloy tested, and also lower than those for Alloy 400 (70–30 nickel–copper alloy) and stainless steel. As the corrosion rate quoted is lower than that encountered in commercial alloys immersed in Atlantic waters, a summary of open sea-water corrosion tests undertaken by the Central Dockyard Laboratory, Portsmouth, England is given in Table 9.4. Typical corrosion rates for aluminium bronzes lie between 51 to 76 µm/year, a range only exceeded if the alloy structure contains the β and γ_2 phases. As much of this corrosion is surface roughening, the above figures provide a very conservative guide for the design of structures subject to sea-water corrosion.

The detrimental effect of sulphide contamination in fresh or sea water was mentioned in Chapter 8. At low flows (< 5 m/sec) and in the absence of dissolved oxygen, sulphides are not particularly detrimental to aluminium bronzes even at concentrations as high as 55g m^{-3}, unless exposure to sulphides is followed by exposure to aerated water.[179] If the sulphide pollution is slight and is dispersed by

Table 9.4 Corrosion of aluminium bronzes in open sea water for 1 year.[127]

% Composition (balance: Cu)				Structure	Plain Corrosion			Crevice Corrosion		Water velocity causing impingement attack†	
Al	Fe	Ni	Other Elements	(w) Wrought (c) Cast	mg dm^{-2} per day	Average depth of attack (mm)	Form of attack	Depth of attack mm	Form of attack	ft s^{-1}	m s^{-1}
3.8	0.1	–	Zn1.2, Fe0.1	(w) α	7.8	0.0508	Surface roughening	0.0025	Surface roughening	12	3.7
5.0*	–	–	–	(w) α	10.6	0.0787	–	–	–	–	–
5.5*	–	–	As 0.29	(w) α	6.5	0.0406	Surface roughening	–	–	–	–
6.5	–	–	Si 2.0	(w) α+κ+γ$_2$	1.2	–	–	0.0508	–	–	–
7.5	3.0	–	–	(w) α	7.4	0.0838	"	2.5400	Layer coppering	14	4.3
3.6	2.4	2.1	Mn 11.4	(w) α	6.6	0.0610	"	0.0508	coppering	10	3.0
7.4	2.9	2.0	Mn 12.5	(c) α+β	2.8	0.0508	"	4.0640	β attacked	20	6.1
9.0	1.0	–	–	(w) α+β	12.0	0.0635		0.0203	β attacked	–	–
9.4	–	–	–	(w) α+γ$_2$	4.3	0.3048	γ$_2$ attacked	3.3528	γ$_2$ attacked	14	4.3
9.3	2.3	–	Mn 0.3	(w) α+β	6.8	0.0559	Surface roughening	0.0254	β attacked	14	4.3
9.7	0.1	–	Mn 1.6	(c) α+β	12.8	0.1626	β attacked	0.5080	β attacked	18	5.5
9.5	4.9	4.8	Mn 0.4	(w) α+β+κ	6.8	0.0508	Surface roughening	0.0610	coppering	15	4.6
9.4	4.8	4.7	–	(w) α+β+κ	9.4	0.0762	Pitting up to 1.143 mm	0.0406	coppering	14	4.3
9.4	5.4	4.1	–	(w) α+κ	9.0	0.0762	Pitting up to 0.889 mm	0.0762	coppering	–	–
8.9	4.5	3.9	Mn 1.0	(c) α+β+κ	18.4	0.1524	Coppering and local deep pitting	–	–	–	–
10.4	4.2	5.5	Mn 0.8	(c) α+κ	8.9	0.0356	Surface roughening	–	–	–	–

* Results reported by Bulow[39]   † Brownsdon-Bannister Test.

currents, the effect on copper-based alloys may be negligible, as is evidenced by the extensive and satisfactory use of copper–aluminium alloys and bronzes in marine applications. If, however, sulphide pollution is contained and if dissolved oxygen is present, it is detrimental to copper-based alloys, including copper–nickel alloys. It is especially detrimental to aluminium bronze. Components, such as pumps and valves, subjected to high flow velocities are particularly vulnerable.

The typical black sulphide film which results from exposure to sulphide-polluted seawater will in time be replaced by a normal oxide film when the component is transferred to clean aerated seawater, although substantially higher corrosion rates persist for some time. This happens when vessels are fitted out in polluted harbours before reverting to the open sea, when the normal protective film replaces the sulphide film in ~9 days. Chemical cleaning with inhibited hydrochloric acid will remove the sulphide film and speed up the formation of the protective film.[179]

Hot Sea Water

Laboratory tests in sea water at 95°C have shown that a 7% aluminium, 2% iron alloy corrodes at only 7.6 μm/year after 5,000 hours immersion.[127] This material corresponds to alloy CuAl7Fe2 (ASTM C61400/B171 Alloy D) which is favoured for condenser tube plates. The corresponding result for alloy CuAl10Ni6Fe3 (ASTM Alloy E), sometimes used for this application, was 45.7 μm/year and suffered from a certain amount of pitting. Hudson[96] reported a loss of 0.025 mg mm^{-2} per day for a 10.6% Al, 3% Fe alloy in aerated sea water at 95°C.

Steam

Steam generated from boilers using distilled or mains water has no significant effect on aluminium bronzes at temperatures up to 400°C and possibly much higher. Hallowes and Voce[85] extended their research on oxidation to include uncontaminated steam and steam containing sulphur dioxide and chlorine. The introduction of these chemically active impurities resulted in some degree of corrosion at 400°C which increased in intensity at higher temperatures. The attack was in the form of dealuminification and the scale was exfoliative. These tests were only carried out for 50 hours, however, and give an indication of the degree of severity of attack which the impurities can cause. Service experience over long periods of time has shown that the more corrosion-resistant aluminium bronzes, free from the γ_2 phase, give good service in clean steam. Valve spindles in alloy CuAl10Fe5Ni5 and silicon–aluminium bronze CuAl7Si2 are satisfactory for temperatures up to 400°C. Silicon–aluminium bronze can however be attacked if the feed water is overheated.

The possibility of stress-corrosion cracking of certain single phase alloys, if used under high stress levels in superheated steam, has been reported by Klement[111]

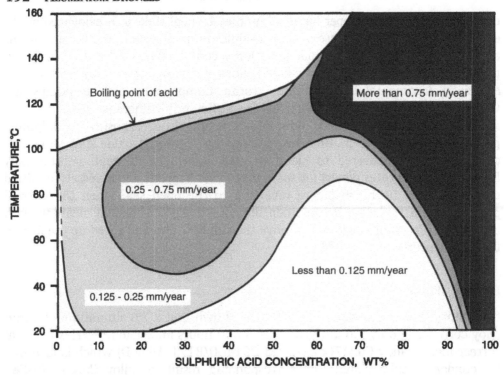

Fig. 9.1   Corrosion rate of complex aluminium bronze CuAl10Ni5Fe4 in sulphuric acid in the presence of oxygen[127]

and his co-workers. An alloying addition of 0.3% tin and/or silver proved completely effective in overcoming the problem encountered with a 7% aluminium, 2% iron alloy (ASTM: B171 Alloy D). As previously mentioned, however, tin in excess of 0.1% can lead to cracking in welding under constraint.

Sulphuric Acid

Sulphuric acid is present to a greater or lesser degree in many industrial corrosive solutions and waters, and the ability of aluminium bronze to resist attack by this acid accounts for an appreciable number of its applications. All aluminium bronzes give satisfactory service under a wide range of conditions but they show an exceptional resistance to concentrations of ~50% sulphuric acid at temperatures up to the boiling point.

A true assessment of all the conditions, under which aluminium bronzes will be suitable, is difficult to obtain as corrosion rates are sensitive to alloy composition and structure, to temperature, and to the concentration and degree of aeration of the acid. Fig. 9.1 shows the complex effect on corrosion rates of varying sulphuric acid concentrations and temperatures in the case of a 9.5% Al, 5% Ni, 4.2% Fe

Table 9.5 48-hour corrosion tests in boiling 10% sulphuric and 2½ wt % hydrochloric acid.[127]

Composition %				Corrosion Rate – mm/year				Condition and Structure
				10% H_2SO_4		2½% HCl		
Cu	Al	Ni	Fe	Aerated	Non-Aerated	Aerated	Non-Aerated	
Rem	9.6	1.9	1.2	0.091	0.081	1.702	1.499	Chill cast. – α, 30% β, trace γ_2 and some κ
				0.132	0.020	–	–	Oil quenched 600°C α, 35% β, little κ
Rem	9.7	1.9	0.1	0.163	0.061	0.991	1.067	Chill cast. – α, 35% β
				0.213	0.041	0.457	0.584	Oil quenched 600°C α, 30%β
Rem	10.2	1.9	0.1	0.046	0.122	1.905	2.108	Chill cast. α, 50% β, trace γ_2
				0.117	0.030	0.787	0.965	Oil quenched 600°C
Rem	9.6	–	0.17	0.254	0.041	0.635	0.889	Chill cast. – α, 35% β, grain boundary γ_2
				0.183	0.051	0.686	0.864	Oil quenched 600°C α, 35%β.
Rem	10.4	–	0.05	0.122	0.102	3.480	4.293	Chill cast. – α, 35% β, 15% eutectoid.
				0.290	0.051	1.194	2.007	Water quench. 650°C α, 60% β.
Rem	7	2	–	1.981	0.239	25.908	0.229	Chill cast. – α.
Rem	10.1	5.4	5.5	0.439	0.173	44.958	0.508	Chill cast. – αβ + fine κ.
				2.743	0.041	34.036	0.381	Water quench 1000°C
				0.305	0.030	77.724	–	Tempered. 1 hr 600°C Air cooled – martensitic
Rem	7	2(Si)	0.5 (Mn)	0.305	–	0.838	0.483	Water quench. 950°C Tempered. 1 hr 600°C Air cooled – Martensitic, with 20% α.
For Comparison Stainless Steel: 18Cr, 10Ni, 2.8Mo, 0.08C				1.524	–	13.462	–	Chill cast and heat treated, 190 VPN.

alloy under normal aeration conditions. Over much of the field covered by the tests the aluminium bronze was superior to Alloy 400 (70–30 nickel–copper alloy). Typical penetration rates for aluminium bronze working in 0%-75% sulphuric acid at room temperature lie in the region of 0.08–0.13 mm/year.

Caney[43] examined in some detail the effect of alloy composition and structure on the corrosion of cast and wrought aluminium bronzes in boiling sulphuric acid under aerated and non-aerated conditions. He concluded that the optimum composition approximated to 9.6% Al, 1% Fe, 2% Ni, although the 90/10-type of alloy was barely inferior under certain conditions. Some of the results he obtained are given in Table 9.5; they are particularly interesting as they show the superiority of the 90/10 alloy over the more expensive CuAl10Fe5Ni5 alloy and stainless steel.

Aeration increases the rate of corrosion of aluminium bronzes in sulphuric acid and accelerated attack occurs when oxidising agents, such as ferric salts or dichromates, contaminate the acid. However, where these are absent, aluminium bronzes appear to offer a possible alternative to lead for sulphuric acid service, especially where mechanical strength is a design consideration.

Acetic Acid

High concentrations of acetic acid have been satisfactorily handled industrially at elevated temperatures.

Hydrochloric Acid

Corrosion rates in hydrochloric acid are appreciably higher than those in sulphuric acid and, except for concentrations below 5%, the maximum temperature for satisfactory service is little above ambient. The optimum alloy composition is not necessarily the same as for sulphuric acid, as Table 9.5 shows. Caney's work[43] suggests that iron should be kept to a low value and the alloy should be free from γ_2. In boiling 2.5% hydrochloric acid, he found that the lowest rate of corrosion occurred with a 9.7% Al, 1.9% Ni alloy which had been quenched from 600°C. Under the same acid conditions a CuAl10Fe5Ni5 alloy suffered severe attack. Other results for this acid are given in Table 9.6 which compares the performance in hydrochloric acid with that in sulphuric acid and in other acids discussed below.

Table 9.6   Corrosion of aluminium bronze (10.2% Al, 0.3% Fe, 0.5% Mn) by dilute mineral acids at ambient temperature.[137]

Acid	Thermal Treatment of Alloy	Loss in Weight after 1000 hrs mg dm^{-2}	Corrosion Rate per day mg dm^{-2}
5% Sulphuric	Quenched from 880°C	1,400	34
	Quenched from 880°C and tempered	1,400	33
	Slowly cooled from 880°C	250	6
	Normalised	250	6
5% Hydrochloric	Quenched from 880°C	1,850	44
	Quenched from 880°C and tempered	1,700	41
	Slowly cooled from 880°C	350	8
	Normalised	350	8
10% Hydrochloric	Quenched from 880°C	2,700	65
	Quenched from 880°C and tempered	2,500	61
	Slowly cooled from 880°C	700	16
	Normalised	1,100	27
5% Nitric	Quenched from 880°C	3,500	84
	Quenched from 880°C and tempered	2,100	51
	Slowly cooled from 880°C	23,700	569
	Normalised	31,800	763

Table 9.7   Corrosion Rates for Cu.Al9 in Hydrochloric Acid.[41]

Temperature	Additions	% HCl	g m² per day
20°C	–	3.6	4–5
100°C	–	15	56
100°C	–	30	14
Room temperature	–	30	14
Room temperature	0.8% chloride	30	42
Room temperature	1% FeCl₃	30	115

An interesting instance of the successful application of aluminium bronze in warm moderately strong hydrochloric acid concerns its use as pickling hooks for steel descaling. The ferrous materials in contact with it provide adequate cathodic protection to reduce corrosion to a very low rate.

Table 9.7 gives the corrosion rates of a 9% Al copper–aluminium in various concentrations of hydrochloric acid. It is taken from E. Rabald's *Corrosion Guide*.[147] Rabald comments: 'The greater attack at 100°C and 15% HCl is caused by the higher air content' and notes that the solubility of oxygen at 100°C is higher in 15% HCl than in 30% HCl.

Phosphoric Acid

Table 9.8 (also from Rabald) gives corrosion rates in pure 20% and 60% phosphoric acid in long term tests.

Table 9.8   Corrosion Rates for Cu 10% Al in Phosphoric Acid.[41]

Temperature	20% H₃PO₄ mm/year	60% H₃PO₄ mm/year
15°C	0.06	0.01
50°C	0.10	0.01
75°C	0.25	0.00
Boiling point	–	0.25

Hydrofluoric Acid

The good resistance of aluminium bronzes to corrosion by hydrofluoric acid is exploited in the 'frosting' of glass bulbs for electric lamps. Nickel-aluminium bronze is used for the nozzles which spray acid into the bulbs and the trays which collect the acid from the process.

Nitric Acid

This strong oxidising acid generally results in an excessive rate of attack which precludes the use of aluminium bronzes to any extent. Corrosion rates for 90/10 alloy are quoted in Table 9.6.

Other Acids

Aluminium bronze gives good results with most other non-oxidising acids at room temperature and resistance to most organic acids is good.

Effect of small alloying additions on corrosion rate in acid

Singh et al.[192] did some experiments on the effect of small additions of tantalum (Ta), lanthanum (La) and neodymium (Nd) on the corrosion rate of a CuAl7Fe2 alloy in HCl, H_2SO_4 and HNO_3. Additions of 0.1% Ta, 0.1% La or 0.05% Nd reduced the rate of corrosion of the alloy in HCl by nearly 40%, in H_2SO_4 by 15–27% and in HNO_3 by 7–14%. It does not follow, of course, that such additions would have a similar effect on other aluminium bronze alloys. The effect of these additions on mechanical properties was not stated.

Alkalis

Apart from ammonia, which attacks all copper-base alloys, most alkalis can be safely handled with aluminium bronze. Excellent results have been obtained in contact with sodium and potassium carbonate and dilute caustic alkaline solutions at all temperatures. Strong solutions of the caustic alkalis, however, attack the protective oxide film, as explained in Chapter 8. Prolonged corrosion tests do not appear to have been carried out to date with other alkaline solutions.

Salts

The small concentrations of hypochlorites and bisulphites found in paper making processes do not significantly attack aluminium bronzes which have given excellent service as beater bars, valves, suction rolls etc. In the chemical industries large quantities of chloride, sulphate and nitrate salts are recovered with the aid of aluminium bronze heat exchangers but salts of iron, copper and mercury, and strongly oxidising permanganates and dichromates under acid conditions should not however be handled in aluminium bronze components.

Aluminium bronze components used in corrosive environments

The high corrosion resistance of aluminium bronzes, combined with high strength and availability in a number of different forms, results in their being used under a wide variety of conditions and by a wide variety of industries. Their principal fields of application are:

Marine service
Water supply

Oil and petrochemical industry
Chemical industry
Building industry

Marine Service

Aluminium bronze is used in a variety of equipment and fittings in ships and in land-based installations that use sea water for cooling. The following are the main types of equipment containing aluminium bronze components:

Marine propellers
Other underwater fittings
Sea water pumps
Valves
Heat exchangers
Pipework

Marine Propellers

The requirements for materials for marine propellers are:

- high resistance to corrosion fatigue, to erosion/corrosion and to cavitation erosion,
- a high strength-to-weight ratio,
- good castability and tolerance of welding and local working for repairing damage sustained in service.

The choice of alloys for the manufacture of large propellers essentially reduces to nickel–aluminium bronze, manganese–aluminium bronze and high tensile brass ('manganese bronze').

Results of corrosion fatigue tests on cast material always show a considerable scatter and the values obtained depend upon such factors as the size of specimen and the frequency of loading. Most published results, however, agree in showing the corrosion fatigue strength of nickel–aluminium bronze in sea water to be approximately twice that of high tensile brass, with manganese–aluminium bronze falling about midway between these two (See Table 9.9). Among the ferrous alloys, spheroidal-graphite cast iron has a corrosion fatigue strength approximately equal to that of high tensile brass. The incorporation of nickel and chromium in the austenitic grade produces no improvement in corrosion fatigue strength although the general corrosion resistance of the material is considerably improved. Cast low alloy austenitic stainless steels are also reported to have a fatigue strength approximately equal to that of high tensile brass.

Note, however, that since results of corrosion fatigue tests are dependent on factors such as test bar design and size, and test frequency, comparisons between results from different sources should be made only with caution.

Table 9.9   Corrosion Fatigue Properties of Marine Propeller Alloys.[41]

Material	Time (days)	Seawater or 3% NaCl	Corrosion Fatigue Strength (108 cycles) N mm^{-2}
Nickel–aluminium bronze CuAl10Fe5Ni5	23*	3% NaCl	± 108
	35	3%NaCl	± 122
	50	Seawater	± 87
Manganese–aluminium bronze CuMn13Al8Fe3Ni3	23*	3% NaCl	± 91
	35	3% NaCl	± 89
	50	Seawater	± 62
High tensile brass	23*	3% NaCl	± 62
	35	3% NaCl	± 74
	50	Seawater	± 42
Spheroidal graphite cast iron (ferritic)	23	3% NaCl	± 46
Spheroidal graphite cast iron (austenitic)	23	3% NaCl	± 46
13% Cr. stainless steel	23	3% NaCl	± 54
19/11 austenitic stainless steel	50	Seawater	± 45

* Indicates samples cut from propellers. All other results were from cast test pieces

Nickel–aluminium bronze shows higher erosion/corrosion resistance than high tensile brass and its resistance to cavitation erosion is greater than that of high tensile brass by a factor of about eight. Manganese–aluminium bronze offers erosion/corrosion resistance approximately equal to that of nickel–aluminium bronze with somewhat inferior resistance to cavitation erosion.

Repair welding of high tensile brass propellers can introduce a corrosion hazard since this alloy is susceptible to stress corrosion cracking in sea water and is, therefore, liable to suffer stress corrosion in the weld and heat-affected zone, where residual stresses remain, unless a stress relief heat treatment is carried out after welding. Manganese–aluminium bronze also shows susceptibility to stress corrosion cracking, although to a considerably smaller extent, and must also be given a stress relief heat treatment after welding. Nickel–aluminium bronze requires more care in welding to avoid formation of cracks in either the weld or the parent metal but, since it is not subject to stress corrosion in sea water, the need for subsequent stress relieving treatment, although always desirable, is not so great. This is a significant advantage in the case of propellers that are difficult to heat-treat without distortion.

The blades, hub body, hub cone and bolts of controllable pitch propellers can be made from nickel–aluminium bronze or from stainless steel. Sound nickel–aluminium bronze components have good resistance to cavitation erosion but duplex stainless steels are more resistant. Austenitic stainless steels, on the other hand, are more susceptible to cavitation damage and are more prone to crevice corrosion but they have higher resistance to erosion/corrosion.

Other Underwater Fittings

Nickel–aluminium bronze, manganese-aluminium bronze and high tensile brass are all used for underwater fittings such as propeller shaft brackets and rudders and are satisfactory in conditions of free exposure to sea water. Under conditions where deposits of silt or mud may form on underwater fittings, high tensile brass and manganese-aluminium bronze are both liable to selective phase attack. Nickel–aluminium bronze may show slight attack of this type but to a very much smaller extent. High tensile brass and manganese-aluminium bronze are not suitable for underwater fasteners because of their liability to stress corrosion cracking. Nickel–aluminium bronze CuAl10Fe5Ni5, silicon-aluminium bronze CuAl7Si2, phosphor bronze or Alloy 400 (70–30 nickel–copper alloy) are used for this purpose. Phosphor bronze and Alloy 400 or K500 are, however, of lower strength than nickel–aluminium bronze.

Sea Water Pumps

Nickel–aluminium bronze is widely used for impellers in centrifugal pumps due to its excellent resistance to both erosion/corrosion and cavitation damage. For the most severe applications or where long life and reliability are particularly important the pump body can also be made of nickel–aluminium bronze together with the shaft and the fasteners. The body is often made of gunmetal, however, even though it is not easily weldable should any repair be required. Gunmetal impellers may be used in pumps operating under relatively low speed conditions. Alloy 400 (70–30 nickel–copper alloy) impellers may be used in high duty pumps, but these do not normally offer any advantage over nickel–aluminium bronze and are usually more expensive. Cast austenitic stainless steel impellers do not provide the same strength or resistance to cavitation damage. These materials are also less reliable for shafts because of their liability to pitting corrosion in the gland area during shut-down periods.

Some sea water pumps have cast iron bodies with impellers of gunmetal, nickel–aluminium bronze, Alloy 400 (70–30 nickel–copper alloy) or K500 or austenitic stainless steel. During the early life of such pumps, the cast iron provides some sacrificial protection to copper alloys or Alloy 400 or K500 impellers but the corrosion of the cast iron takes the form of selective phase corrosion leaving a surface that is essentially graphite. A heavily graphitized pump body is strongly cathodic to non-ferrous impeller materials and can cause accelerated attack on them. Consequently the use of aluminium bronze or other non-ferrous impellers is not recommended for sea water pumps with cast iron bodies if long service is expected. A cast austenitic stainless steel impeller with a Ni-resist body or a super-duplex impeller and body are preferable in those circumstances.

Valves:

Valves in salt water systems with steel or galvanised steel pipework are usually also of ferrous material but are protected internally by non-metallic coatings. The discs and seats are usually of cast nickel–aluminium bronze or Alloy 400 (70–30 nickel–copper alloy) and the stems of wrought nickel–aluminium bronze or phosphor bronze but sometimes of Alloy 400 (70–30 nickel–copper alloy) or 70/30 copper–nickel. High tensile brass is not suitable for the valve stems because of its liability to dezincification nor are low alloy austenitic stainless steels because of their liability to pitting in crevices.

For systems using copper alloy pipework, gunmetal valves are most commonly used – often with nickel–aluminium bronze stems – but for high integrity systems valves made entirely from nickel–aluminium bronze or from copper–nickel are used. Nickel–aluminium bronze has the advantage of greater strength and is usually less expensive than copper–nickel for this purpose.

Heat Exchangers:

Steam condensers, oil coolers and other heat exchangers operating on sea water usually have tubes of aluminium brass or of copper–nickel. Aluminium bronze CuAl7 is manufactured in tube form but is not very often used in heat exchangers. There is an increasing tendency to use titanium tubes for condensers and heat exchangers where conditions are very severe or where extended trouble-free life is essential – for example, drain coolers in ships and main condensers in electricity generating stations.

The 'traditional' material for heat exchanger tubeplates is rolled naval brass and this is usually satisfactory with tubes of coated steel or aluminium brass. Experience of deep dezincification in naval brass tubeplates used with 70/30 copper–nickel tubes in main condensers led the British Navy, several years ago, to change to aluminium bronze CuAl9Ni6Fe3 (alloy E) tubeplates which have proved quite satisfactory. CuAl8Fe3 (alloy D) is, however, very widely used with copper–nickel tubes in sea water cooled condensers and heat exchangers without problems.

Apart from titanium-clad steel, aluminium bronzes CuAl9Ni6Fe3 (alloy E) or CuAl8Fe3 (Alloy D) are the only satisfactory materials for tubeplates in heat exchangers using titanium tubes. CuAl9Ni6Fe3 (alloy E) is preferred since its reliability from the corrosion point of view is higher, but it is harder and was somewhat more difficult to drill (with modern tooling this should not be a problem). Waterboxes and covers for heat exchangers, condensers and desalination plant are historically of rubber-coated cast iron or steel. CuAl8Fe3 (Alloy D) or the corresponding alloy with addition of tin (UNS Alloy 61300) are suitable provided that appropriate care is taken over welding (see 'Corrosion associated with welds' and 'Stress corrosion cracking' in Chapter 8).

Pipework:

Sea water piping is often made of steel or galvanised steel if first cost is a ruling factor and corrosion failures will not result in serious loss or damage. For higher quality systems 90–10 copper–nickel is used, 70–30 copper–nickel being occasionally employed where maximum strength and corrosion resistance are required. These properties can be obtained also from aluminium bronze plates and there is consequently interest in the use of pipes made from aluminium bronze by seam welding which show resistance to corrosion and erosion/corrosion equal to or better than 90–10 copper–nickel but with a better strength/weight ratio. Seam welded aluminium bronze pipes should be heat treated to ensure resistance to corrosion (see Chapter 7).

Fresh Water Supply

The principal application of aluminium bronzes in the fresh water supply industry is for pumps of the centrifugal and axial flow types but they are also employed for valve trims, especially for valve spindles.

Pumps
The corrosive conditions for pumps handling fresh water are obviously less severe than for sea water systems but impeller tip velocities are nevertheless usually too high for gunmetals, and aluminium bronze is, therefore, specified. CuAl9Fe5Ni5 is the most suitable alloy because of its higher resistance to erosion/corrosion. Very long working lives are expected from pumps in the water supply industry and the service conditions often change, usually making greater demands on the pumps.

Centrifugal pump bodies are sometimes of cast iron but more frequently of gunmetal or aluminium bronze to avoid contamination of the water by corrosion of the cast iron. The shrouds of axial flow pumps can be made of cast iron since they can be effectively protected by non-metallic coatings. Pump spindles are usually of nickel aluminium bronze or stainless steel. Stainless steel rarely suffers pitting attack in this type of pump since the pumps are normally operated continuously and chloride concentrations are low (typically ≤ 100 mg/l).

Valves
The larger size valves used in the water supply industry are of coated cast iron or steel but with internal trim in non-ferrous material or stainless steel. Alloy 400 (70–30 nickel–copper alloy) and aluminium bronze CuAl10Fe3 or CuAl10Fe5Ni5 are often used, with aluminium bronze CuAl10Fe3 or CuAl10Fe5Ni5 for spindles; either will give satisfactory corrosion resistance but CuAl10Fe5Ni5 has higher strength. High tensile brass is sometimes used for valve spindles but aluminium bronze is much to be preferred because of the liability of high tensile brass to selective phase dealloying (dezincification) in some waters.

Aluminium bronze is not generally used in small stopvalves for domestic water installations except for valves that are to be installed underground. The British Standard Specification BS 5433 requires these to be made from materials immune to dezincification and permits forged aluminium bronze CuAl10Fe5Ni5 for all parts of the valve except the body, which is cast in gunmetal.

Oil and Petrochemical Industries

The use of aluminium bronzes in the oil and petrochemical industries is largely restricted to pumps, heat exchangers and valves in cooling water systems and the comments under 'Marine Service' above generally apply.

Aluminium bronze CuAl7 tubes are sometimes used with tubeplates of CuAl9Ni6Fe3 or CuAl8Fe3 in heat exchangers especially when these are operating under relatively high pressure and it is desired to weld the tubes to the tubeplates. For most purposes, however, heat exchangers are tubed with aluminium brass or, for particularly severe service conditions, with titanium and the tubeplates are of naval brass or aluminium bronze CuAl9Ni6Fe3 or CuAl8Fe3.

For coolers dealing with streams of high pressure products, tube-and-shell condensers are used with sea water on the shell side and with the product passing through the tubes. The shells and baffles are fabricated from CuAl8Fe3 ('Alloy D') plate 12 to 15 mm thick, using a duplex (9 to 10% Al) alloy as the weld filler but with a capping run of the parent CuAl8Fe3. The tubeplates are also of CuAl8Fe3; with tubes of 70/30 copper–nickel or titanium.

Aluminium bronze bolts are usually employed on submersible cooling water pumps made in copper alloys. Since the pipes themselves are of steel and are cathodically protected, an alternative procedure, which is often used, is to fit steel bolts and flanges with a 'bolt protector' filled with grease fitted round the bolt shanks, and to rely upon the cathodic protection to take care of the exposed ends. In critical cases or in cases where there is no cathodic protection, a corrosion resistant alloy is used such as superduplex stainless steel, nickel aluminium bronze, high strength copper–nickel, or nickel alloys such as Alloy 625 and 925. Alloy K-500 (65–35 Ni-Cu) bolts are not used in offshore installations because of the problem of hydrogen embrittlement.

Sea water piping ranging from approximately 50 to 600 mm diameter is used on offshore oil platforms to convey cooling water and water for injection back into the well. The first generation of North Sea platforms used cement-lined steel for these but super austenitic and superduplex stainless steels are the most commonly used alloys for new installations. Welded aluminium bronze tubes would also give good corrosion resistance.

One important application of aluminium bronze in the oil industry is its use for fans in the inert gas protection systems on board oil tankers. These fans are used to maintain the flow and pressure of the inert gas blanket over the oil cargo so as to avoid the danger of explosion or fire. The inert gas used is produced from the exhaust gas of the main engines, auxiliary engines or sometimes from a special generator, and is 'scrubbed' with sea water. The fan operating conditions can be very corrosive and far from easy to predict, involving salt-laden water vapour,

Table 9.10   Uses of aluminium bronzes in corrosive chemical environments.[41]

Environment	Alloy	Details
Acetic acid 33% at bp*.	CuAl8Fe	Pumps, fittings, pipelines, condensers.
Acetic anhydride at bp*.	—	Pipelines, pumps, caps for copper stills.
Aluminium fluoride solutions at 100°C.	—	Linings for reaction vessels producing AlF_3 from $Al(OH)_3$ and HF.
Aluminium sulphate all concentrations up to b_2*. Free H_2SO_4 present.	—	Valves, pipes, fittings.
Boric acid saturated solutions 105°C.	—	Fittings, filter closures.
Citric acid 33–36% solution up to bp*.	—	Fittings, piping.
Dyestuffs, acid dyes.	—	Stirrers, fittings.
Ethylene dibromide up to bp*.	—	Pumps.
Fatty acids up to 200°C.	—	Vessels, valves, pumps, vacuum stills.
Fluorine −196 to +200°C.	—	Valve seats.
Fluorosilicates MgSiF6 and $ZnSiF_6$ at pH 2 120°C.	—	Vacuum evaporators.
Formaldehyde 200 to 400°C.	—	Parts of apparatus for oxidation of MeOH and manufacture of acrolein from HCHO and CH_3CHO.
Formic acid room temperature to bp*.	CuAl8Fe or CuAl10Fe	Fittings pumps.
Furfural.	CuAl5	Autoclaves for manufacture from corn-cobs or straw by acid hydrolysis.
Gelatin up to 100°C.	—	Boilers for acid and alkaline digestion of skins, bones, etc.
Isobutyl chloride room temperature to bp*.	—	Valves, pumps, fittings.
Limonene: mixtures of turpentine with benzoic and salicylic acid 105°C.	—	Valves, pumps, fittings.
Linseed oil.	—	Bottoms of varnish boilers.
Mixed acids 55% H_2SO_4 40% HNO_3, 5% SO_3, room temperature.	CuAl10	Bolts for pumps, stuffing boxes.
Molasses room temperature to 180°C.	—	Fittings, valves, piping.
Potassium sulphate all concentrations room temperature to bp*.	—	Heating tubes, stirrers, pumps, valves.
Quinine sulphate.	—	Fittings, valves.
Sodium bisulphate all concentrations room temperature to bp*.	—	Evaporators, heating coils.
Sodium bisulphites all concentrations room temperature.	—	Blades for beaters in paper industry.
Sodium fluorosilicate 100°C.	—	Evaporators.
Sodium hypochlorite 2% solution room temperature.	—	Knives for hollanders.
Sodium sulphate all concentrations room temperature to bp*.	CuAl8	Valves, centrifuges, pumps.
Stearic acid room temperature to 250°C.	—	Vacuum still pans, condensers.
Sulphur dioxide dry or moist bp to 100°C.	—	Valves, pumps.
Sulphurous acid 6% solution room temperature.	—	Piping, valves.
Zinc chloride all concentrations room temperature to 140°C.	—	Vessels, stirrers.
Zinc sulphate all concentrations room temperature to bp*.	—	Pumps, piping.

*bp = boiling point

sulphurous gases and traces of carbon. Several materials have been used to construct the fans, many of which are large and run at high speeds. Only titanium and aluminium bronze have been found to give reliable service and, of these, aluminium bronze is far less expensive. The smaller fans may use cast impellers but the larger ones are fabricated by welding.

Chemical Industry

The general policy in the chemical manufacturing industries is to construct equipment of mild steel wherever possible and to use stainless steel for those parts where the corrosion resistance of mild steel is inadequate. Copper-base alloys tend to be used mainly in cooling water pumps and valves. Aluminium bronzes are, however, also used for small items in a very wide variety of chemical environments where high resistance to corrosion and erosion are required.

Table 9.10, compiled by E. Rabald,[147] lists a number of chemical environments in which aluminium bronzes, have been successfully used. It lists industrial uses of aluminium bronzes in contact with corrosive chemicals where corrosion rates of < 2.4 g m^{-2} per day (0.008 mm per year) were recorded. The list by no means covers all industrial uses of aluminium bronzes in chemical industry and excludes cases where aluminium bronzes have been used successfully but with corrosion rates somewhat higher than those specified above. The information provided by Rabald does not always include details of the particular aluminium bronzes concerned. Where this is the case, a dash has been inserted under the heading 'Alloy' in Table 9.10.

Building Industry

The combination of high strength, weldability, good general corrosion resistance and low susceptibility to stress corrosion cracking presented by aluminium bronzes make them a preferred material for load bearing structural features and masonry fixings. The extensive use of aluminium bronze castings and wrought parts in the new British Parliamentary Building is the most outstanding example to-date of the suitability of aluminium bronze for structural members in a building intended to last for many years. These structural members not only provide strength and durability but, being the most visible features of this impressive building, they give it a unique and pleasing appearance. All aluminium bronze castings in this building are in the CuAl10Fe5Ni5 alloy and the wrought parts in the corresponding wrought alloy. A number of lightly loaded features or more intricate shapes than can be produced in copper–aluminum, are made in brass.

The appearance and tarnish resistance of aluminium bronzes make them suitable for decorative architectural features but their relatively high price compared with brasses and high tensile brasses limits their use for this purpose.

A specialised application for aluminium bronze in building is as reinforcement bars or clamps used in the repair of old stonework. Restorations of stonework carried out in the nineteenth century were commonly made using cast or wrought iron for reinforcement and subsequent rusting of the reinforcement has resulted in severe further damage to the stonework. Aluminium bronze is the most suitable of the copper alloys for this purpose, not only because of its high resistance to general corrosion and to stress corrosion cracking but because its high strength enables bars and clamps of relatively small section to be used. Hence correspondingly less damage is done to the old stonework during installation of the reinforcement.

10
RESISTANCE TO WEAR

Aluminium bronze as a wear resisting material

Tribology, the study of friction, wear and lubrication, is relatively new and it is only in the last 25 years that the understanding of wear has developed most rapidly.[97] Although aluminium bronze has found increasing recognition for a wide variety of applications requiring resistance to mechanical wear, and although some very valuable research has been done in recent years, it is still not possible to obtain comprehensive published information on acceptable combinations of load, speed, temperature and lubrication for this range of alloys. It is to be hoped that further research will be done to establish these parameters for the benefit of designers.

Nevertheless, some individuals and companies have found by experience that aluminium bronze provides a valuable alternative to more conventional materials for a number of specialised purposes and it has become well established for high stress gears and bearings applications, notably in earth-moving equipment. But it is also used for a variety of less arduous applications such as: gears, wear strips, bushings, valve seats, plungers, pump rods, sleeves and nuts.

Wear

Wear is liable to occur when two surfaces, in contact with each other and usually under load, move relative to each other. In many cases, one surface is stationary. The relative movement is either,

(a) a sliding action as in the case of plain rotary bearings or of various types of linear reciprocating machinery; or
(b) a rolling action as in the case of wheels running along a track or of ball or roller bearings; or
(c) a combination of both, as in gears.

Another type of wear is known as 'fretting'. It results from two surfaces rubbing against each other with a reciprocating or oscillatory motion of very small amplitude (e.g.: typically less than 0.1 mm) and high frequency (e.g.: typically 200 cycles/sec). This oscillatory motion is not normally intended but is, more or less, the inevitable consequence of some factor such as vibration.

Wear also occurs in dies, rollers and tools used to shape materials in various wrought processes or in equipment used to handle loose materials. Threaded assemblies are examples of sliding friction.

Erosion and cavitation erosion which are caused by flowing fluids on metal parts under certain circumstances (propellers, pumps etc.) are forms of wear although normally dealt with under the heading of corrosion (see Chapters 8 and 9). In the case of cavitation erosion, there is a hammering effect which can cause fatigue. Under certain conditions involving high local flow velocities of the lubricant, bearing failures have resulted from cavitation damage at the surface of contact. Aluminium bronze has an exceptional resistance to this form of attack (see Chapter 9).

Wear may be relatively slight, in which case it does not impede the working of a machine but will in time limit its life, or it may be severe, as in galling (also known as scoring or scuffing) which causes deep scratches or grooves in a surface and can lead to a rapid break-down.

Mechanism of wear

When two surfaces slide or roll against each other under load, two forces come into play:

(1) The load which acts normal to the surfaces in contact. It exerts a compressive force on the materials and is usually more concentrated in the case of a rolling contact.
(2) A force exerted by the machine in the direction of motion which overcomes the following types of resistance:

- The friction force which is the product of the load and of the coefficient of friction of the combination of materials in contact. The coefficient of friction is higher at the start of motion than in a dynamic situation and is different for sliding motion than for rolling motion. It is significantly reduced by lubrication.
- Adhesion: the tendency of the two mating metals to adhere to each other when not separated by an insulating film, such as a lubricant (see 'adhesive wear' below). An oxide film can reduce or even eliminate adhesion. The coefficient of adhesion is the ratio of the force required to overcome the adhesion to the applied load normally. Adhesion may result in the surfaces being locally bonded together: this is known as a 'junction'.
- In extreme cases, resistance to motion is caused by abrasive material (see abrasive wear below)

These two forces (the load and the force overcoming friction or adhesion) combine to submit the surface and the sub-surface of the mating materials to stresses. This may have the following effects:

(a) to work-harden the softer surface or perhaps both surfaces,
(b) to cause plastic deformation of the softer of the two materials, particularly when overcoming adhesion,

(c) when junctions occurs, to dislodge particles from the more wear-vulnerable of the two surfaces.

(d) in the presence of abrasive material, grooves are ploughed into the softer material.

It has been observed[163] that both the surface and subsurface deformation is non-uniform due to the difference in subsurface structures and to the different level of stresses acting on them. The highly deformed areas consequently form raised areas or 'plateaux' on the worn surfaces and are of higher hardness.

Z Shi et al.[163] carried out rolling-sliding unlubricated wear test on a nickel–aluminium bronze CuAl10Ni5Fe4 to BS 1400 CA104 against hardened En19 steel. They found that two types of wear took place:

(a) adhesive wear and
(b) delamination wear.
    To these two types of wear, must also be added:
(c) abrasive wear

Adhesive wear

Adhesive wear is caused by the strong adhesive force that develops between mating materials. Prior to the surfaces beginning to move relative to each other, minute areas of contact between the mating surfaces become joined together (these are known as 'junctions'). If, when the machine applies a force to break these junctions, the resulting stresses in the metals are small, only small fragments of the metals become detached. In the case of aluminium bronze (and some other metals), these fragments or particles are quickly transferred from the softer metal (aluminium bronze) to the harder metal (steel).[163] They adhere firmly to the steel in the form of a thin layer and are work-hardened. Thereafter, newly transferred particles agglomerate with the existing transferred layer. Some transferred particles may transfer back to the aluminium bronze.[163] Provided adhesive wear is moderate, no debris form and the resultant small degree of wear may be acceptable, depending on the desired service life. On the other hand, metals which adhere strongly are more liable to cause debris and are therefore more susceptible to galling.[97]

Delamination wear

Delamination wear is the result of cracks forming below the surface of the aluminium bronze and propagating to link up with other cracks. They are the result of the subsurface strain gradient caused by the load and the anti-adhesion force and are aggravated by fatigue or defective material. As a result, subsurface deformation occurs and material becomes detached as wear debris of a platelet or laminated form. The structures of the debris therefore reflect that of the subsurface structures from which they originated.[163] If the subsurface structure of the alloy is itself of a laminar type, as in the case of some aluminium bronze structures (see Chapters 11–13), it is more vulnerable to this kind of wear. Debris, resulting from delamination wear, may become part of the transferred layer and be work-hardened in the same way as the adhesive wear transferred particles. In a lubricated bearing, the debris may combine with constituents in the lubricant to form a gel structure.[76] If there is no lubrication and if the debris do not become part of the transferred layer, they may lead to galling.

Given bearing design appropriate to the conditions, the likelihood of this kind of wear occurring with aluminium bronze is very slight provided the material is sound and of the right microstructure (see Chapters 11–13).

Abrasive wear

Abrasive wear is the result of one very hard material cutting or ploughing grooves into a softer material.[159] The harder material may be one of the rubbing surfaces or hard particles that have found their way between the mating surfaces. These may be 'foreign' particles or particles resulting from adhesive or delamination wear. Due to the build up of elastic energy in the transferred layer, some of this layer may eventually, become detached and form tiny debris.[163] These debris have undergone considerable deformation and work hardening and are therefore liable to have an abrasive effect on the softer surface and cause severe galling. It may be possible to arrest this effect by removing the debris. Otherwise, they may lead to rapid deterioration and to machine break-down. Aluminium bronze has however very good galling resistance (see below).

It is advisable to give the harder of the two surfaces a finer finish to eliminate asperities that can plough into the softer material and steps need to be taken to prevent the ingress of hard foreign particles.

Factors affecting wear

The degree of wear that occurs is the result of the inter-play of a number of factors that apply in a given situation. The correlation between these factors has been the subject of much research with results that are not always applicable to all material combinations, particularly the relationship of the wear rate and the load, the speed, the coefficients of friction and of adhesion, hardness and tensile and yield strength.[149] An approximate indication of how load (W) and hardness (H) affect the wear rate (Q) is given by the following formula by Archard[97] in which K is a 'wear coefficient' of the system and is dependent on many of the factors described below:

$$Q = KW/H$$

The factors affecting wear have been grouped under the following headings:

- Operating conditions
- Material structure and properties
- Environmental conditions

Operating conditions

Loading

Loading may be anything from low to high, depending on the application. It may be unidirectional or reversing, continuous or intermittent. It governs the friction and adhesion resistance and consequently the rate of wear of the oxide film. It has therefore a paramount influence on wear. The resistance of metal to severe wear under high load conditions does not always correlate with their wear resistance under less severe conditions.[97] In a sliding wear situation, wear rate increases with load and sliding distance although not necessarily linearly. This indicates that there can be more than one wear mechanism operative.[169]

Velocity

Velocity, like loading, can be anything from low to high, unidirectional or reversing, continuous or intermittent. It is one of the factors that affect the erosion of the oxide film although, in some cases, speed has little effect on wear. In other cases it increases the rate of wear and in yet other cases it reduces it. This is because the effect of speed is related to other factors such as lubrication and the temperature it generates by friction (see 'inter-face temperature' below). In the case of fluid erosion (propellers, pumps etc) there is a velocity above which the shear stresses it induces in the metal surface, begins to strip off the oxide film. For nickel–aluminium bronze, this velocity is 22.9 m s^{-1} and for duplex aluminium bronze 15.2 m s^{-1}.

Fatigue

Reversing or intermittent loading result in repeated stressing and un-stressing which give rise to fatigue. It is particularly prevalent in rolling contact as in ball bearings and gears and may also be caused by the hammering action of cavitation. Fatigue may in time lead to the formation of cracks at or below the surface and hence ultimately to spalling (chips or fragments of metal breaking off) and de-lamination wear. Aluminium bronze is reputed for its excellent fatigue resistant properties. Fatigue is greatly affected by surface conditions such as hardness and finish, by the structure of the alloy, by residual stresses and by freedom from internal defects. Generous fillets and fine finish reduce the high notch or stress-concentration factors that can lead to accelerated fatigue failure.[159]

Lubrication

The object of lubrication is to reduce friction and the tendency to adhesion and to mitigate their effects. There are five types of lubrication.[50–97]

- *hydrodynamic* lubrication in which the mating surfaces are separated by a fluid film resulting from the movement of one surface relative to the other; adhesion is prevented and little surface distortion occurs;
- *hydrostatic* lubrication in which the lubricant is supplied under pressure and is able to sustain higher load without contact taking place between the surfaces.
- *elasto-hydrodynamic* lubrication in which the pressure between the surfaces are so high and the lubricant film so thin that elastic deformation of the surfaces is likely to occur and is a feature of this kind of lubrication;
- *boundary* lubrication in which an oil or grease, containing a suitable boundary lubricant, separates the surfaces by what is known as 'adsorbed molecular films'; appreciable contact between asperities and formation of junctions may occur;
- *solid lubricants* which provide a solid low shear strength film between the surfaces.

It may not always be possible to lubricate in a given wear situation and there are many demanding unlubricated sliding systems in various industries. In other cases, it may be necessary to adapt to a lubricant dictated by circumstances, such as water.

Surface finish
Surface finish affects wear: A well-polished surface finish – say less than about 0.25 μm rms (root mean square distance from peak to trough) – provides more intimate contact between the surfaces.[159] This results in more interaction between them and may lead to local weld junctions forming and therefore a greater susceptibility to galling. Lubricants also tend to be swept away between smooth surfaces whereas shot peening a surface helps to retain a lubricant. If, on the other hand, the surfaces are too rough – say 2 μm rms – the asperities will tend to interlock resulting in severe tearing and galling. Most machined finish, however, fall within an intermediate range of surface finish. It is advisable to give the harder of the two surfaces a finer finish to eliminate asperities that can plough into the softer material.

Material structure and properties

Among the most important factors affecting wear are those relating to the structure and properties of the mating materials themselves.

Microstructure and space lattice structure
Yuanyuan li et al.[190] have carried out wear tests on nickel–aluminium bronzes within the following ranges of wt % compositions:

Cu	Al	Fe	Ni	Mn
Bal	8–13	2–5	1–3	0.5–3

They found that the microstructure of this range of aluminium bronze alloys, both at its surface and at its subsurface, determines its wear behaviour (see Chapter 13). By adjusting the structure of the alloy, a balance is struck between plasticity and hardness. A 'soft' structure is more plastic and more prone to adhesion and distortions. Consequently it results in a high wear rate. A hard structure is likely to be abrasive and to lead to rapid deterioration of at least one of the surfaces in contact. An intermediate structure results in the lowest wear rate which also correspond with the lowest coefficient of friction and the most favourable tensile and yield strength.

The softness or hardness of a phase in a metallurgical structure is a function of its space lattice structure. Hexagonal close-packed structures are less ductile than face-centred or body-centred structures and generally show lower wear rates and less galling tendencies.[97] Most phases in nickel–aluminium bronze have cubic structures, the exception being the martensitic beta phase which has an hexagonal close-packed structure and is less ductile (see Chapter 13).

Adhesion also seems to be related to the energy stored in a distorted crystalline structure which is known as its stacking fault energy: the lower this energy, the lower generally is the coefficient of adhesion.[149] This is because a low stacking fault energy inhibits dislocation cross-slip and hence favours a high work-hardening rate which in turn results in lower adhesion and friction,[98] but this correlation does not apply in every case.

Oxide film

As explained in Chapter 8, the film of oxides that forms on aluminium bronze consists of a copper-oxide-rich (Cu_2O) outer layer and of an alumina-rich (Al_2O_3) inner layer.[161–16] Sullivan and Wong[168] report that alumina (Al_2O_3) is easily removed from nickel aluminium bronze at the initial stages and adheres very strongly to a hard steel mating material (known as the 'counter-face'), forming a stable aluminium-rich transfer layer on the steel and leaving a stable wear resistant copper-oxide-rich (Cu_2O) film on the aluminium bronze. It is this combination of a strongly adhesive alumina-rich transfer film on the counter-face and of a stable copper-oxide-rich film strongly bonded to the aluminium bronze which gives aluminium bronze its excellent wear resistance. It is widely recognised that a stable oxide film, such as copper oxide (Cu_2O), is an essential feature for wear resistance because it reduces or prevents adhesion. The rate at which the oxide film is eroded is a function of load, speed and temperature. It is vital that oxidation should constantly renew this film as it wears in service (it is oxygen in solution in the lubricant which causes oxidation). Indeed, if the load and speed conditions are too severe, then the rate of growth of the copper oxide is less than the rate of surface removal and Cu_2O debris form and cause severe galling or even seizure. This is known as 'oxidation wear'.[169]

According to Poggie et al.,[142] the copper-oxide-rich layer has mechanical properties similar to those of the parent aluminium bronze and is resistant to mechanical disruption during sliding. It results in a very low coefficient of friction in the boundary lubrication (see below) condition. The alumina-rich inner layer, on the other hand, has poor mechanical strength. Poggie et al. found that, in the case of binary copper–aluminium alloys having aluminium contents of less than 6 wt%, if the aluminium content is increased and the alumina-rich (Al_2O_3) inner layer is disrupted, the chances of a bond forming between the aluminium bronze and the counter-face is increased. Since the shear strength of this bond is greater than the shear strength between the alumina-rich film and its parent metal, the process of adhesive wear explained above takes place. Hence, the higher the aluminium content of the binary copper–aluminium alloy, the greater the degree of transfer to the counter-face.

It has also been observed[142] that, at a temperature of 600K (327°C), aluminium segregates towards the surface and displaces the oxygen bonded to copper to form alumina, thus making the alloy more prone to adhesion wear for the reasons just given.

Tribological compatibility and adhesion

As has been shown above, the tendency of materials to adhere to one another is the major cause of ordinary wear. It is thought to be usually related to the degree of mutual solubility in the solid state of the mating materials: the more soluble they are in each other the higher their tendency to adhesion and therefore the less tribologically compatible they are. The less tribologically compatible two materials are the higher the strain hardening of the softer material and the less their suitability as a mating pair. A pair of identical metals are completely mutually soluble and have therefore poor compatibility. As has already been seen, the oxide film affects tribological compatibility. According to Reid et al.,[149] compatibility also seems to determine whether metal transfer occurs, but is no guide to subsequent surface damage which is more likely to be a function of the mechanical properties of the adhered surfaces. Tribological compatibility is not to be confused with metallurgical compatibility which, being the degree of mutual solubility of two materials, is the opposite of tribological compatibility.

Coefficient of friction

Since friction opposes motion, it determines the efficiency of a machine. A designer will therefore aim to use the lowest friction combination of materials consonant with other design considerations. It is not clear, however, how significant is the part played by friction in the wear mechanism. Yuanyuan Li and Ngai,[190] have demonstrated that, in the case of aluminium bronze, the effect of changes in microstructure on the coefficient of friction follows the same trend as its effect on the rate of wear (see Chapter 13). The metallurgical structure and tribological compatibility of mating pairs of materials govern the magnitude of the friction between them with the lowest friction being obtained with the most tribologically compatible materials.[190–150]

There is no general correlation between wear rate and the coefficient of friction.[97] Some metals experience high friction and low wear and others are the reverse.[159] This inconsistency between friction and wear of different materials may however be accounted for by the fact that any effects that friction may have on wear rate, would not only be dependant on the magnitude of the load and the friction force, but also on the nature of the materials in contact. As we have seen, however, lubrication has the effect of reducing both friction and wear rate.

Friction can also have an indirect effect on wear by causing inter-face heating (see below).

Tensile properties

As mentioned above, the load and anti-adhesion force together subject the subsurface of the mating materials to a strain gradient. It is the mechanical properties of the material that resist this strain and governs the amount of deformation that will occur. Yuanyuan Li and Ngai[190] found that, in the case of aluminium bronze, wear rates for different microstructures are inversely proportional to the corresponding yield strength and, less markedly, to tensile strength.

Since machinery that is subject to wear may also be subjected to bending and other loads, as in the case of gear teeth, it is an attractive feature of aluminium bronze that the structure that gives the best wear resistance should also have the best tensile properties.

Elastic property

The elastic properties of the softer of two mating materials ensures that deformation can take place under stress without rupture occurring which leads to delamination and galling.

Hardness

When comparing the wear resistance of different materials, the harder materials are often found to be the most wear resistant. There is considerable service experience to show that an aluminium bronze with a hard surface has excellent galling resistance (see below). It was thought therefore at one time that wear was inversely proportional to the hardness of the surface being worn away.[159] The relationship between wear and hardness is not so clear cut, however, as more recent researchers have found. Harder material do not imply lower adhesion and metal transfer, nor lower galling resistance.[97]

According to Reid and Schey,[149–150] there is no correlation either between the coefficient of friction and overall hardness. Yuanyuan Li and Ngai[190] have come to a similar conclusion.

Although hardness is undoubtedly an important factor in wear performance, its role is more complex than was once thought and, as explained above, is closely linked to the structure of the materials involved. It is evident that the combination of one hard and one less-hard material is an important feature of a successful matching pair. The hard surface controls the interaction and the softer surface conforms. The softer material is able to embed hard abrasive particles thereby minimising damage to the surfaces. Its lower shear strength means that, should contact occur in a lubricated bearing, seizure is less likely to happen. The softer material, being the one that experiences most wear, can be designed to be the cheaper and more easily replaced component.

It has been found, in the case of aluminium bronze, that the presence of hard intermetallic particles in a soft constituent of the microstructure is an advantageous feature in resisting wear[190] (see Chapter 13).

As explained above, surface hardness is increased by the work hardening that occurs during sliding or rolling, but higher strain-hardening does not necessarily imply lower friction or lower adhesion.[190–150] Although there is evidence that high-strain-hardening alloys, such as austenitic stainless steel, outwear harder alloys like the precipitation-hardening stainless steel,[159] austenitic stainless steels are notoriously susceptible to galling.[97] It is possible however that the excellent wear performance of aluminium bronze may be due in part to the fact that, it too, is a high-strain-hardening alloy, because a high working rate in a metal usually gives good resistance to severe wear and galling.[97]

Metal defects
Gas porosity, inclusions or shrinkage defects are all liable to have a very detrimental effect on wear resistance.

Thermal conductivity
The thermal conductivity of at least one of the materials in a mating pair determines the rate at which the heat generated by friction is dissipated and therefore helps to control the inter-face temperature (see below) to an acceptable level.

Environmental conditions

Inter-face temperature
Inter-face temperature also influences wear performance. It may result either from ambient conditions or from frictional heating caused by heavy load and high speed.[159] As explained above, high temperature has an effect on the oxide film which adversely affects wear performance. It also affects mechanical properties, reduces hardness and increases the tendency to galling and to surface deformation due to plastic flow. It is possible however to use aluminium bronze as a bearing material at up to 260°C.[77]

Corrosion
In many cases, the apparent 'wear' of a metal surface is the result of corrosion followed by mechanical wear of the corrosion product. The corroding agent varies widely, from sulphuric acid (originating from products of combustion) to

Table 10.1 Comparison of mechanical, physical and tribological properties of bearing alloys.[50]

Material designation		Mechanical and physical properties						Guide to operating limits		Tribological properties		
Material Category	CEN/ISO** designation	0.2% proof strength MNm^{-2}	Modulus of elasticity GNm^{-2}	Thermal conductivity WmK^{-1}	Coefficient of thermal expansion $10^{-6}\,K^{-1}$	Hardness H_v or HB	Elongation %	Fatigue Resistance	Maximum recommended operating temp. °C	Resistance to seizure	Embeddability and conformability	Recommended min. journal hardness H_v or HB
Tin bronze	CuSn12Ni2-G	130–160	–	50	18	70–90	20–9	High	170	Moderate	Moderate	300
Phosphor bronze	CuSn10-G	160–180	–	50	18	75–110	20–12	Very High	220	Moderate	Poor	350
	CuSn11P-G	130–170	95	50	18	70–80	20–10	Very High	220	Moderate	Poor	350
	CuSn12-G	140–150	–	50	18	60–85	22–8	Very High	220	Moderate	Poor	350
	CuSn11Pb2-G	130–150	95	50	18	80–90	15–7	Very High	220	Moderate	Poor	350
	–	80–200	–	50	18	80–90	20–12	Very High	220	Moderate	Poor	350
	CuSn8	260–550	115	59	17	95–200	18–3	Very High	220	Moderate	Poor	350
Leaded bronze	CuP9Sn5-G	60–100	85	71	18	55–60	40–5	Very High	170	Good	Good	250
	CuSn10Pb10-G	80–110	90	47	18	60–70	20–12	Mod/High	170	Good	Good	250
	CuPb15Sn7-G	80–90	85	47	18	60–65	15–6	Mod/High	170	Good	Good	250
	CuPb20Sn5-G	70–90	75	59	19	45–50	10–8	Mod/High	170	Very Good	Good	200
Aluminium bronze	CuAl10Fe5Ni5-G	250–280	120	38–42	16	140–150	16–5	Very High	300	Moderate	Poor	350
	CuAl10Fe4Ni5	480–530	118	33–46	16	180–220	20–13	Very High	300	Moderate	Poor	350
Gunmetal	–	130–140	105	51	18	70–95	25–8	Mod/High	200	Moderate	Good	300
	CuZn8Pb5Sn3-G	85–100	100	75	18	60–70	25–12	Mod/High	200	Moderate	Good	300
	CuPb5Sn5Zn5-G	90–100	90	71	18	60–65	15–12	Mod/High	200	Moderate	Good	300
	CuSn7Pb3Zn2-G	130	105	65	18	65–70	25–13	Mod/High	200	Moderate	Good	300
	CuPb78Sn7Zn4-G	100–120	85	59	18	60–70	14–12	Mod/High	200	Moderate	Good	300
Brass	CuZn33Pb2Si-G	170–280	105	95	21	110–120	35–20	Moderate	200	Moderate	Poor	300
	CuZn37Mn3Al2PbSi	280–350	100	65	19	150–170	15–8	Moderate	200	Moderate	Poor	300
	CuZn31Si1	250–330	105	67	18	120–150	18–10	Moderate	200	Moderate	Poor	300
	CuZn38Pb2	250	96	109	20	120	15	Moderate	200	Moderate	Poor	300
Copper–beryllium	CuBe2	1260	130	100	17	400	2	High	260	Good	Poor	450
Whitemetal* tin based	–	–	52	55	23	32	–	Moderate	120	Excellent	Excellent	140
lead based	–	–	29	24	25	25	–	Moderate	120	Excellent	Excellent	140
Alum.–tin low tin	–	–	70	200	24	45	–	High	170	Mod/Good	Good	250
high tin	–	–	70	200	24	45	–	Mod/High	160	Good	Good	250

NB. Where a range of values is indicated for cast alloys, the lower number relates to sand casting and the higher to either continuous or centrifugal casting. Chill casting lies between the two values.
* These materials are normally only used in layers between 0.5mm and 1.5mm on a steel backing. The actual strength of the bearing depends on the lining thickness.
** In ISO/CEN designations, the suffix 'G' indicates a cast material.

atmospheric contamination in industrial or marine environments. The proportion of wear attributable to corrosion is impossible to assess, but it is advisable to use a corrosion-resistant material, such as aluminium bronze. Detailed information on the degree of resistance to corrosion of aluminium bronzes is given in Chapter 9.

Because corrosion is liable to attack both the surface and subsurface of an alloy, it is liable to undermine its wear performance.

Foreign particles

Hard foreign particles finding their way between the mating surfaces can plough grooves into the softer surface and cause severe abrasive wear. Steps need to be taken, therefore, to prevent the ingress of hard foreign particles. Filtering systems normally only remove the coarser particles, and the resistance of the material to abrasion therefore assumes considerable importance for most bearing applications.

Wear performance of aluminium bronzes

Properties of copper alloys used in wear applications

A comparison of the fundamental properties of the more popular alternatives for sliding contact with steel is made in Table 10.1. Aluminium bronze has superior mechanical properties to phosphor bronze; in this respect it closely approaches medium carbon steel, and it may therefore be subjected to considerably heavier loading. Its high proof and fatigue resistance, in particular, represent the major advantages which it offers over phosphor bronze. The design stress is significantly greater than that of the most popular grade of phosphor bronze and this allows a considerable reduction in the dimensions of certain components such as gears. Its resistance to impact and shock loading is also exceptional, and has led to its use in plant such as earth-moving equipment, which involve heavy loads of this type.

It will be seen that the coefficient of friction of aluminium bronze is higher than that of phosphor bronze, and this limits its use for applications involving continuous rubbing contact, particularly at high speeds. As we have seen, a high frictional resistance leads to higher running temperatures, with a consequent increase in the tendency to gall. With components subjected to discontinuous surface loading, e.g. gears and worm wheels, the surface temperature does not build up in the same way and the effect of friction is of less consequence.

Comparison of wear performance of copper alloys

Table 10.2 gives a comparison of wear rate of a grease lubricated cylindrical plain bearing in some copper-base alloys.[76] In heavily loaded, boundary lubricated conditions, frictional heating is often the limiting factor.

Table 10.2 Comparison of wear rate of a grease lubricated cylindrical plain bearing in some copper-base alloys.[77]

Alloy	Brinell hardness	Bearing* pressure range N mm^2	Wear rate** 10^{12} mm^3 m^{-1}
Leaded tin bronze UNS C93200	65	0–14	6.4
		14–40	33.3
Tin bronze UNS C90500	75	0–40	2.7
		14–40	13.4
Heat treated aluminium bronze CuAl11Fe4	170	0–100	1.3
		100–200	6.7
Beryllium Copper UNS C82500	380	0–550	1.1

*Bearing pressure = radial load ÷ (length × dia of bearing)
**Wear rate = volume of wear at slow speed over a given number of cycles

Table 10.3 Comparison of adhesion of copper and its alloys mated with two different hard materials, by Reid et al.[149]

*Copper or copper alloy (in annealed condition unless marked 'H')	Surface damage to copper or copper alloy specimen	Metal transfer to hard specimen
Mated to 16% Al copper–aluminium (Ampco 25)		
Cupro-nickel	Cu–Ni (H) Severe	Thick and accumulative (more
	Cu–Ni	transfer than to D2 below)
Copper	Cu (H) Severe	Thick and accumulative
	Cu	
Copper–Zinc	Cu–Zn (H) Moderate	Accumulative but self limiting
Copper–aluminium	Cu–6.5Al Moderate	Accumulative but self limiting
	Cu–4Al	
Copper–tin	Cu–5Sn Moderate	Thin burnished transfer layer
	Cu–9Sn	
	Cu–13Sn	
Copper–aluminium	Cu–8Al Burnished surface	Accumulative and self limiting but smaller area
Mated to tool steel D2		
Cupro-nickel	Cu–Ni (H) Severe	Thick and accumulative
	Cu–Ni	
Copper	Cu (H) Severe	Thick and accumulative but not
	Cu Moderate	continuous
Copper–Zinc	Cu–Zn (H) Moderate	Accumulative but self limiting
	Cu–Zn	
Copper–aluminium	Cu–6.5Al Moderate	Accumulative but self limiting
	Cu–4Al	
Copper–tin	Cu–5Sn Burnished surface	No visible transfer
	Cu–9Sn	
	Cu–13Sn	
Copper–aluminium	Cu–8Al Burnished surface	Accumulative and self limiting but smaller area

(H) signifies work-hardened condition

Table 10.4 Properties of alloys mated with aluminium bronze.

Alloy	Tensile Strength N mm⁻²	0.2% Proof Strength N mm⁻²	Elongation %	Hardness Vickers HV	Brinell HB	Form (annealed)
Austenitic s.s.						
Type 301	758	276	60	B85	–	Sheet
303	621	241	50	–	153	Bar
304	579	290	55	B80	140	Sheet
310	655	310	45	B85	–	Sheet
316	579	290	50	B79	150	Sheet
Nitronic 50	827	414	50	B98	–	Bar (1121°C)†
	862	448	45	C23	–	Bar (1066°C)†
Austenitic type s.s. galling resistant Nitronic 60	–	–	–	B95	205	
Ferritic s.s.						
Type 430	517	345	25	B85	159	Sheet
Martensitic s.s.						
Type 410	483	310	25	B80	352	Sheet
416	517	276	30	B82	342	Bar
440C	758	448	14	B97	560	Bar
Precipitation hardening s.s.						
17–4PH*	1000	862	13.0	C32–39	302–375	available in
17–4PH**	931	724	16.0	C28–37	277–352	most forms
Cobalt-based						
Stellite 6B††	935–1000	590–621	10–12	C36–37	–	Sheet and plate
Cast Iron						
BS 1452 Grade 17	540	278	18		180	
Cast steel						
BS 592 Grade C	278	–	0		250	
Wrought steel						
En8 Normalised	540	216	20		170	
En8 Heat treated	726	355	19		200	

†annealing temperature
††Solution heat treated at 1232°C, air cooled

*Hardened at 579°C
**Hardened at 621°C

Adhesion comparison of aluminium bronze with copper and its alloys

Reid et al.[149] carried out research into the adhesion of copper and its alloys. Table 10.3 compares the adhesion of copper aluminium alloys to that of copper and of some copper-based alloys when mated with two very different hard alloys, both used for dies: D2 tool steel and Ampco 25, of the following compositions:

Alloy	Cu	Al	Fe	C	Cr	Mo	Co	V
D2 tool steel	–	–	Bal	1.5	12.0	1.0	<1.0	<1.1
Ampco 25 alum. bronze.	79.25	16.0	5.8	–	–	–	–	–

The load applied to the wear specimens was sufficient to cause plastic deformation of the copper or copper alloy. It varied between 20 to 40 kN. The tests were done without lubrication at a relative velocity of 1 cm s^{-1}.

It will be seen that 8% Al copper–aluminium is the copper alloy least prone to adhesion, but if the aluminium content is reduced, the alloy becomes more adhesive than copper–tin alloys. It will also be noted that the order of adhesiveness of copper and of copper alloys is the same for both the hard mating materials used in the experiments.

Wear performance of aluminium bronze mated with other alloys

Comparison of compositions and properties
Tables 10.4 and 10.5 give the properties and compositions respectively of alloys most commonly mated with aluminium bronzes.

Self-mated

Table 10.6 shows that the wear performance of aluminium bronze compares favourably with a number of other alloys, when self-mated and unlubricated at relatively low RPM and low loading. The aluminium bronze alloy used in these tests

Table 10.5   Composition of alloys mated with aluminium bronze.

Alloy	C	Mn	P	S	Si	Cr	Ni	Mo	Others
Austenitic s.s									see
Type 301	0.15	2.00	0.045	0.030	1.00	17.0–19.0	6.0–8.0	–	note
303	0.15	2.00	0.20	>0.15	1.00	17.0–19.0	8.0–10.0	0.60*	–
304	0.08	2.00	0.045	0.030	1.00	18.0–20.0	8.0–10.5	–	–
310	0.25	2.00	0.045	0.030	1.50	24.0–26.0	19.0–22.0	–	–
316	0.08	2.00	0.20	0.030	1.00	16.0–18.0	10.0–14.0	2.0–3.0	–
Nitronic 50	0.06	4–6	0.040	0.030	1.00	20.5–23.5	11.5–13.5	1.5–3.0	(1)
Nitronic 60	0.10	7–9	–	–	3.5–4.5	16.0–18.0	8.0–9.0	–	(2)
Ferritic									
Type 430	0.12	1.00	0.040	0.030	1.00	16–18	0.75	–	–
Martensitic s.s.									
Type 410	0.15	1.00	0.040	0.030	1.00	11.5–13.5	–	–	–
416	0.15	1.25	0.060	>0.15	1.00	12.0–14.0	–	0.60*	–
440C	0.95–1.2	1.00	0.040	0.030	1.00	16.0–18.0	–	0.75	–
Precipitation hardening s.s.									
17–4PH	0.07	1.00	0.04	0.03	1.00	15.0–17.5	3.0–5.0	–	(3)
Cobalt-based									
Stellite 6B	0.9–1.4	2.0	–	–	2.0	28.0–32.0	3.0	1.50	(4)

Above figures are max. unless otherwise stated	(1) **N:** 020–0.40 **Cb:** 0.10–0.30	**V:** 0.10–0.30
* May be added at manufacturer's option	(2) **N:** 0.08–0.18	
	(3) **Cu:** 3.0–5.0 **Cb+Ta:** 0.15–0.45	
	(4) **Co:** Bal. **Fe:** 3.0	**W:** 3.50–5.50

Table 10.6   Comparison of the self-mated and unlubricated wear performance of aluminium bronze and stainless steels under a 7.26 kg load by Schumacker.[160]

Alloy	Rockwell Hardness	Weight Loss mg/1000 cycles		
		105 RPM over 10^4 cycles	415 RPM over 10^4 cycles	415 RPM over 4×10^4 cycles
Nickel Aluminium Bronze	B87	2.21	1.52	1.70
Nitronic 60 austenitic	B95	2.79	1.58	0.75
Type 301 austenitic	B90	5.47	5.70	–
Type 304 austenitic	B99	12.77	7.59	–
Type 310 austenitic	B72	10.40	6.49	–
Type 316 austenitic	B91	12.50	7.32	–
17–4 PH precipitation hardening	C43	52.80	12.13	–
CA 6 NM	C26	130.00	57.00	–
Type 410 martensitic	C40	192.79	22.50	–
Stellite 6B		–	1.27	1.16
Chrome Plate		–	–	0.68

may not have had the optimum grain size or combination of constituents in its microstructure for best wear performance established by Yuanyuan Li et al.[190] (see above and Chapter 13). It is possible therefore that lower weight loss could be achieved than indicated.

Z. Shi et al[163] have found that electron beam surface melting (see Chapter 7) of nickel–aluminium bronze results in an increase of the martensitic beta phase (see Chapter 13) at the surface of the alloy thereby increasing its hardness. In certain circumstances, this may improve wear resistance. However, in the light of what has been said above on the effect of hardness on wear, such a procedure may render the surface of the alloy more brittle and give rise to debris and lead to galling.

Sliding pairs
It is standard engineering practice, that steel surfaces are only allowed to slide on one another when complete dependence can be placed on the lubricant film. Copper alloys, however, are selected when lubrication is not ideal, phosphor bronze or aluminium bronze being the most popular for moderate and heavy loading.

Table 10.7 compares the rates of wear of a number of sliding pairs of aluminium bronze and stainless steels with the self-mated rates of wear of the individual alloys. It shows that the pairs containing aluminium bronze perform best. It will also be seen that the rate of wear of aluminium bronze reduces when it is paired with another alloy, whereas the rates of wear of other pairs of alloys generally lie between their individual self-mated values.

Abrasion or galling resistance
Whereas wear limits the life of a component over a period of time, galling has an immediate and potentially devastating effect on a piece of machinery.

Table 10.7 Comparison of the rates of wear of various sliding pairs of stainless steels and aluminium bronze under a 7.26 kg load, with their individual self-mated rates of wear for comparison, by Schumacker.[159]

Sliding pairs	Rockwell Hardness	Weight Loss mg/1000 cycles		
		Self-mated	Paired	
		105 RPM over 10^4 cycles	105 RPM over 10^4 cycles	415 RPM over 10^4 cycles
Nickel Aluminium Bronze	B87	2.21	1.36	–
17–4 PH precipitation hardening s.s.	C43	52.8		
Nickel Aluminium Bronze	B87	2.21	1.64	–
Nitronic 60 austenitic stainless steel	B95	2.79		
Nickel Aluminium Bronze	B87	2.21	1.49	1.24
Type 301 austenitic stainless steel	B90	5.47		
Nitronic 60 austenitic stainless steel	B95	2.79	5.04	2.83
17–4 PH precipitation hardening s.s.	C43	52.8		
Nitronic 60 austenitic stainless steel	B95	2.79	2.74	–
Type 301 precipitation hardened ss	B90	5.47		
Nitronic 60 austenitic stainless steel	B95	2.79	5.95	–
Type 304 austenitic stainless steel	B99	12.77		
17–4 PH precipitation hardening s.s.	C43	52.8	25.0	–
Type 304 austenitic stainless steel.	B99	12.77		

Table 10.8 Unlubricated galling resistance of various combinations of aluminium bronze and stainless steels, by Schumacker.[160]

Alloy	THRESHOLD GALLING STRESS (N/mm²)									
	Type 440C	17-4 PH	Type 410	Type 416	Nitronic 60	Type 430	Type 303	Type 316	Type 304	Nickel* Alum Bronze
Brinell hardness:	560	415	352	342	205	159	153	150	140	140–180
Type 440C (martensitic)	108	29	29	206	490	20	49	363	29	500
17-4 PH (precipitation hardened)	29	20	29	20	490	29	20	20	20	500
Type 410 (martensitic)	29	29	29	39	490	29	39	20	20	500
Type 416 (martensitic)	206	20	39	128	490	29	88	412	235	500
Nitronic 60 (austenitic)	490	490	490	490	490	355	490	373	490	500
Type 430 (ferritic)	20	29	29	29	355	20	20	20	20	500
Type 303 (austenitic)	49	20	39	88	490	20	20	29	20	500
Type 316 (martensitic)	363	20	20	412	373	20	29	20	20	500
Type 304 (austenitic)	29	20	20	235	490	20	20	20	20	500
Nickel* Alum Bronze	500	500	500	500	500	500	500	500	500	500

* to ASTM C95400

Shaded figures denote: did not gall.
Framed figures are self-mated.

Table 10.9 Unlubricated galling resistance of various combinations of aluminium bronze and stainless steels under reversing load condition, by Schumacker.[160]

Alloy	THRESHOLD GALLING STRESS UNDER REVERSING LOAD (N/mm^{-2})						
	Type 410	Type 430	Type 316	17-4 PH	20 Cr–80 Ni	Nitronic 50	Nitronic 60
Nickel* Alum. Bronze	332	332	275	385	332	275	384
Nitronic 60	<231	–	88	416	–	147	<167
Stellite 6B	346	–	<35	416	–	<165	502

* to ASTM C95400 Shaded figures denote: did not gall

Table 10.8 by Schumacker[160] gives the threshold galling stress (lowest load at which galling damage occurs) of various unlubricated combinations of aluminium bronze and stainless steels. The table shows that:

- hardness has no noticeable influence on galling resistance (note that the steels are arranged in descending order of hardness),
- nickel aluminium bronze and Nitronic 60 have the best galling resistance and nickel aluminium bronze did not gall under test in combination with any of the other alloys – they both performed well when self-mated,
- there is no detectable difference in the wear performance of aluminium bronze against martensitic, austenitic or ferritic stainless steels.

Schumacker[160] also carried out threshold galling stress tests involving three consecutive reversals of load for a better simulation of operating conditions. The results are given in Table 10.9. It will be seen that aluminium bronze was outstanding under these very severe test conditions: no galling occurred with any of the mating pairs involving aluminium bronze. Nitronic 60 and Stellite 6B, which is a Cobalt-based alloy widely used for wear and galling resistance, did not fare well except in a few mating combinations.

Fretting comparison of aluminium bronze with other alloys

We have seen above that fretting is the type of wear that results from two surfaces rubbing against each other under load with a reciprocating motion of very small amplitude and high frequency. It might be the result of vibration in a machine causing two surfaces to rub against each other under load.

Cronin and Warburton[60] compared the fretting performance of six materials: mild steel (EN3), 12% Chrome steel (EN56), 18/8 steel (EN58), copper, titanium and nickel–aluminium bronze (BS 1400 AB2) under a load of 1000 N and at a frequency of 190 Hz (cycles/sec). The tests were carried out at two amplitudes: 6.5μm and 65μm. The total sliding distance of each test was 2 km which gave 10

Table 10.10   Comparison of fretting performance of various alloys by Cronin and Warburton.[60]

Alloy	Weight changes per pair of specimen × 10³ g⁻¹		Specific wear rate × 10⁸ mm³ J⁻¹		Average machine finish: μm
	A	B	A	B	
6.5 μm fretting amplitude					
Nickel aluminium bronze BS 1400 AB2	−0.26 to −0.14	−0.68 to −0.66	1.24	4.17	0.25
Copper 99.9%	−0.74 to +0.09	−1.47 to −0.62	1.9	6.6	0.35
Mild Steel EN3	+0.38 to +0.26	–	wt gain	–	0.35
Stainless steel EN58	−1.23 to −1.63	−2.20 to −2.13	0.94	1.42	0.39
12% Chrome steel EN56	−0.10 to −0.23	−0.25 to −0.30	1.12	1.85	0.29
Titanium	−0.07 to −0.02	–	0.57	–	0.26
65 μm fretting amplitude					
Nickel alum. bronze BS 1400 AB2	+0.35 to −0.15	−0.202 to −0.61	wt gain	1.99	0.25
Copper 99.9%	−1.44 to −3.40	−2.2 to −3.92	13.9	17.6	0.48
Mild Steel EN3	−0.68 to −1.74	–	8	–	0.72
Stainless steel EN58	−4.99 to −18.37	−9.77 to −23.47	76.6	10.9	0.47
12% Chrome steel EN56	−23.84 to −39.44	−33.22 to −48.56	209	309	0.5
Titanium	−0.31 to +0.56	–	wt gain	–	0.26

A = as fretted   B = oxide stripped

days fretting at the smaller amplitude and one day at the larger amplitude. The results are given in Table 10.10. They show that whereas, at the higher amplitude of 65μm, aluminium bronze performs better than other materials with the exception of titanium, it is only better than pure copper at the low amplitude of 6.5μm (if the oxide has been removed). In the 'as fretted' condition, however, it is better than mild steel and stainless steel. The wear of all the materials at the 6.5μm amplitude is low, in any case, and aluminium bronze is much less affected by changes of amplitude than other materials with the exception of titanium. The latter gained weight due to the formation of a cohesive oxide which could not be removed.

Galling resistance of aluminium bronze with high-aluminium content

The degree of galling resistance which a material possesses, is related to the shear strength and hardness. Standard aluminium bronzes are among the most highly rated of the copper alloys in both these respects, but, for those applications where abrasion resistance is of prime importance, the composition may be modified to give even better properties. Copper–aluminium–iron alloys with aluminium content of up to 16% have exceptional hardness and have been found to be advantageous in very high load and very low speed applications not subject to a corrosive environment (see Chapter 12).

In sheet metal forming, lubrication is not always sufficient to prevent adhesion between the sheet and the die and this results in severe galling of the sheet and even damage to the die. To overcome this problem, aluminium bronze inserts are used where the conditions are most severe. These aluminium bronze inserts have a high aluminium content of about 14–15%. They have a high compressive strength but low tensile strength and are very brittle.

According to Roucka et al.,[154] the optimum hardness required in aluminium bronze alloys used in tooling for sheet drawing is in the range of Brinell Hardness 390–400 HB. If hardness drops below 360–370 HB, particles of aluminium bronze adhere to the drawn sheet and the tool life is considerably reduced; and if hardness is above ~ 420 HB, the cast aluminium bronze is too brittle and difficult to work. The desired hardness can be achieved with an alloy of the following range of composition:

Cu	Al	Fe	Ni	Mn
Bal	14.9–15.1%	3.3–3.5%	0.9–1.2%	~ 1%

Table 10.11 shows the effect of heat treatment on hardness and tensile strength for a range of aluminium bronze alloys which all have high aluminium contents. It would seem that a Rockwell hardness of 40 HRC is approximately equivalent to the desired Brinell hardness figure of 390–400 HB and that a Rockwell hardness of 43–44 HRC is approximately equivalent to a Brinell hardness of 410–420 HB. Alloy A has a slightly lower aluminium content than the above alloy range but otherwise falls within it. Alloys B to D have substantial additions of nickel and iron in various combinations. For an understanding of the effect of heat treatment on the metallurgical structure, see Chapters 12 and 13.

Table 10.11   Effect of heat treatment on tensile strength and hardness of various aluminium bronze alloys with high aluminium content – by Roucka et al.[154]

Heat Treatment	Tensile Strength N mm^{-2}				Rockwell Hardness (HRC)			
Alloy:	A	B	C	D	A	B	C	D
Annealed at 960°C for 1 h, air cooled	83	171	100	228	36.5	32	36	32
Annealed at 960°C for 1 h, air cooled, annealed at 550°C for 6–8 h and furnace cooled	105	63	40	71	37.0	41	39.5	38.5
Annealed at 960°C for 1 h, air cooled, annealed at 620°C for 5 h and furnace cooled	141				40.0			
Annealed at 960°C for 1 h, furnace cooled at 1.8°K min from 960 to 650°C and at 1.0° K mm^{-1} from 650 to 500°C	155	154	109	165	40.5	43.5	43.5	34

	Alloy composition wt%				
	Cu	Al	Fe	Ni	Mn
Alloy A	Bal	14.6	3.3–3.5	0.9–1.2	~1
Alloy B	Bal	14.9	4.9	5.2	
Alloy C	Bal	15.1	7.2	5.8	
Alloy D	Bal	14.9	4.8	7.1	

The following conclusions can be drawn from Table 10.11:

- a high nickel figure of 7.1% (alloy D) gives the highest tensile figures but lower hardness figures than alloys with 5–6% nickel contents (alloys B and C);
- slow cooling from 960°C gives the highest hardness figures for all alloys;
- with the exception of alloy A, the best tensile figures are obtained by air cooling from 960°C;
- the best combination of hardness and tensile strength is given by alloy B, but the hardness is only marginally higher than that obtained with alloy A with the low nickel content. If the aluminium content of alloy A was increased to 15%, there would probably be little difference between alloys A and B when cooled slowly from 960°C. The evidence suggest that the aluminium content combined with slow cooling are the overriding factors in achieving the highest hardness. As explained in Chapter 13 however, alloy B would have a much less corrosive structure than alloy A and would therefore be a better choice in a corrosive application.

Roucka et al.[154] experimented with a higher iron content than in alloy A but with no increase in nickel. They found that, provided the alloy was slowly cooled, increasing the iron content to 7.2–9.0% resulted in slightly higher tensile and comparable hardness figures to those obtained with a 3% iron content. There was however an undesirable tendency for some fine grains to break out during machining resulting in poor surface finish.

Roucka et al.[154] also experimented with a titanium addition of 0.3–0.45% to an alloy similar to alloy A but containing 15.2% aluminium. They found that, unlike alloy A, the titanium containing alloy benefited from being cooled in air from 960°C: a considerably higher Brinell Hardness of 440–455 HB was obtained and the tensile strength was 30–50% higher than with a titanium-free alloy. Slow cooling, on the other hand, resulted in properties similar to those of the titanium-free alloy. It would appear therefore that a titanium addition to a type A alloy, combined with relatively rapid air cooling, provides the best combination of strength and hardness, but the extra cost may not be justified if titanium-free alloys perform adequately.

Summary of comparative wear performance of aluminium bronzes

- Aluminium bronzes have higher mechanical properties than phosphor bronzes and can therefore sustain higher loads associated with wear conditions (see Table 10.1).
- They have however a higher coefficient of friction than phosphor bronzes which limits their use in continuous rubbing conditions (see Table 10.1).
- Their rate of wear in lubricated conditions is significantly less than that of

leaded bronze or tin bronze and only slightly higher than Beryllium bronze (see Table 10.2).

- Copper–aluminium, with 8% Al, is less prone to adhesion at 1cm s^{-1} under non-lubricated conditions than other copper alloys when paired with hard steels (see Table 10.3).
- When aluminium bronze is paired with a variety of ferrous alloys, the resulting wear performance is better than that of these alloys paired between themselves (see Table 10.7).
- The wear performance of unlubricated self-mated aluminium bronzes at low RPM and low loading compares favourably with that of various ferrous and other alloys (see Table 10.6).
- The fretting resistance of nickel–aluminium bronze at low amplitude (6.5μm) is only slightly better than that of pure copper, but it performs better than other materials, except Titanium, at high amplitude (65μm) – see Table 10.10.
- The non-lubricated galling resistance of aluminium bronze with high Al, when mated with a variety of alloys, compares favourably with that of various pairs of these alloys (see Table 10.8).

Aluminium bronze coatings

Aluminium bronze sprayed coatings

Aluminium bronze sprayed coatings on various ferrous and non-ferrous bases combine the excellent wear resistance of aluminium bronze with the lower initial cost of the base metal. Sprayed coatings of approximately 0.15 mm can be applied to components such as clutch plates, lathe guide-rails, press ram sleeves, push-pull rods and a wide variety of parts involving mechanical wear against steel surfaces. The porosity of the sprayed coating has only a slight effect upon its mechanical properties, and has the advantage of retaining a lubricant film under conditions of imperfect lubrication.

Ion-plated aluminium bronze coatings on steel

Sundquist et al.[170] experimented with ion-plated aluminium bronze coatings on steel, using an alloy of approximately 14% Al, 4½ Fe, 1% Ni and bal. Cu. The process involved melting and evaporating the aluminium bronze in a vacuum chamber and depositing it on a steel work-piece. Work-pieces of both carbon tool steel and of mild steel were used in the experiments. They were coated with films of different thicknesses, as shown in Table 10.12.

Coating composition
Because of the different evaporation rates of the constituent elements of aluminium bronze (nickel has a very slow evaporation rate), the coatings were not fully

Table 10.12 Details of ion-plated aluminium bronze coatings on steel by Sundquist et al.[170]

	Original aluminium bronze	Coating A	B	C
Thickness (μm)		4.9	5.2	10
Evaporation rate (g min^{-1})		0.38	0.43	1.04
Coating time (min)		55	48	20
Aluminium content %	14	11.7	12.4	14.2
Knoop Micro-hardness Number (KHN)	380	320	380	380
Pin-on-disc test:				
Sliding distance to penetration of steel pin through the coating (m)		34	60	105

homogeneous. To reduce this effect, the coatings were applied in layers of about 0.4 μm thickness by melting and evaporating only a small slug of metal at a time. The evaporation rate was increased approximately in line with the coating thickness as indicated in Table 10.12. It will be seen that the faster the evaporation rate, the nearer is the aluminium content of the coating to that of the original aluminium bronze. The nickel content of all the coatings was less than 1% and the iron content could not be reliably measured because of the proximity of the steel and the high iron content on the surface of the coatings.

Hardness
The micro-hardness figures of the coatings obtained by Sundquist et al.,[170] using a Knoop indenter and a load of 25 gf (~0.245N) , are given in Table 10.12. Coatings B and C, with the high aluminium contents, had similar microstructures and the same hardness as the original aluminium bronze. The differences in microstructure between the coatings are discussed in Chapter 12.

Strip drawing test
This test, which simulates a sheet drawing operation, consisted in drawing a mild steel strip through two flat aluminium bronze-coated steel dies of dimension 25 mm × 25 mm which exerted a force of 6.6 kN. The strip surfaces were cleaned with a solvent and there was no lubrication. The resultant coefficient of friction was 0.2–0.25. The surfaces of the drawn strips were smooth and free from scratches. With non-coated steel dies the coefficient of friction was 0.5–0.6, the surface of the strip was severely galled and seizure and tensile fracture of the strip occurred at a drawing distance of 150 mm.

Pin-on-disc test
This test measured the coating's resistance to penetration by a hard steel pin and is an indication of galling resistance. It consisted in a hard steel pin, with a tip radius of 3.175 mm, sliding with a force of 6.6 kN against an aluminium bronze coated

rotating disc. The sliding velocity of the pin on the disc was 53 mm s^{-1}. In all the pin-on-disc tests, the coefficient of friction was initially 0.18–0.2 and this coincided with a penetration rate of the coating of 0.1 μm m^{-1}. It then increased to 0.25–0.35 when the penetration rate increased sharply to 0.25 μm m^{-1}, corresponding with the point at which the coating was worn through. The sliding distance at which this point was reached for each coating is given in Table 10.12. The longer sliding distance of coating B compared with that of coating A is due to the harder gamma$_2$ microstructure (see Chapters 11–13); whereas the longer sliding distance of coating C compared with that of coating B is apparently due to the greater coating thickness of the former, since both coatings have a similar microstructure.

Advantage of aluminium bronze coated steel

The advantage of using a high-aluminium aluminium bronze-coated die as against using a solid aluminium bronze insert of the same composition is that it partly overcomes the problem of the brittleness of the high aluminium alloy. The tough steel to which the coating is applied gives resilience to the coated die.

There are no doubt many other applications where an aluminium bronze coated steel would have significant advantages.

Applications and alloy selection

Applications

Aluminium bronze finds many applications where wear resistance is of prime importance, e.g.: gear selector forks, synchronising rings, friction discs, cams, lead-screw nuts, wear plates and a wide range of bearings, bushes, gears, pinions and worm wheels. Table 10.13 compares the suitability of various copper alloys for gear applications. Aluminium bronze alloys with high aluminium content have been found particularly advantageous as dies and other tools used in metal drawing. They have a longer life, are less liable to seizure, they reduce spoilage and, in some cases, the number of forming operations can be reduced.[154]

Table 10.13   Comparison of suitability of various copper alloys for gear applications.[50]

Material	CEN/ISO designation	Typical application
Leaded brass	CuZn33Pb2	Lightly loaded small gears
	CuZn39PbAl	
Leaded gunmetal	CuPb5Sn5Zn5	Lightly loaded small gears
High tensile brass	CuZn33Pb2Si	Heavy duty low speed gears
Aluminium bronze	CuAl10Fe5Ni5	Heavy duty low speed gears
Phosphor bronze	CuSn12	Heavy duty gears
Gunmetal	CuSn7Ni5Zn3	Very heavy duty gears
	CuSn10Zn2	Heavy duty gears

Alloy selection

Light loading

For applications involving light loading, the choice of materials is very wide. As aluminium bronze is suitable for gravity diecasting it is often the most economic for large quantity batch-production when a material superior to brass is required. Examples of aluminium bronze components running satisfactorily against parts of the same alloy composition have been shown to have a wear rate of only one-tenth of that experienced with brass against brass.

Heavy loading

Since the majority of applications involve heavy loads, large masses of material are required which are normally cast or hot-forged. A material of inherent high strength is therefore desirable; the most popular being the CuAl10Fe5Ni5 type of alloy. However, if the component is to be die-cast, the CuAl10Fe3 alloy will provide a more economical substitute for most applications. The silicon containing alloy CuAl7Si2 has good wear resistance, especially against steel pins in pintle bearings.

For bushes and wear plates, thin gauge material may be produced by cold rolling or drawing processes. It is therefore possible to choose a lower strength alloy containing less than 8% aluminium and to obtain the desired hardness by cold-working. Very thin gauge material can in fact be obtained far more readily in this work-hardened type of aluminium bronze.

Highly abrasive conditions

Alloys with higher aluminium contents have been found to be particularly suitable for heavily abrasive conditions, e.g. the cutting blades of a refuse pulveriser. They have been produced successfully from an alloy containing 11–11.5% aluminium with 5% each of nickel and iron which has a hardness of up to 300 HV.

Tooling for sheet drawing

Alloys with aluminium in excess of 12 per cent have a low elongation value (below 5%) and are unsuitable for applications involving severe impact. They have, however, very high hardness and wear resistance and an alloy containing 15% aluminium is successfully used for deep-drawing dies handling stainless steel and other sheet materials. This alloy is very brittle and can fracture when subjected to only mild impact loads, but for deep-drawing dies and similar applications this is not a serious handicap. As explained above, the practice of ion-plating a high-aluminium aluminium bronze on steel would overcome the disadvantage of brittle-ness of the tool whilst providing a very hard surface.

Part 2
MICROSTRUCTURE OF ALUMINIUM BRONZES

INTRODUCTION TO PART 2

Alloy systems

In order to explain the development of the microstructure of aluminium bronzes, from those with the simplest to those with the most complex composition, it is convenient to divide them into the following three systems:

(1) The Binary System which consists of only two elements, copper and aluminium, in varying proportions.
(2) Ternary Systems which consist of three elements: copper, aluminium and a third element (iron, manganese, nickel, silicon etc.), also in varying proportions.
(3) Complex Systems which consist of four or more elements: copper, aluminium and two or more other elements, likewise in varying proportions.

These systems will be considered in the following chapters:

Chapter 11: Binary Systems
Chapter 12: Ternary Systems
Chapter 13: Copper/Aluminium/Nickel/Iron System
Chapter 14: Copper/Manganese/Aluminium/Nickel/Iron System

It will be seen that the ternary and complex systems are modifications of the basic binary system. An understanding of the binary system is therefore essential to the understanding of the more alloyed systems.

The structures of existing alloys are considered rather than their past development. The rational for the composition of these alloys will, however, become evident, at least in part, from the effects of the various elements on mechanical and corrosion properties.

The effect of heat treatment on microstructure will also be considered. This should be considered in conjunction with Chapter 6 where the standard forms of heat treatment are explained.

The structure of aluminium bronzes is explained in these chapters in a way that, it is hoped, will be understandable to readers who may have little or no knowledge of metallurgy.

Crystalline structure

As previously explained in Chapter 4, the structure of a metal consists of crystals bonded together. These crystals are themselves made up of atoms or, more properly speaking of ions, which were previously dispersed randomly in the liquid state, and

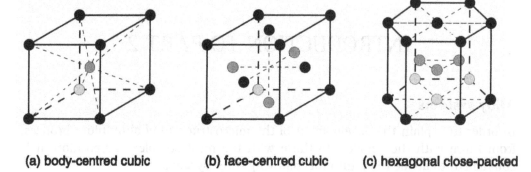

(a) body-centred cubic   (b) face-centred cubic   (c) hexagonal close-packed

Fig. 11.1   Three principal types of space lattices.

which assumed, on solidification, an orderly geometrical pattern, known as a 'space lattice'. There are several types of space lattice but the three most common are as follows and are illustrated in Fig. 11.1:

(a) body-centred cubic (bcc), see Fig. 11.1a,
(b) face-centred cubic (fcc), see Fig. 11.1b,
(d) hexagonal close-packed, see Fig. 11.1c.

The space lattices shown in Fig. 11.1 are the simplest units and a crystal is made up of a continuous series of these units in which adjoining units share a common face. This assembly of space lattices constitutes therefore the structure of the crystal.

As we shall see in the following chapters, different chemical constituents of the alloy may have different space lattices.

The type of space lattice has a bearing on the ease or difficulty with which a wrought alloy can be worked and on the choice of working temperature. For example, a face-centred cubic structure is far more malleable and ductile than a hexagonal close-packed structure.

Growth of crystals

As a liquid metal approaches its solidification temperature, a number of 'nuclei' are formed simultaneously in the melt, a 'nucleus' being a single unit of a given type of space lattice. Other atoms then attach themselves to these nuclei, building up a crystal of the same type of space lattice as the nucleus. The crystal initially grows into a dendrite (see Fig. 11.2) which conforms to a rigid geometrical pattern. Eventually the outward growth of the dendrite is impeded by other growing dendrites in the vicinity. The crystal then grows in thickness as the liquid metal remaining between the arms of the dendrite solidifies. This results in irregularly shaped crystals. This growth process of crystals is illustrated in Fig. 11.3. The more nuclei appear in the melt the sooner the growth of the crystal is halted by neigh-

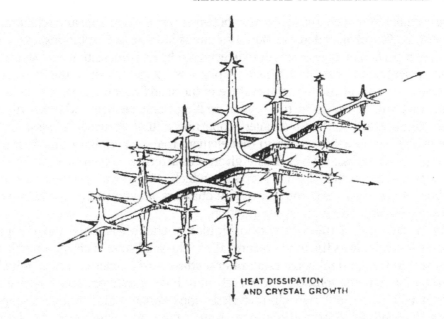

HEAT DISSIPATION
AND CRYSTAL GROWTH

Fig. 11.2   Early stage in the growth of a dendrite.[92]

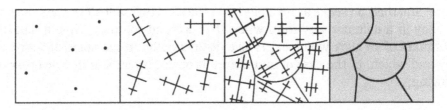

Fig. 11.3   Formation of crystals or grains by dendritic growth[19]

bouring crystals and therefore the smaller the 'grain' structure of the alloy. The word 'grain' is effectively interchangeable with the word crystal and is more commonly used when referring to the microstructure. The grain size has very important effects on alloy properties as will be discussed later.

If the alloy cools slowly below the solidification temperature, the crystals keep growing, but at each other's expense. Thus a sand casting will have a coarser grain than a die casting and the thick sections of a sand casting will have a coarser grain than its thin sections.

Chemical constitution of an aluminium bronze alloys

An aluminium bronze alloy in the liquid state consists of a solution of its various elements in each other. It may also contain some intermetallic compounds which have resulted from a chemical reaction between certain elements and which are also in solution. On solidification, the liquid solutions become solid solutions, each crystal being of a particular type of solution determined by its composition and space lattice arrangement and known as a 'phase'. A phase is not necessarily a uniform solution, however, because, at the top temperature of the solidification range, the metal solidifying first will be richer in the higher melting point element, whereas the metal solidifying last will contain a smaller proportion of that element. It follows that the core of any crystal will be richer in that element than its periphery. This is known as 'coring'. Nevertheless, a phase has a given characteristic appearance under the microscope and has certain specific properties which affect the properties of the alloy as a whole. As we shall see in the subsequent chapters, an alloy may solidify into one or more phases, depending on its composition.

As in the case of the solubility of liquids, metallic elements and compounds become less soluble as the temperature falls in the solid state. This may result, as we shall see, in the gradual conversion of one phase into a different phase which will consist of a different solution and which may have a different space lattice structure. It will have a different characteristic appearance under the microscope and different properties. Intermetallic compounds may also come out of solution as precipitates. They then become visible under the microscope and will appear either within a crystal or at the boundary between two crystals. They too constitute a 'phase' and their presence as precipitates in the structure will affect the properties of the alloy in a different way than when they were in solution. Thus a 'phase' is a constituent of an alloy which exists as a distinct entity in the microstructure of the alloy and which, in the case of aluminium bronzes, consists in one or other of the following:

(a) a solid solution of one or more elements in another, or
(b) an intermetallic compound which has formed by chemical reaction and has come out of solution before or after solidification.
(c) a combination of two or more individual phases to form duplex or complex phases.

Heat treatment

If the alloy is re-heated, phase changes are reversed provided sufficient time is allowed. By controlling the time of exposure to a higher temperature and the rate of cooling thereafter, the nature of the alloy, both in its grain size and phase constitution, can be adjusted to achieve a desired combination of properties. This is the object of heat treatment.

11
BINARY ALLOY SYSTEMS

This chapter will consider the microstructure of alloys within the binary system and its effects on properties. This and the next two chapters should be read in conjunction with chapter 1 which dealt with the effects of alloying elements on mechanical properties and with chapter 8 in which the mechanism of corrosion was discussed.

Copper-aluminium equilibrium diagram

Aluminium is the main alloying element in all aluminium bronzes except those with high manganese content. It is primarily responsible for the mechanical and corrosion resisting properties of these alloys.

When an alloy of a given composition is allowed to cool very slowly from the solidification temperature to room temperature it undergoes changes in its crystalline structure. The change may simply be a matter of grain growth within the structure which otherwise remains fundamentally the same. Alternatively the alloy may experience, as it cools, several successive transformations of its structure. The temperature at which each new crystalline structure arises is particular to the specific composition of the alloy. The equilibrium diagram, shown in Fig. 11.4, indicates the temperatures at which each change in structure occurs for any given alloy of copper and aluminium.

Single phase alloys

Considering, for example, an alloy consisting of 7% aluminium and 93% copper: above 1060°C, this alloy is an homogenous molten solution of aluminium and copper. At 1060°C the alloy begins to solidify in the form of numerous crystals which grow as the liquid metal solidifies. This solidification process takes place over a relatively small temperature drop from 1060°C to around 1045°C. Below 1045°C the alloy is therefore fully solid and consists entirely of a copper-rich solid solution of aluminium in copper, known as the 'α-phase'.

If the alloy is allowed to cool very slowly to room temperature, the crystals will grow in size at each other's expense. In practice, cooling to room temperature may occur fairly rapidly thereby restricting this grain growth.

In the case of an alloy containing 8% aluminium, it will be seen from Fig. 11.4 that the alloy will solidify at 1040°C into a mixture of two solutions: one a copper-rich solid solution (the α-phase) and the other a high temperature solid solution which is richer in aluminium, known as the β-phase. If allowed to cool slowly, it

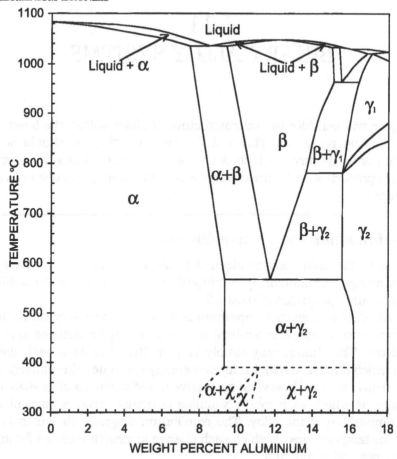

Fig. 11.4   Binary copper–aluminium equilibrium diagram.[127]

will change at about 900°C into a single α-phase alloy. Although the equilibrium diagram indicates that single α-phase alloys are obtainable up to 9.4% aluminium, in practice the transformation from β to α is so retarded, that the limit for a homogeneous α alloy is 7.5–8%.[33]

An alloy which ends up consisting of only one phase is known as a 'single phase' alloy, although it may have gone through changes of crystalline structure during which it may have consisted of more than one phase.

Single α-phase alloys have excellent ductility but low mechanical properties. An alloy consisting entirely of the α-phase also has very good resistance to corrosion provided it is not subject to internal stresses.[151] There is a progressive improvement in the corrosion resisting properties of the α alloys as the aluminium content is increased up to about 8%. It should be noted that these characteristics of single α-phase alloys apply to the single phase alloys of other aluminium bronze alloy systems (containing other alloying elements).

Duplex (two-phase) alloys

Let us now consider an alloy of 10% aluminium and 90% copper. Solidification begins at about 1045°C and is complete at about 1040°C – an even shorter freezing range than for the previous alloy. In this case, however the crystals that are formed are high temperature solid solutions, known as the β-phase which is richer in aluminium than the α-phase. If the alloy is allowed to cool very slowly, a further change in crystalline structure occurs at about 900°C. Below this temperature, an α-phase begins to form and to grow at the expense of the β-phase and the alloy becomes a combination of both α- and β-phases. The rate of cooling is rarely slow enough, however, for the phase transformation to begin at the temperature indicated by the equilibrium diagram. An alloy which ends up with a structure consisting of two phases is known as a 'duplex' alloy.

The β-phase, shown in the equilibrium diagram (Fig. 11.4), is an intermediate high temperature solid solution and cannot exist at room temperature but goes through a number of successive intermediate phases (see below) before becoming β' at room temperature (also known as 'martensitic β' because of its similarity to a quenched steel structure known as 'martensite').

If the rate of cooling remains very slow, the proportion of the α-phase will increase by comparison with that of the β'-phase.

In the case of an alloy with 10% aluminium, the next change of structure from α+β to the highly corrodible α+γ₂ will not occur, unless it cools very slowly from any temperature between 900°C and 565°C to room temperature. The alloy will end up as an α+β' duplex alloy. Although a 10% aluminium alloy has been taken as an example, alloys with aluminium contents within the range of about 8% to 11 % may end up with an α+β' structure if the cooling rate is sufficiently fast.

For the γ₂ phase to begin to appear depends on the combination of cooling rate and aluminium content as shown on Fig. 11.5. For example, it will be seen that, at a 9.0% aluminium content, the rate of cooling has to be less than 70°K min⁻¹ for γ₂ to begin to appear. In warm sand, the cooling rate is ~65°K min⁻¹ and therefore most cast structures contain little, if any, α+γ₂ eutectoid. Air cooling is sufficient to prevent this change taking place in most wrought sections, although, for heavy sections, it may be advisable to quench from 600°C. It will be seen in Chapter 12 that alloying additions of nickel, iron and manganese also tend to stabilise and permit lower cooling rates to be used without the change to α+γ₂ occurring.

Duplex α+β' alloys generally offer higher strength than the single phase α alloys but are susceptible to corrosion by de-aluminification. This is because β' is more anodic than α and is therefore potentially vulnerable to inter-phase corrosion. Weill-Couly and Arnaud[183] are of the opinion that an excess of β' is more a matter for concern than the likelihood of the γ₂ phase occurring. They point out that 'very slow cooling rates are required for the β phase to be totally transformed and that the structure of cast binary alloys were practically always α+β' or α+β'+γ₂. Furthermore they report that 'systematic trials, carried out on samples at different stages of phase changes, have clearly shown that the β' phase, although less anodic than the γ₂ phase, was also strongly attacked if the aluminium content was sufficiently high to make β' appear in sufficient proportion'. They go on to point out that 'it is probable that the susceptibility to corrosion of the β' phase is aggravated by internal stresses, which would explain the micro-ruptures in de-aluminised zones, whereas γ₂ is attacked uniformly' and is therefore not similarly vulnerable. This is another reason why it is advisable to limit the aluminium content of binary alloys to around 9.0%, even when rapidly cooled, if vulnerability to de-aluminification of the β' phase is to be minimised. If the β' and γ₂ phases are successfully broken up by heat treatment so that they are not continuous through the structure, corrosion will not penetrate and its effect will remain superficial. The treatment normally adopted is to soak at a temperature between 600–800°C for 2 to 6 hours, depending on section thickness and on the properties required, followed by rapid air cooling or quenching. Its success in breaking up the corrodible phases cannot however be assured.

Binary alloys show a distinct advantage over the α alloys at high water velocities where erosion and cavitation damage become important. Again it should be noted that these characteristics of α and of α+β' structures apply to aluminium bronze alloy systems containing other alloying elements.

Eutectoid formation

If, however, the alloy is allowed to cool very slowly past 565°C, a third change of structure will occur. Below this temperature, the β phase breaks up into an intimate mixture of α and γ₂ (gamma-two) phases, forming what is known as a *eutectoid*. Although this eutectoid has good wear properties and is sometimes encouraged for this reason, it has poor corrosion resisting properties and is therefore normally avoided. The reason for its poor resistance to corrosion is that, as the γ₂ phase is richer in aluminium than the α-phase and is consequently more anodic, it is liable to corrode at a greater rate under certain conditions (see chapters 8 and 9). The γ₂ phase was referred to as the δ-phase in earlier literature.

Fig. 11.5 shows the influence of the aluminium content and of cooling rate on the formation of the corrosion-prone γ₂ phase in a binary alloy. In alloys containing 8.5–9.5% aluminium, it has been found that, where γ₂ forms a continuous network in the alloy structure, penetration rates were five or six times greater than with the normal α+β' structure and under crevice conditions the effect was further accentuated. Isolated areas of γ₂ were not as dangerous and resulted only in minor pitting.

As the proportion of eutectoid depends on both the aluminium content of the alloy and the cooling rate, it is possible to indicate the conditions under which an unacceptable structure will form, as shown in Fig. 11.5. Thus with a 9.2% aluminium content, the rate of cooling would need to be slower than approximately 35°K min⁻¹ for the danger of continuous eutectoid to arise and, at 9.0% Al, it is unlikely to arise at all.

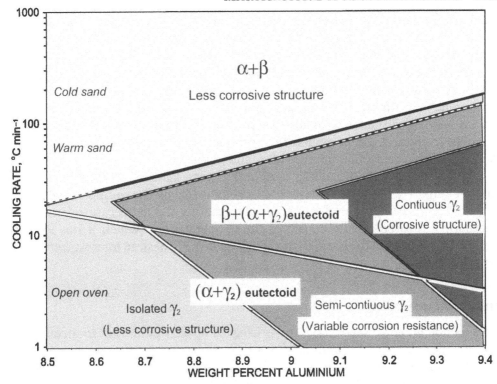

Fig. 11.5   Influence of aluminium content and cooling rate on the corrosion
resistance of binary copper–aluminium alloys.[127]

The shapes of the different areas are governed by the speed of cooling. At very
slow cooling rates, the proportion of β available to transform to α+γ₂ diminishes as
equilibrium conditions are approached, whereas at high cooling rates decomposition is suppressed. Continuous eutectoid and its attendant danger can be readily
avoided by limiting the aluminium content to 9.0% in binary alloys.

At high aluminium contents the γ₂ phase can be avoided by faster cooling but
this further increases the proportion of β'. In addition to its susceptibility to corrosion, the α+γ₂ eutectoid renders copper-aluminium alloys more brittle, particularly
if it is continuous. One of the main purposes of other alloying additions is to reduce
or eliminate the possibility of this eutectoid forming and still take advantage of the
enhanced mechanical properties that higher aluminium content offer.

Eutectic composition

A third combination of copper and aluminium is of special interest. At about 8.5%
aluminium content, the molten alloy solidifies at 1040°C into an intimate mixture
of two solid solutions: one a copper-rich solution (α-phase) and the other a solution

richer in aluminium (β-phase), this mixture being known as a *eutectic*. Solidification occurs at this single temperature and there is therefore no intermediate stage between liquid and solid, as with the above two alloys. If cooling continues at a very slow rate, the α constituent of the eutectic will gradually increase at the expense of the β constituent and, at about 750°C, the β constituent will disappear altogether as may be seen from the equilibrium diagram. If, on the other hand, the alloy is cooled very rapidly after solidification, the eutectic formation will appear in the microstructure and will be similar in appearance to the eutectoid previously mentioned but will consist of a mixture of α- and β-phases.

In practice the rate of cooling is almost always faster than that required for phase changes to occur at the temperatures indicated by the equilibrium diagram, but the latter serves the useful purpose of indicating the temperature from which the alloy has to be rapidly cooled or the temperature to which it has to be reheated for a period of time and then rapidly cooled, if a desired microstructure is to be achieved.

Intermediate phases

This section is rather specialised and is likely to be only of interest to metallurgists and manufacturers of wrought products.

The β-phase, shown in the equilibrium diagram (Fig. 11.4), is an intermediate high temperature solution and has, according to many researchers, a random or disordered body-centred cubic structure. As mentioned above, it cannot exist at room temperature. If the rate of cooling is too great for β to transform to α+γ$_2$, the β-phase goes through a number of successive intermediate phases before becoming β' at room temperature. The β'-phase has a martensitic structure and is often referred to as 'martensitic β'. These changes are shown on Fig. 11.6 by Jellison and Klier[103]

- Above the eutectoid temperature, and over a range of temperature around 500°C, β changes into an ordered structure, known as β$_1$. It has twice the lattice parameter of β[175] and is based on a Cu$_3$ Al superlattice.[132] Even at the very high cooling rate of 2000°K sec^{-1}, this transformation cannot be suppressed.

- On further cooling and passing through another range of temperature (M_S to M_F see Fig. 11.6), β$_1$ changes into a martensitic structured phase. If the aluminium content is below 13.1%, this phase is designated β' (with an approximately closed-packed hexagonal structure), and if the aluminium content is above 13.1%, it is designated γ'' (with a closed-packed hexagonal structure). At around 13.1% aluminium it will be a combination of both β' and γ'.[103–175] Since most commercial copper–aluminium alloys have less than 13.1% aluminium, they will be free of γ'. Some researchers[135] designate β' as β$_1$' and γ' as γ$_1$'.

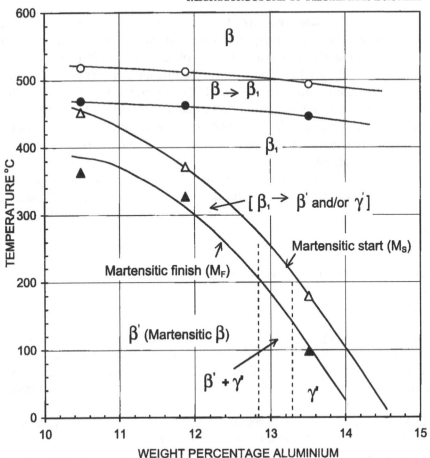

Fig. 11.6 Transformation of the β phase into (martensitic) β' at room temperature, through a number of successive intermediate phases.[103]

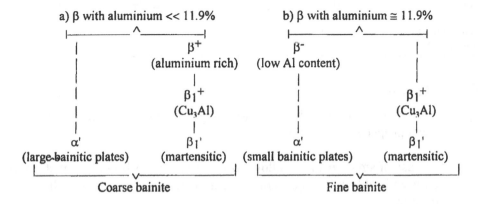

Fig. 11.7 Intermediate Phases by Brezina.[31]

Brezina and others[132] have reported the existence of an α' phase with a bainitic type structure, which forms at temperatures between 500°C and 475°C and depends on the rate of cooling and composition. This α' phase is said to form from areas that are low in aluminium content. The transformation for two alloys – one with an aluminium content significantly less than 11.9% and the other with approximately 11.9% aluminium – are shown in diagrammatical form in Fig. 11.7 by Brezina.

Although only the formation of the α' is bainitic, being due to shear and diffusion, Brezina describes the whole structure as coarse or fine bainite because of the predominance of the α' phase.

Other researchers[132] have also reported the existence of two sizes of α' plates in an alloy with 11.2% aluminium. As in the above diagram, the large α' plates were precipitated from disordered β, whereas the small α' plates were formed after ordering.

Summary of effects of structure on properties

Corrosion resistance

In the case of binary alloys, it is only single α-phase alloys that are fully resistant to corrosion by de-aluminification, provided, as previously mentioned, that the alloy is not subject to internal stresses.[151] Table 11.1 summarises the various structures obtainable with binary alloys and indicates which are fully resistant to corrosion.

Table 11.1   Binary alloy structures and their vulnerability to de-aluminification by Weill Couly and Arnaud.[183]

Aluminium content:	< 8.2%	> 8.2%		
Cooling rate:	any	Rapid (quenched)	Medium (cold sand or air cooled)	Slow (oven cooled)
Structure:	100% α	α+β'	α+β+γ$_2$	α+γ$_2$
Protection against de-aluminification:	yes	no	no	no

Mechanical properties

The proportion of aluminium in binary alloys affect the microstructure and hence the mechanical properties as follows:

(1) In the case of single α-phase alloys, the tensile strength and proof strength increase with aluminium content, with a decrease in elongation above 7% aluminium.

(2) In the case of duplex (α+β') alloys, the tensile strength and proof strength are increased in proportion to the amount of β' in the structure, but this is

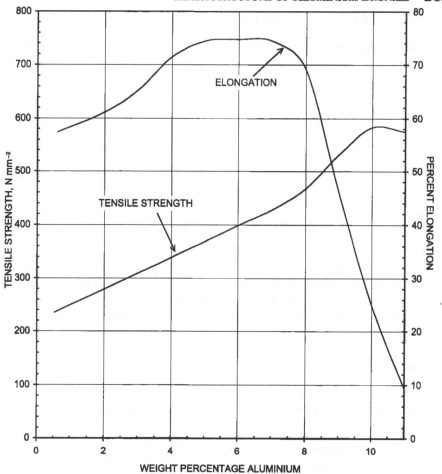

Fig. 11.8   Effect of aluminium content on the elongation and tensile strength of binary alloys at ambient temperature.[127]

accompanied by a progressive drop in ductility. The mechanical properties are also influenced by the distribution of α and β'.

(3) Hardness rises with tensile strength.
(4) The finer the structure, the greater the toughness (see 'Effect of heat treatment on structure of duplex alloys' below).
(5) Dissociation of β' into the α+γ₂ eutectoid lowers the elongation value and tensile strength. It renders the alloy hard and brittle.

Fig. 11.8 indicates the general relationship between mechanical properties and aluminium content. In single α-phase alloys (approximately up to 8% aluminium), the strength increases with aluminium content. In alloys of higher aluminium content, the presence of the β' phase modifies the mechanical properties markedly in proportion to the volume of this phase present.

As cast and hot-worked microstructure

We have seen above that, within the normal commercial range of alloy compositions (namely 5–12% aluminium), the basic structures at room temperature can
be:

- either single α phase solid solution (below about 9.4% Al)
- or duplex $\alpha+\beta'$
- or three-phase $\alpha+\beta'+\gamma_2$ (above 9.4% Al)
- or $\alpha+\gamma_2$, with very slow cooling and at higher aluminium contents.

The all-β alloys are found at room temperature, only at the extreme upper limit of
this range of aluminium content, provided the β-phase is retained artificially by
very rapid cooling, e.g. in quenching during heat treatment.

The solubility of aluminium in copper increases considerably with decrease in
temperature, as may be seen from the slope of the $\alpha/\alpha+\beta$ boundary in Fig. 11.4.
Consequently, the proportion of α and β in the structure at room temperature is as
much affected by the rate of cooling as by the aluminium content.

As-cast structures

The effect of the cooling rate on the structure of a binary alloy with 9.3% Al is
shown in Fig. 11.9, from the work of Weill-Couly and Arnaud.[183]

- Fig. 11.9a: Water cooled (in salt water) at 2000°K min-1.
  This reveals a martensitic structure containing $\alpha+\beta'$.
- Fig. 11.9b: Cooled in sand at 30 – 40°K min-1.
  This shows an $\alpha+\beta'$ martensitic structure in an α matrix (grey), with traces of
  the γ_2 phase (black).
- Fig. 11.9c: Cooled in hot (800°C) refractory mould at 8°K min-1
  The eutectoid transformation is more advanced and the γ_2 phase progressively
  invades the martensitic areas.
- Fig. 11.9d: 20 mm thick section re-heated to 970°C and cooled very slowly at
  0.5°K min-1
  The grain is very coarse and the β' phase has disappeared.

Hot-worked structures

The hot-worked alloy of the nominal 90Cu/10Al type has a high proportion of β in
the structure at the temperature of working and thus the majority of the α is
formed during cooling to room temperature. However, it is more common for alloys
to contain less than 10% aluminium and these can contain a high proportion of α

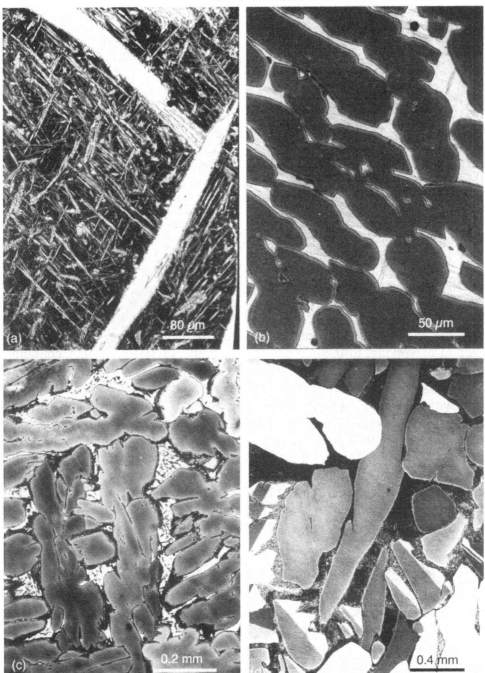

Fig. 11.9   Effect of cooling rate on a binary cast alloy containing 9.3% Al.[183] (a) Water cooled (in salt water) at 2000 K min⁻¹; (b) Cooled in sand at 30–40 K min⁻¹; (c) Cooled in hot (800°) refractory mould at 8 K min⁻¹; (d) Cooled very slowly at 0.5 K min⁻¹, by Weill Couly and Arnaud.[183]

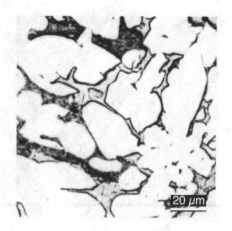

Fig. 11.10   Binary alloy containing 9.6% Al, showing elongation of the β phase due to hot working.[127]

Fig. 11.11   Binary alloy containing 9.8% Al, as extruded and cooled slowly in air from 900°C.[127]

at the hot-working temperature. The β' phase therefore, becomes elongated in the direction of working, in the manner shown in Fig. 11.10 for an alloy containing 9.6% Al (the β' constituents are the darkest lines or streaks).

The amount of β' in the structure varies considerably with the rate of cooling as explained above and, in practice, the proportion indicated by the equilibrium diagram is always exceeded. For example an alloy containing 9.8% aluminium, cooled slowly in air from 900°C, has a structure containing approximately 70% α, as shown in Fig. 11.11. This specimen has a microstructure comparable to that of an alloy in equilibrium at approximately 550°C. For comparison reference may be made to Fig. 11.12d which shows the microstructure of the same alloy after slow cooling to 500°C followed by quenching. It will be noticed that these two treatments yield similar proportions of α in the microstructure.

Re-crystallisation

It is possible to reverse the process by which the crystalline structure is transformed at critical temperatures, as an alloy slowly cools. Reheating any alloy to some high temperature for sufficient time (as happens during hot-working), will recreate the structure appropriate to that temperature. This may be necessary simply for the purposes of hot-working. If, on the other hand, it is desired for some reason to retain this high temperature structure at room temperature, the alloy must then be rapidly cooled by quenching in water. This will create internal stresses in the component which will need to be relieved by temper annealing (see below: 'Effect of heat treatment on microstructure and properties').

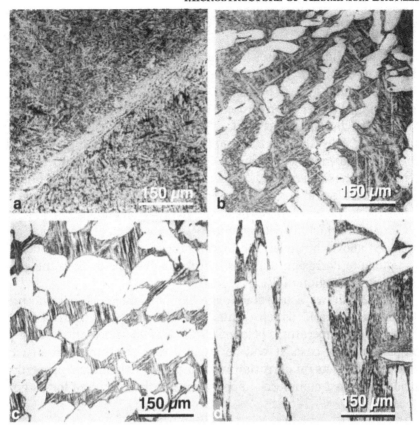

Fig. 11.12 Effects of quenching from different temperatures on a binary alloy[127] containing 9.8% Al (see Table 11.2 for associated properties: (a) Soaked at 900°C for 1 hour and water quenched; (b) Slowly cooled to 800°C and water quenched; (c) Slowly cooled to 650°C and water quenched; (d) Slowly cooled to 500°C and water quenched.

Effect of heat treatment on structure of duplex alloys

The heat treatment described in this section aims at adjusting the mechanical properties of binary alloys with a duplex structure. As mentioned above, these alloys are unsuitable for corrosive environments (see Table 11.1).

Effect of quenching from different temperatures

The effect of soaking a 90/10 alloy at different temperatures followed by quenching is illustrated in Fig. 11.12a–d, the structures are shown approaching equilibrium conditions at the temperature concerned. It will be seen that the α phase is progressively increasing from condition 'a' to 'd'. These are the kind of structures which would be obtained with the same alloy cooled at progressively slower rates after hot working.

The following should be noted in the photomicrographs shown in Fig. 11.12a–c:

(1) The white α phase increases in area from 'a' to 'c' as well as in grain size.

2) The darker-etching β' phase reveals an acicular (needle-like) structure in 'a', 'b' and 'c' similar in form to martensite in steels. Although the crystals appear needle-shaped under the microscope, these 'needles' are in fact cross-sections through flat disc-shaped crystals. This martensitic β-phase or β' is an unstable form of α in β and is the result of a cooling rate too great for the α+γ$_2$ eutectoid to form. Since normal β cannot exist at room temperature, all the so-called β-phase at normal temperatures is in fact β'. The β'-phase is very hard and brittle and has high tensile properties. An alloy composed of α and β' phases, however, can result in an attractive combination of mechanical properties and ductility (elongation) – see 'Mechanical properties' above. As explained above, however, too high a proportion of β' may render the alloy susceptible to corrosion and it is therefore advisable to keep the aluminium content in the region of 9%.

(3) Fig. 11.12d shows a lamellar (plate-like) structure. This is the previously mentioned eutectoid which results from the transformation of the β phase into α+γ$_2$ at temperatures below 565°C – this transformation being virtually complete in this case. It will be recalled that this eutectoid structure has undesirable effects on corrosion resistance and mechanical properties and is avoided for most commercial applications, unless required for its good wear property.

Table 11.2   Effect of quenching from different temperatures on the mechanical properties (at ambient temperature) of a duplex alloy, containing 9.8% aluminium, (see Fig. 11.12 for corresponding photomicrographs).[127]

Heat Treatment	0.1% Proof Strength N mm^{-2}	Tensile Strength N mm^{-2}	Elongation % on 50 mm	Hardness HB
Heated to 900°C and quenched	322	671	42	55
Heated to 900°C, slowly cooled to 800°C and quenched	297	592	9	216
Heated to 900°C, slowly cooled to 650°C and quenched	148	425	17	138
Heated to 900°C, slowly cooled to 500°C and quenched	136	297	5	136

A high-power photomicrograph is given in Fig. 11.13 which illustrates partially dissociated β which clearly shows the form of the α+γ$_2$ eutectoid (this type of structure is sometimes called 'perlite' because of its similarity to a structure found in ferrous alloys). The breakdown of β into α+γ$_2$ can be avoided in practice by ensuring that the material is rapidly cooled through the range 600–400°C. In special cases involving localised heating, e.g. welding, this cannot be achieved and heat

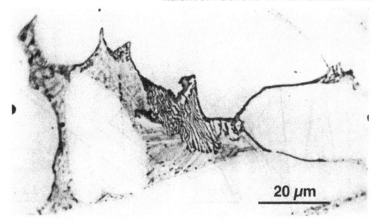

Fig. 11.13 Binary alloy containing 10% aluminium showing partial
transformation of the β phase to α + γ$_2$, due to slow cooling to below 650°C.[127]

treatment may be necessary after fabrication of the assembly to remove the eutectoid.
The treatment normally adopted is to soak at a temperature between 600–800°C,
depending on the properties required, followed by rapid air cooling or quenching.

It will also be noted from Fig. 11.13 that an alloy consisting only of the α phase
would be entirely made up of white grains. It will be seen that where two α grains
are in direct contact with each other, the boundary between them is hardly visible.
The microstructure of a single α-phase alloy would therefore be barely discernible
under the microscope. Another point to note is the large size of the α grains
resulting from the slow cooling which led to the formation of the α+γ$_2$ eutectoid
(slow cooling allows time and energy for the grains to grow).

The effect on mechanical properties of the above treatment is shown in Table
11.2. It will be seen that properties are altered as dramatically as the structure. The
lower the temperature at which the alloy is quenched, the greater the proportion of
α in the structure and the lower the tensile strength, proof strength and hardness.
The elongation increases until the quenching temperature is below 565°C, when γ$_2$
is produced and elongation is reduced.

Table 11.3 Effects of different tempering temperatures on rod, in a binary alloy
containing 9.4% aluminium, following quenching from 900°C.[127]

Heat Treatment	0.1% Proof Strength N mm⁻²	Tensile Strength N mm⁻²	Elongation % on 50 mm	Hardness HV
Heated 1 hr at 900°C and quenched	195	751	29	187
Quenched from 900°C and tempered at 400°C for 1 hr	212	750	29	185
Quenched from 900°C and tempered at 600°C for 1 hr	238	699	34	168
Quenched from 900°C and tempered at 650°C for 1 hr	223	646	48	150

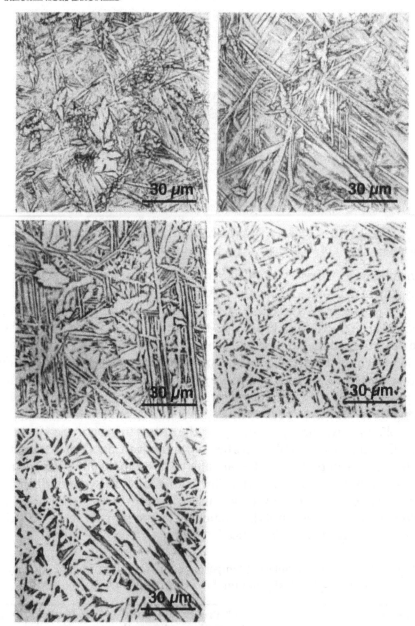

Fig. 11.14 Effect on a binary alloy containing 10% Al of soaking for 1 hour at 900°C and quenching followed by tempering at different temperatures:[123] (a) Soaked for 1 hour at 900°C and quenched; then (b) Tempered for 1 hour at 400°C; (c) Tempered 1 hour at 500°C; (d) Tempered 1 hour at 600°C.

Effects of quenching followed by tempering at different temperatures

The most usual heat treatment of a binary alloy containing around 10% aluminium involves heating to about 900°C or above in order to bring the alloy into the all-β field, soaking for a time at that temperature and quenching. The alloy will then consist of a martensitic β or β' structure. In order to achieve different forms and degrees of α precipitation, the alloy is then tempered which consists in reheating to and soaking at a pre-selected temperature for the desired α precipitation to take place and then cooling rapidly to retain this structure. Tempering also relieves internal stresses which can render the alloy brittle.

In Fig. 11.14 microstructures of alloys thus heat treated are illustrated and properties of similar treated alloys are given in Table 11.3. With this method of heat treatment, the α is precipitated along crystallographic planes and results in a much finer precipitate than that formed during continuous cooling with a consequent improvement in toughness. It is, therefore, possible by this means to control the mechanical properties to within much closer limits.

These figures should be compared with the figures given in Table 11.2 for a 90/10 alloy which was cooled slowly from 900°C to a temperature below the eutectoid temperature of 565°C. They show the drop in mechanical properties resulting from the formation of the eutectoid. The tensile strength can fall by approximately 155 N mm^{-2} and the elongation figure to about 5%. This means that, if the soaking temperature is reduced from 650°C to 500°C, an alloy containing a high proportion of eutectoid is produced which has a much poorer ductility than the all-β' material as quenched, and only half the strength.

Binary alloys in use

The only truly binary alloy in use is the wrought CuAl5 alloy. Other alloys which are nominally binary such as CuAl7 and CuAl8, may contain deliberate additions of iron, nickel or manganese.

Although lead is normally an undesirable impurity in aluminium bronzes, with very detrimental effects on mechanical properties and on welding, 1–2% lead is sometimes added in special wrought alloys to improve machinability. From a structural point of view, lead may be considered insoluble in aluminium bronze. It is therefore not an alloying element and does not alter the basic alloy system. It appears as isolated particles dispersed throughout the structure in a fashion depending on the previous history of the material. Normally leaded aluminium bronze is only manufactured as extruded rod, and the structure has the appearance of a 60/40-type of leaded brass. The microstructure of a 90/10 duplex copper-aluminium alloy containing l% lead is shown in Fig. 11.15. Most foundries would however prefer to steer clear of such an alloy for fear that the use of lead in one alloy might contaminate other alloys.

Applications of wrought binary alloys, both single phase and duplex, are given in Chapter 5.

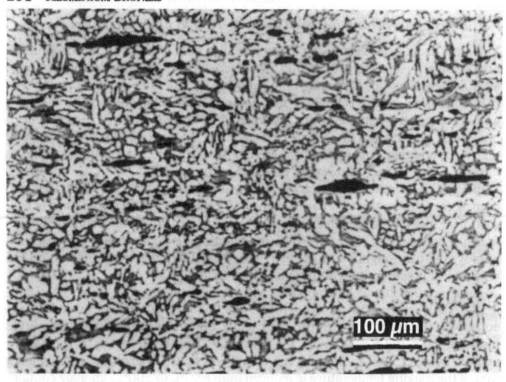

Fig. 11.15   Duplex copper–aluminium containing 10% lead, 90% Copper, 10% Al, 1% Pb alloy, as extended.[127]

12
TERNARY ALLOY SYSTEMS

Ternary systems consist of copper and aluminium plus one other alloying element: most commonly iron, nickel, manganese or silicon.

The copper-aluminium-iron system

Small additions of iron are made to copper-aluminium alloys primarily as a grain refiner in order to improve toughness and to prevent the formation of the corrosion-prone γ_2 phase. The way that iron refines the grain will be discussed below. The effects of iron addition on mechanical properties has already been dealt with in Chapter 1 (Tables 1.1 and 1.2). According to Goldspiel et al.,[78] iron narrows the solidification range.

Equilibrium diagram

Iron additions of 3% and 5% only slightly modify the binary diagram. The influence of iron on the copper-aluminium system[191] is shown in Figs 12.1a and b. Arnaud[10] reports that 3% iron moves the α–β boundary to higher aluminium contents, which would account for the increase in tensile strength caused by iron. As in the case of the binary system, aluminium content determines, under equilibrium conditions, whether an alloy is basically single α phase or duplex $\alpha+\beta$. Intermetallic Fe(δ) particles precipitate in both single and duplex alloys provided iron is above 1%. There are iron-containing wrought alloys in both these categories, as may be seen from Chapter 5. There is only one common cast alloy: the CuAl10Fe2-C CEN alloy. It has a range of aluminium content of 8.5–10.5% and an iron content of 1.5–3.5%. Even at the lowest aluminium content of 8.5% and at the relatively slow cooling rate of a sand mould, there is likely to be a partial retention of the β phase.

Development of microstructure

Hasan et al.[89] carried out research into the development of the microstructure of an alloy of composition 8.6% Al, 3.2% Fe, bal. Cu, as it cooled continuously from 1000°C at the rate of 25–30°K min^{-1}. The micrographs of the structure at successive quenching temperatures are shown on Fig. 12.2.

- Quenched at 1000°C, the microstructure (Fig. 12.2a) consists only of the β phase which transformed to a martensitic structure on quenching.

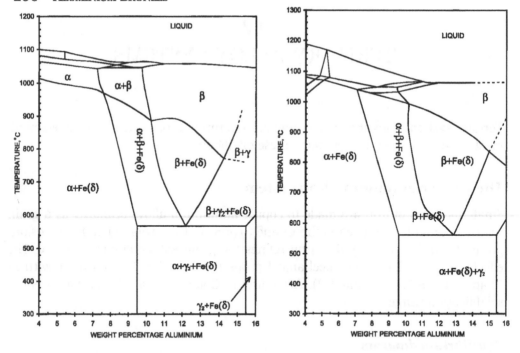

Fig 12.1   Vertical section of the Cu–Al–Fe system at 3 and 5% Fe, by Yutaka.[191]

- Quenched at approximately 900°C, the microstructure (Fig. 12.2b) shows the Fe(δ) particles (based on Fe_3Al) precipitated both within the β phase and at its boundaries.
- Quenched at 860°C, the microstructure (Fig. 12.2c) shows that the α phase has nucleated at both the β grain boundaries and around the Fe(δ) particles within the β phase.
- Quenched at 800°C, the microstructure (Fig. 12.2d) shows that the α grains have grown further but that their growth appears to have been impeded by the Fe(δ) particles remaining in the β phase.
- Quenched at 550°C, the microstructure (Fig. 12.2e) is similar to the as-cast structure (Fig. 12.3) indicating that the alloy did not undergo any appreciable changes below this temperature.

It follows from the above that the as-cast rate of cooling was not sufficiently slow for the $\alpha+\gamma_2$ eutectoid to form below 575°C as indicated by the equilibrium diagram.

As cast microstructure of a Cu–Al–Fe alloy
The microstructure of an alloy containing 8.6% Al, 3.2% Fe, bal. Cu, is shown in Fig. 12.3 by Hasan et al.[89] It will be noted that, except for the presence of the iron precipitate, the main constituent phases of this alloy have the same appearance

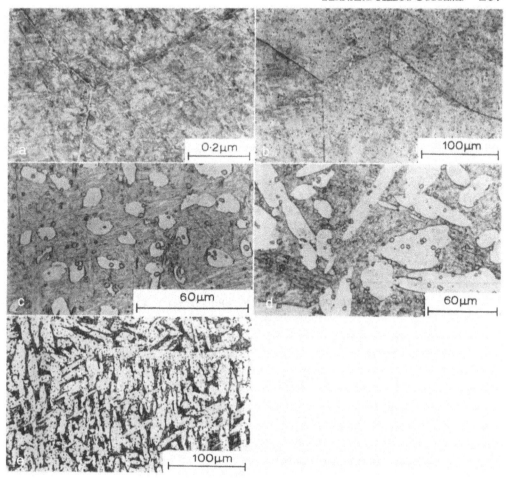

Fig. 12.2 Micrographs showing the development of the structure of an alloy containing 8.6% Al, 3.2% Fe, bal. Cu, during continuous cooling from 1000°C. by Hasan et al.[89]: (a) Quenched at 1000°C; (b) Quenched at 900°C; (c) Quenched at 860°C; (d) Quenched at 800°C; (e) Quenched ay 550°C.

under the microscope as the corresponding phases of the binary copper-aluminium alloys (Chapter 11). The proportion of iron which is in solution does not affect the appearance of the main phases. The composition of the phases by various researchers is given in Table 12.1. The light etching α phase Fig. 12.3b is a fcc copper rich solid solution. The dark etching β phase is martensitic.

Although the aluminium content of the alloy is less than 9.5%, the rate of cooling results in the retention of the β phase, but in a modified form. The retained β phase is martensitic whereas the high temperature β phase is an intermediate solid solution with a bcc structure. The smaller magnification of Fig. 12.3a shows the white α forming a boundary around the former high temperature β grains in which the α phase has precipitated in the typical needle-like form known as a

Table 12.1   Composition and crystallography of phases of a CuAl10Fe3 alloy.

Phases	wt % composition					Crystallography		
	Cu	Fe	Al	Si	Particles size (μm)	grain structure	Lattice structure	Based on
[1] α	Rem	1.6	9.2	–		Solid solution	fcc	
High temp. β						Intermed. solid solut.	bcc	
[1] Low temp. β	Rem	1.5	12.5	–		Martensite	9R	
[1] Fe(δ)	6.8	80.5	9.9	2.8	~2	Globular	DO_3	Fe_3Al
[2] Fe(δ)	4.4–6.3	79.7–82.3	12.4–14.0	–		or dendritic		
[3] Fe(δ)	9–26	69–77	7–12	–	4			
"	5–14	66–80	–	–	10–20			

[1] by Hasan et al.[89]   [2] by Mullendore & Mack[134]   [3] by Le Thomas et al.[171]

Table 12.2   Solubility of Fe in various phases as a function of temperature by Brezina.[33]

Temperature °C	%Fe (by weight) soluble in:		
	α (7.5% Al)	β (12.5% Al)	γ_2 (17.5% Al)
500	0.6*	–	0.9
600	0.9	1.1	1.3
800	1.5	2.6	–
1000	2.6	6.0	–

* 0.2% according to Weill-Couly.[185]

Widmanstätten structure. It should be noted that the grain size of the β phase, prior to decomposition, was relatively large: 0.5–1.0 mm.[89]

The uniform spread of intermetallic Fe(δ) particles in both α and β is clearly visible in Fig. 12.3b at the higher magnification. These particles are based on Fe_3Al.[89] Most are globular and of approximately 2μm diameter but some are in the form of 'rosettes' or dendritic.[89] There is an interesting line of closely spaced Fe(δ) particles going diagonally across the micrograph. They are thought to follow the line of a pre-existing high temperature β grain boundary at which the particles had precipitated. This is evidence that the alloy initially solidified into an all-β phase from which the Fe(δ) particles precipitated (see Fig. 12.1a).

Belkin[23] claims that the increase in hardness caused by the addition of iron is attributable to the presence of these iron precipitates.

Solubility of iron

The solubility of iron in the various phases is given in Table 12.2, as a function of temperature. It will be seen that, at any given temperature, it is highest in the phases

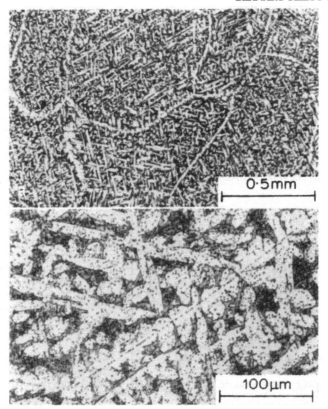

Fig. 12.3   As-cast microstructure of an alloy containing 8.6% Al and 3.2% Fe by
Hasan et al.[89]: (a) Low magnification showing α at former β grain boundaries and
forming a needle-like pattern, known as a Widmanstätten structure, within the β grain;
(b) High magnification showing intermetallic precipitates in α and β.

richest in aluminium. Opinions are divided regarding the solubility of iron at room
temperature; it is not likely to be more than 1%. According to Yutaka,[191] the solubility
of iron below 550°C under equilibrium conditions is less than 1%, but moderately
rapid cooling conditions can retain up to 2% iron in solid solution.

As shown in Table 12.2, iron is soluble in all α and β phases below the freezing
temperature of the alloys.

Iron in excess of 3% begins to precipitate in single α-phase alloys when the
temperature drops below about 1000°C, whilst in alloys with greater amounts of
aluminium, iron does not precipitate until the temperature has fallen at around
900°C (see below). The chemical composition of these iron-rich precipitates is given
in Table 12.1.

Grain refining action of iron

It was thought at one time that iron-rich particles came out of solution in the melt,
creating nuclei which initiated the formation of crystals and that this was the

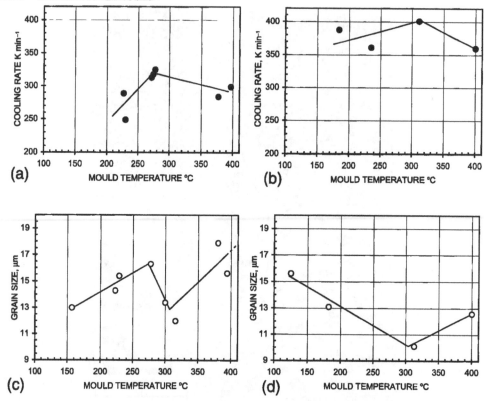

Fig. 12.4 Effect of pre-heated temperature of mould on cooling rate and grain size, by Gozlan et al.[79] (a) Effect of mould temperature on cooling rate with 0.01 mm graphite coating; (b) Effect of mould temperature on cooling rate with 0.10 mm graphite coating; (c) Effect of mould temperature on grain size with 0.01 mm graphite coating; (d) Effect of mould temperature on grain size with 0.10 mm graphite coating.

explanation for the grain refining action of iron. This explanation could still be valid in the case of any iron-rich particles which had not been properly dissolved in the melt. A more likely explanation is put forward by Hasan et al.[89] They point out that the β grains can undergo substantial growth before iron-rich particles are precipitated (as one would expect from the high solubility of iron in β at high temperatures – see Table 12.2). The iron-rich particles precipitate within β and at its grain boundaries (see above) prior to the formation of the α phase which it helps to nucleate. The resultant number of nucleation sites is large and the growth of the numerous α grains is consequently mutually restrictive. This grain refining action of iron is one of the most important advantages provided by iron additions. The fineness of the as-cast structure can be seen in Fig. 12.5a. It contrasts with the large size of the prior β phase grain boundary which clearly shows that the refinement of the microstructure, brought about by iron, has not resulted in a smaller β grain

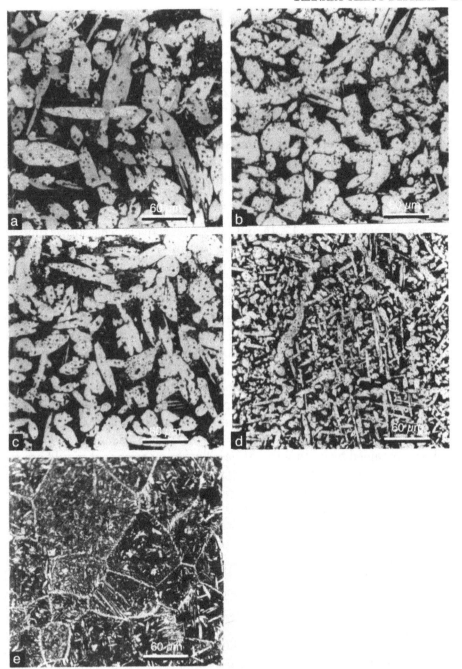

Fig. 12.5 Effects of various pre-heated mould temperatures on the microstructure of a CuAl10Fe3 alloy by Gozlan et al.[79]: (a) Pre-heated to 400°C – graphite coating: 0.01 mm.; (b) Pre-heated to 280°C – graphite coating: 0.01 mm.; (c) Pre-heated to 150°C – graphite coating: 0.01 mm.; (d) Not pre-heated – graphite coating: 0.10 mm.; (e) Water-quenched.

size, as would be the case if particles of iron had precipitated in the melt and nucleated the β phase.

Microstructure of a die-cast Cu–Al–Fe alloy

Cu–Al–Fe alloys with 3–4% Fe are used extensively in die-casting. Gozlan et al.[79] carried out experiments on the effects of various pre-heated mould temperatures on the microstructure of a CuAl10Fe4 alloy.

The mould was made of heat-resistant steel and coated with colloidal graphite. Surprisingly, the experiments showed that, up to certain mould temperature, the higher this temperature the faster the cooling rate. Only beyond this mould temperature did the cooling rate decrease as the mould temperature increased. The variation of cooling rate with mould temperature for graphite coatings of 0.01 mm and 0.10 mm is shown on Figs 12.4a and 12.4b respectively. This shows that, with a 0.01 mm graphite coating, a peak cooling rate of ~320°K min^{-1} is reached at a pre-heated mould temperature of 280°C, and with a 0.10 mm graphite coating, a peak cooling rate of ~400°K min^{-1} is attained at a pre-heated mould temperature of ~315°C. In each case, below this peak cooling rate, the higher the mould temperature the faster the cooling rate. The reason proposed by Gozlan and Bamberger[79] for this apparent anomaly is that, at mould temperatures below that corresponding to the peak cooling rate, the hot metal coming into contact with the mould causes it to deflect. The higher the temperature gradient between incoming metal and the mould, the greater the deflection of the mould and the consequent interruption in heat flow. Above the mould temperature corresponding to the peak cooling rate, the increase in mould temperature becomes the predominant factor in reducing the cooling rate.

The phases obtained by casting Cu/Al/Fe alloys in a permanent mould are α+γ$_2$ and the martensitic β-phase which have formed on cooling from the high temperature β-phase. Gozlan and Bamberger[79] report that the martensitic β-phase is in two forms: β' in which the aluminium content is less than 13.1% and γ' where it is higher than 13.1%. The latter only occurs at rapid rates of solidification. There are also intermetallic Fe(δ) particles in both α and β phases. They are of the following compositions: Fe$_3$Al, Al$_5$Fe2 and Al$_{13}$Fe$_4$. There are also some Fe particles which did not react with aluminium.

The effect of the above variations of cooling rate on the α-grain size is shown on Figs 12.3c and 12.3d for graphite coatings of 0.01 mm and 0.10 mm respectively. In the case of the thicker graphite coating (Fig. 12.3d), the α-grain size varies, as one would expect, inversely with cooling rate. With the thinner coating (Fig. 12.3c) however, the correlation is rather different: up to a pre-heated mould temperature of 280°C, the grain size surprisingly increases as the cooling rate increases and then decreases up to a pre-heated mould temperature of ~310°C, as the cooling rate decreases. This apparent anomaly is thought to be due, according to Gozlan et al[79], to the different ways in which the α-grains form. With the 0.10 mm graphite coating, which results in high cooling rates at all pre-heated mould temperatures,

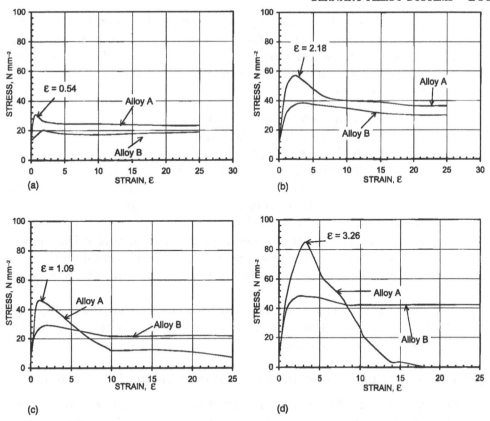

Fig. 12.6 Effect of hot-working and strain rate on the flow stress/strain relationship in two Cu–Al–Fe alloys, by Gronostajski and Ziemba.[81]: (a) Hot-worked at 830°C and at strain rate of 0.95/sec.; (b) Hot-worked at 830°C and at strain rate of 18.15/sec.; (c) Hot-worked at 750°C and at strain rate of 0.95/sec.; (d) Hot-worked at 750°C and at strain rate of 18.15/sec. Alloy A: 9.90 Al – 3.64 Fe – 1.63 Mn – bal Cu. Alloy B: 10.74 Al – 4.00 Fe – 1.60 Mn – bal Cu.

the α-phase forms by nucleation around the Fe(δ) particles and grow to needle-shaped grains, as shown in Fig. 12.5d for a non pre-heated mould. By contrast, with a thin graphite coating, it is only at the lowest cooling rates, associated with the pre-heated temperatures of 150°C and 400°C, that the α-phase has these needle-shaped grains (see Fig. 12.5a and 12.5c). At the highest cooling rate, associated with a pre-heated mould temperature of 280°C, the α-phase is formed by a process of massive transformation which results in spherical α grains (see Fig. 12.5b).

Gozlan et al.[79] report that it is the segregation of the γ'-phase, associated with rapid solidification, which impedes massive transformation and which allows the nucleation and growth mechanism of the α-phase to occur.

Finally, the effect of water-quenching on the alloy is shown on Fig. 12.5e which illustrates the nucleation of the α-phase at the β-phase grain boundaries and around the Fe(δ) particles, mentioned at the beginning of this section on the development of microstructure of Cu/Al/Fe alloys. The rapid rate of cooling impedes the growth of the α-phase.

Summary of the effects of iron on the microstructure of a Cu/Al/Fe alloy

1. Iron has a refining effect on microstructure.
2. It narrows the solidification range.
3. It increases hardness due to the iron-rich Fe(δ) precipitates
4. It moves the α–β boundary to higher aluminium contents which means that, in an equilibrium condition, an α alloy can have higher mechanical properties without entering a corrodible β-phase.
5. In practice, however, even at a low aluminium content of 8.5% and at the relatively slow cooling rate of a sand mould, an alloy containing 1.5–3.5% iron, is likely to experience a partial retention of the corrodible martensitic β phase.
6. Even at the rapid cooling rate associated with die-casting, the highly corrodible $\alpha+\gamma_2$ eutectoid is likely to form at an aluminium of 10% (see below: 'Vulnerability to corrosion').

Effect of hot-working temperature on structure and mechanical properties

Gronostajski and Ziemba[81] carried out research on two wrought Cu/Al/Fe alloys of the following compositions to determine the effects of hot-working temperatures on wrought microstructure and mechanical properties.

Alloy	Al	Fe	Mn	Impurities	Cu
A	9.90	3.64	1.63	0.5	bal
B	10.74	4.00	1.60	0.5	bal

The relationship of flow stress to strain when hot-working the above alloys at 750°C and at 830°C and at two strain rates of $0.95s^{-1}$ and $18.15s^{-1}$ is shown on Figs 12.6a–c. The flow stress is the stress measured in the direction in which the grains are elongated. It will be seen that, in the case of alloy B with the higher aluminium content of 10.74%, both hot-working temperatures of 750°C and 830°C come within the $\beta+$Fe(δ) zone (see Fig. 12.1a). The resultant effect is that, at both strain rates, the flow stress increases initially to a certain figure and thereafter remains fairly constant. This is an indication that work-hardening is taking place and that a banded structure (Fig. 12.7a) is retained as recovery from the effects of hot working takes place. It is in fact retained even after annealing.

In the case of alloy A, on the other hand, with its lower aluminium content of 9.90%, both hot-working temperatures are in the $\alpha+\beta+$ Fe(δ) zone (see Fig. 12.1a). The flow stress increases initially to a higher value due to the presence of the α-phase which is less ductile than the β-phase at high temperature. This results in a

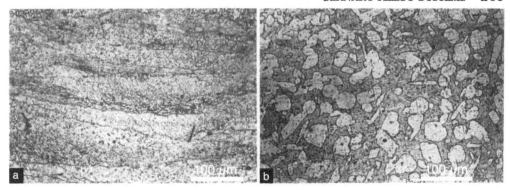

Fig. 12.7 Microstructure of two Cu–Alz–Fe alloys after hot-working at 830°C and at a strain rate of 18.15/sec, by Gronostajski and Ziemba.[81]: (a) Alloy B: banded structure after deformation and quenching – Strain $\varepsilon = 3$; (b) Alloy B: globular and elongated precipitates of α-phase in matrix of β-phase after deformation – $\varepsilon = 2$

build-up of stored energy, known as the stacking fault energy (SFE), which initiates recrystallisation once the strain has reached a critical value. This critical strain corresponds to the maximum value of the flow stress (see Fig. 12.6a–d) since re-crystallisation has a softening effect and the stress value consequently falls beyond the critical strain. The α-phase recrystallises before the β-phase.

If the strain is less than the critical value, recrystallisation of both phases does not occur unless the alloy is subsequently annealed. The drop in mechanical properties following re-crystallisation is due to the grain structure becoming equi-axial and the mechanical properties consequently becoming isotropic (equal in all direction) and significantly inferior to the anisotropic (unidirectional) properties of a banded grain structure. The highest stress value for alloy A (Fig. 12.6d) occurs at the combination of the lower temperature of 750°C (at which the proportion of α-phase would be greatest) and the higher strain rate of $18.5s^{-1}$. At the higher temperature of 830°C, on the other hand, and specially at the lower strain rate of $0.95s^{-1}$ (Fig. 12.6a), there is very little re-crystallisation occurring, due to the low proportion of α-phase at that temperature, and the banded structure is retained.

Conclusion on choice of hot-working temperature for Cu–Al–Fe alloys
It follows from the above that, for the wrought alloy to have a banded fibrous structure and therefore the highest tensile properties, it is necessary to hot-work at a temperature that falls within the $\beta+$ Fe(δ) zone or just below this zone where there is little or no α-phase. At a lower temperature, where there will be a significant proportion of the α-phase present, the alloy recrystallises as it recovers from hot-working and the mechanical properties cease to be anisotropic.

Vulnerability to corrosion

As shown in Fig. 12.3, the standard Cu–Al–Fe alloys have a duplex structure of α and β phases with iron-rich precipitates in both. The β phase is anodic to α and is

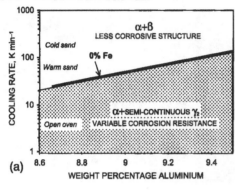

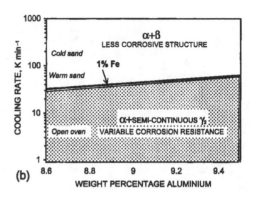

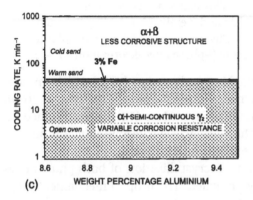

Fig. 12.8 Effect of aluminium content and cooling rate on the formation of the corrosion-prone γ_2 phase at various iron additions.[127] (a) 0% iron; (b) 1% iron; (c) 3% iron.

therefore liable to corrode preferentially. Under equilibrium conditions (see Fig. 12.1), there would be no β phase at an aluminium content below 9.5%. In practice the rate of cooling in sand and especially in die casting, results in the retention of some β phase at an aluminium content as low as 8.5%, the usually specified bottom limit of this kind of alloy. Furthermore, the tendency is to aim towards the upper limit of the specification in order to achieve the higher mechanical properties of a duplex (two phase) alloy. This means that, not only will the β phase be present, but it is potentially liable to transform to the more corrodible $\alpha+\gamma_2$ eutectoid.

An important advantage of iron additions, however, is that of slowing down the rate of breakdown of the β phase into $\alpha+\gamma_2$. Even when γ_2 is formed by very slow cooling, the grain-refining action of the iron helps to disrupt the continuity of the γ_2 phase and makes it less prone to corrosion. Moreover, P. Aaltonen et al.[1] report that, if the alloy is heat treated between 600–700°C for three hours and air cooled, although the amount of aluminium-rich phases is greater than in the as-cast condition, these phases are more uniformly distributed and do not form a contin-uous path for the selective dissolution of the eutectoid phases. With isolated areas of

γ_2, only pitting corrosion occurs.[62] The effect of aluminium content and cooling rate on the formation of a semi-continuous γ_2 phase for different percentage iron additions, is shown on Fig. 12.8.

It will be seen from Fig. 12.8 that:

(a) the higher the aluminium content the greater the risk of γ_2 forming as the cooling rate increases,

(b) the greater the iron additions up to 3%, the smaller the risk of γ_2 forming as the cooling rate increases. Furthermore, as the iron content increases its beneficial influence becomes progressively independent of the aluminium content.

It is unlikely however that γ_2 will be in a continuous form because the grain-refining action of iron helps to disrupt the continuity of the γ_2 phase, making it less corrosion prone. If care is taken therefore in the choice of aluminium content and cooling rates the concentration and corrodible nature of the γ_2 phase can be minimised.

Although significantly less anodic than γ_2, the martensitic β phase is anodic both to the α phase and to the Fe_3Al particles and has been shown by Lorimer et al.[122] to be vulnerable to corrosion in sea water. They point out however that the Fe_3Al particles are not affected and are cathodic to the α phase since they would otherwise be severely attacked. For this reason, it will be seen that the complex alloys (Chapter 13) are preferable to copper-aluminium-iron alloys in sea water and have largely superseded them for marine service. See also below 'Effects of tin and nickel additions'.

A light and widespread rust 'staining' occasionally forms on iron-containing aluminium bronze components exposed to a corrosive atmosphere, such as a marine environment. If this rust staining is superficial, it may be due to the presence of precipitates and is likely to be of no consequence.[183] If, on the other hand, localised 'rust spots' form, which reveal the presence of large iron particles, caused by poor foundry melting techniques, they are likely to be corroded areas and have extremely harmful consequences. They have been found to initiate cavitation on impellers, propellers and other components and to lead to their early failure.

Applications involving contact with hydrochloric acid requires an aluminium bronze alloy free of iron because the formation of ferric chloride accelerates the corrosion rate.

Summary of the factors affecting the corrosion resistance of a Cu/Al/Fe alloy

1. The rate of cooling in sand and especially in die casting, results in the retention of some β-phase at an aluminium content as low as 8.5%. These alloys are therefore duplex alloys, containing the corrosion-prone martensitic β-phase.

2. The martensitic β-phase is potentially liable to transform to the more corrodible $\alpha+\gamma_2$ eutectoid. The higher the aluminium content the greater the risk of γ_2 forming as the cooling rate increases. However, the greater the iron additions up to 3%, the smaller the risk of γ_2 forming as the cooling rate increases.

Furthermore, as the iron content increases its beneficial influence becomes progressively independent of the aluminium content.

3. The grain-refining action of iron helps to disrupt the continuity of the γ_2 phase, making it less corrosion prone.
4. Alloys containing iron are unsuitable for use in contact with hydrochloric acid.

Effects of tin and nickel additions

Research in the USA has shown that susceptibility of α phase alloys to intergranular stress corrosion cracking in high pressure steam service can be eliminated by an addition of 0.25% tin (American specification UNS 61300).

Weill Couly has found that a 0.2% tin addition can also improve the corrosion resistance of a duplex alloy (with or without iron) and that of a copper-aluminium-iron alloy containing 2% nickel.[185]

Tin may however lead to weld cracking and, for this reason may have to be restricted to 0.1%, particularly if the weld metal is restrained from shrinking on cooling (see Chapter 7 – Ductility dip).

Copper-aluminium-iron alloys with small additions of nickel and manganese

Although copper-aluminium-iron alloys with up to 1% each of nickel and manganese are strictly speaking complex alloys, their metallurgy is very similar to that of ternary copper-aluminium-iron alloys. Manganese remains in solution and acts as an aluminium equivalent: 1% manganese being equivalent to 0.25% aluminium. A nickel content as low as 1% also remains in solution.

Copper-aluminium-iron alloys with high aluminium content

Copper-aluminium-iron alloys containing 12–14% aluminium have an exceptionally high hardness level and consequent low ductility; these properties are useful in special applications involving resistance to heavy wear and galling, provided the loading is wholly compressive. As these alloys contain little or no α phase, they are very susceptible to grain growth at high temperature. For this reason iron is invariably added, and this is supplemented occasionally with a special addition to improve further the grain refinement. In this connection, the grain refining effect of small addition of Titanium is disputed: Roucka et al.[154] observed no grain refinement effect on adding 0.5% Titanium. A small addition of nickel may also be made.

Microstructure

Roucka et al.[154] did some research on aluminium bronze alloys suitable for dies used in sheet drawing. They found that the alloy which had the required Brinell Hardness of 390–400 HN had the following wt % range of composition:

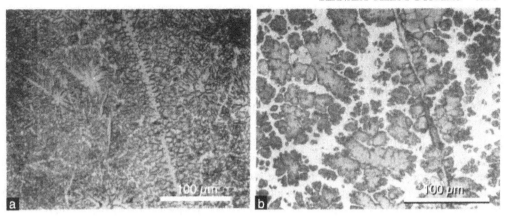

Fig. 12.9 High aluminium Cu–Al–Fe alloy cooled from 960°C by Roucka et al.[154]: (a) with 14.6% Al, cooled in air; (b) with 15.2% Al, cooled slowly.

Cu	Al	Fe	Ni	Mn
Bal	14.9–15.1%	3.3–3.5%	0.9–1.2%	~ 1%

Although the alloy contains a small amount of nickel and manganese, it resembles more closely a Cu–Al–Fe alloy than a complex alloy.

The structures may be understood from the 3% iron section of the Cu-Al-Fe diagram, Fig. 12.1a. If the alloy is cooled relatively fast in air from a high temperature (e.g. 960°C), the structure (see Fig. 12.9a) takes the form of loose aggregates of the acicular β' phase which appear throughout the structure. There may be some iron precipitation but probably little or no γ_2. The result is a very brittle alloy with a tensile strength of only ~83 N mm^{-2}. The brittleness of the alloy will not allow quenching in water or oil, and air cooling is the fastest practicable method of cooling.

Table 12.3 Analysis of typical structure of aluminium bronze alloy containing 15.2% aluminium, by Roucka et al.[154]

Phase	Composition wt%			
	Al	Fe	Ni	Mn
Field	18.7	1.5	1.0	1.3
γ_2 phase	18.4	2.3	0.7	0.8
$\alpha+\gamma_2$ eutectoid	14.3	0.5	0.15	2.0
Iron-rich precipitate Fe(δ)	16.3	24.5	1.5	4.4

If the alloy is cooled very slowly in a sand mould or in a furnace, the structure will consist of comparatively compact rosette-shaped particles of γ_2 surrounded by a continuous $\alpha+\gamma_2$ eutectoid (light areas on Fig. 12.9b), the volume of these two constituents being approximately equal. The details of the eutectoid cannot be seen distinctly even at high magnification. It contains iron-based precipitates that include aluminium and manganese. The γ_2 phase is rich in aluminium and contains some iron in solution. Its presence results in improved hardness and improved wear properties, especially at high loads and low speed and where high galling resistance is required (see Chapter 10). This structure also results in a significantly increased tensile strength of 140–155 N mm^{-2}. An analysis of the various phases of this structure is given in Table 12.3.

It should be noted that in all high aluminium alloys that have been slowly cooled, the γ_2 phase also occurs at the primary grain boundaries where it creates lines of weakness along which fracture would tend to occur.

If the aluminium content is increased, the γ_2 phase increases at the expense of the $\alpha+\gamma_2$ eutectoid and, when it reaches 15.6% or more, the eutectoid forms only envelopes around the γ_2 phase and does not provide sufficient strength to the alloy. It creates lines of weakness across the grain along which fracture is more likely to occur than along the γ_2 primary grain boundary. The alloy consequently becomes very brittle and trans-crystaline fracture predominates.

If, on the other hand, the aluminium content is reduced to 14.0–14.5%, the $\alpha+\gamma_2$ eutectoid becomes visible as a granular structure and, if further reduced to 12.7–13.5%, it becomes clearly lamellar.

Ion-plated aluminium bronze coatings on steel
As described in detail in Chapter 10, Sundquist et al.[170] experimented with ion-plated aluminium bronze coatings on steel, using an alloy of approximately 14% Al, 4.5 Fe, 1% Ni and balance Cu. Work-pieces of both carbon tool steel and of mild steel were coated with films of different thicknesses, as shown in Table 12.4.

The coatings were applied in layers of about 0.4 μm thickness by melting and evaporating only a small slug of metal at a time. The aluminium content of the coatings are given in Table 12.4. The nickel content of all the coatings was less than 1% and the iron content could not be reliably measured because of the proximity of the steel and the high iron content on the surface of the coatings.

Table 12.4   Details of ion-plated aluminium bronze coatings on steel by Sundquist et al.[170]

	Original aluminium bronze	Coating		
		A	B	C
Thickness (μm)		4.9	5.2	10
Coating time (min)		55	48	20
Aluminium content %	14	11.7	12.4	14.2
Micro-hardness (Hm K)	380	320	380	380
Microstructure		mainly martensitic β'	mainly γ_2	mainly γ_2

The micro-hardness figures of the coatings, using a Knoop indenter with a 25gf (~0.245N) load, are given in Table 12.4. Coatings B and C, with the high aluminium contents had the same hardness as the aluminium bronze used to supply the coating material.

The type of microstructure of the coatings are given in Table 12.4. The grain size was approximately 0.4 μm (similar to the thickness of the layers that made up the coatings). Coating A contained a high proportion of the martensitic β' phase which indicates that the long interval of time between the application of each layer of coating allowed time for the latter to cool before the next layer was applied with the result that successive layers were rapidly cooled, thereby retaining the β-phase. There was consequently only a small proportion of α and γ_2 in the structure. In the case of sample B, the faster deposition rate allowed less time for each layer to cool. Consequently the α + γ_2 eutectoid was formed although the rest of the structure was similar to that of the A coating. This effect was even more pronounced in the case of coating C which had a much faster deposition rate combined with greater coating thickness. As a result, there was no β' phase in this coating and the structure approached that of the equilibrium condition. Surprisingly, there was little difference in hardness between coating B and C and the longer sliding distance of the latter, prior to full penetration, was due to its greater thickness.

The use of a high-aluminium aluminium bronze-coated die in metal forming instead of a solid aluminium bronze insert of the same composition offers the advantage that it partly overcomes the problem of the brittleness of the high aluminium alloy. The tough steel to which the coating is applied gives resilience to the coated die.

Summary of effects of high aluminium content in Cu/Al/Fe alloys
1. Alloys containing 12–14% aluminium have exceptionally high hardness but low ductility. At higher aluminium content, the strength of the alloy reduces.
2. They have excellent wear and galling resistance.
3. The presence of iron reduces the grain growth
4. Slow cooling in sand or in a furnace produces the best combination of hardness and tensile strength.
5. Steel-forming dies, ion-plated with a high-aluminium Cu/Al/Fe alloy, offer an ideal combination of a hard surface and a strong and resilient sub-structure.

Standard copper–aluminium–iron alloys

The only common copper-aluminium-iron cast alloy is CuAl10Fe3 which is used mainly in die casting. Although it may contain as much as 1% each of nickel and manganese it is effectively a ternary alloy with a duplex α+β structure. The corresponding wrought alloy is CuAl8Fe3 which is a single α-phase alloy. It too may contain less than 1% each of nickel and manganese. An American version of this alloy, C61300, contains 0.25% tin to eliminate susceptibility of this α phase alloy to intergranular stress corrosion cracking in high pressure steam service.

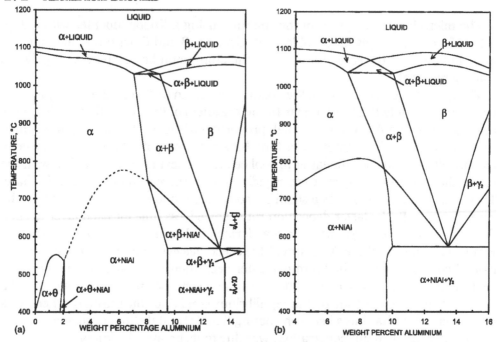

Fig. 12.10 Vertical section of the Cu–Al–Ni system at 3% and 6% Ni by Alexander.[5]: (a) 3% Nickel; (b) 6% Nickel.

The copper-aluminium-nickel system

Effects of nickel

There are no ternary copper–aluminium–nickel alloys in common use because nickel is almost always associated with iron. The only known copper–aluminium–nickel alloy, is the low aluminium content alloy CuAl7Ni2 which cools as an all-α phase, although particles of NiAl will precipitate under slower cooling.[5] A study of the copper–aluminium–nickel system is nevertheless of interest to show the effect of nickel as an allowing element. Its principal effect is to improve the corrosion resistance of aluminium bronzes. In conjunction with iron, it improves tensile strength and especially proof strength. It also improves hardness and therefore resistance to erosion, but lowers elongation.

In single α-phase alloys, minor additions of nickel probably have a slightly beneficial influence, especially in improving the resistance to erosion by high velocity water-flow. Nickel contents higher than 2% are known to give good, though not necessarily better results and the main reason for nickel additions to single α-phase alloys is to improve mechanical properties.

Equilibrium diagram

Fig. 12.10 shows the copper-aluminium-nickel equilibrium diagram at 3% and 6% nickel. This shows that, on slow cooling, a precipitate of nickel aluminate NiAl is

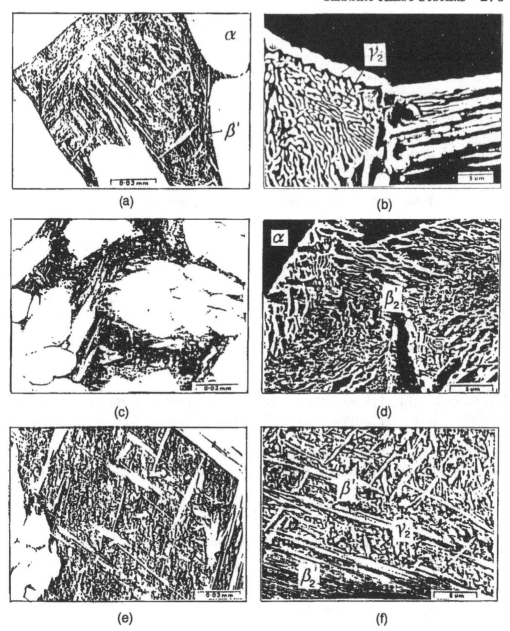

Fig. 12.11 As-cast microstructure of Cu–Al–Ni alloys of various Al and Ni contents: (a) Optical micrograph and (b) SEM micrograph of as-cast alloy 1. Alloy 2, similar structure; (c) Optical micrograph and (d) SEM micrograph of as-cast alloy 4. Alloy 3, similar structure; (e) Optical micrograph and (f) SEM micrograph of as-cast alloy 5—See Table 12.5 for alloy compositions, by Sun et al.[194]

formed in all phases up to about 13% aluminium.[5] Under fairly rapid cooling however, this may not occur.

Phase θ, shown at very low aluminium content consists of the compound Ni_3 Al.

Microstructure of copper-aluminium-nickel alloys

Sun et al.[194] carried out research into the microstructure of five Cu/Al/Ni alloys of compositions given in Table 12.5. They reported on the as-cast structure of these alloys and on the development of their structure as they cooled continuously from 1020°C at about 5°K min^{-1}.

Table 12.5   Composition of alloys investigated by Sun et al.[194]

Element	Alloy				
	1	2	3	4	5
Cu	90.7	90.0	87.9	86.4	87.6
Al	9.2	9.2	9.2	9.1	9.7
Ni	0.1	0.8	2.9	4.5	2.7

As-cast structure

The as-cast structures of Alloys 1 to 5 are shown in Figs 12.11a to f.

Alloys 1 and 2

Figs 12.11a and 12.11b show the as-cast microstructure (at different magnifications) of Alloys 1 and 2, containing just over 9% Al and less than 1% Ni. It consists of:

- dark etching regions of α phase (top of Fig. 12.11b),
- a fine lamellar mixture of $\alpha+\gamma_2$ forming an eutectoid (left hand side of Fig. 12.11b),
- a band of light etching γ_2 phase along the former α/β boundaries (Fig. 12.11b) and
- a martensitic β phase, sometimes designated β', (right hand side of Fig. 12.11b).

This means that, at less than 2.5% Ni, the structure is similar to a very slowly cooled binary Cu/Al alloy of similar Al content (see Fig. 11.9d).

Alloys 3 and 4

Figs 12.11c and 12.11d show the as-cast microstructure of Alloys 3 and 4 containing just over 9% Al and 2.9%–4.5% Ni. It consists of:

- a dark etching α phase (Fig. 12.11d) – light etched in Fig. 12.11c – and
- a lamellar $\alpha+NiAl$ eutectoid (Fig. 12.11d) – NiAl is here designated β'_2. There is no martensitic β phase.

Alloy 5

Figs 12.11e and f show the effect on the microstructure of increasing the aluminium content to 9.7% with a Ni content of 2.7%, as in the case of Alloy 5 by comparison with Alloy 3. It consists of:

- a light etching α phase (Fig. 12.11e),
- darker etching α+NiAl and α+γ₂ eutectoids (Fig. 12.11e) together with martensitic β (designated β') needles. The small NiAl particles (designated β'₂) are located mainly on or near the former α/β boundaries, whereas the much larger γ₂ particles (0.5 to 1 μm in size) are within the former β regions.

Comparison of as-cast structures

Comparing the microstructure of Alloys 1 and 2 with that of Alloys 3 and 4, it will be seen that, at 9.1%–9.2% Al, the effect of increasing the nickel content from 0.1%–0.8% to 2.9%–4.5% is to eliminate both the γ₂ and martensitic β phases. As explained below, this is very important from the point of view of corrosion resistance.

Comparing the microstructures of Alloys 3 and 5 which have similar nickel contents (2.9% and 2.7% respectively), it will be seen that, increasing the aluminium content from 9.2% to 9.7%, has the effect of reintroducing both the γ₂ phase and the martensitic β phase in the as-cast microstructure.

Development of structure

Alloys 1 and 2

These two alloys had virtually the same alloy development.

- Quenched at 1020°C, both alloys consisted of the β phase (see Fig. 12.12a). It will be noted that, according to the equilibrium diagram (Fig. 11.4), these two alloys with 9.2% Al should be in the α+β field at 1020°C and not in the single phase β field. This applies also to Alloys 3 and 4 below, which have likewise solidified into the β phase (see Fig 12.10a). This indicates that the boundary between these two fields should be slightly more to the left in the equilibrium diagram.
- Quenched at 1000°C, the α phase had started to precipitate (see Fig. 12.12b) and continued to precipitate as the temperature decreased. There were no significant change in microstructure until the temperature reached 520°C.
- Quenched at 520°C, the β phase began to decompose into the α+γ₂ phase (see Figs 12.12c and d). It will be noted that this transformation occurred at a significantly lower temperature than at the higher nickel content of 3% shown in the equilibrium diagram (Fig. 12.10a). This is evident also in the case of Alloy 3 below. It can be seen from Fig. 12.12d that γ₂ formed initially at the α/β boundaries before developing into the α+γ₂ eutectoid.
- Quenched at 400°C (Fig 12.10e), the microstructure was similar to the as-cast structure.

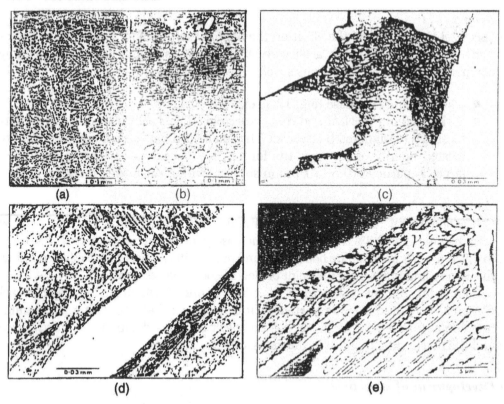

Fig. 12.12 Development of microstructure of Alloy 1: 9.2% Al, 0.1% Ni; Alloy 2: 9.2% Al, 0.8% Ni, similar structure: (a) Quenched at 1020°; (b) Quenched at 1000°C; (c) Quenched at 520°C; (d) Quenched at 520°C (SEM micrograph); (e) Quenched at 400°C, by Sun et al.[194]

Alloys 3 and 4

- Quenched at 1020°C, both alloys were in the β phase (see Fig. 12.13a), as in the case of the previous alloys.
- Quenched at 1000°C, the α phase began to form at the β phase boundaries (see Fig. 12.13b).
- Quenched at about 700°C (Alloy 3) and 790°C (Alloy 4), the β phase began to decompose into the α+NiAl eutectoid at the α/β boundaries (see Figs 12.13c and d). It will be noted that, as mentioned above, the higher the nickel content, the higher the temperature at which this decomposition begins to occur. This can be seen also by comparing Fig. 12.10a with Fig. 12.10b. In the case of Alloy 4, NiAl particles, designated $β'_2$ by Sun et al[194], began to precipitate at about the same time in the α phase. In the case of Alloy 3, however, they did not begin to precipitate until the much lower temperature of about 550°C.
- As the quenching temperature was further lowered, the α+NiAl eutectoid reaction products grew into the former β phase.

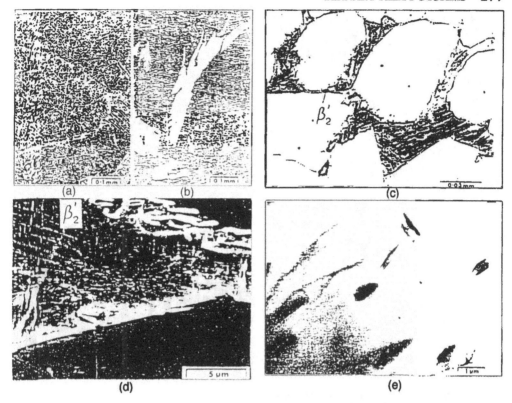

Fig. 12.13 Development of microstructure of Alloy 4: 9.1% Al, 4.5% Ni; Alloy 3: 9.2% Al, 2.9% Ni, similar structure. (a) Quenched at 1020°C; (b) Quenched at 1000°C; (c) Quenched at 790°C; (d) Quenched at 790°C (SEM micrograph; (e) Quenched at 600°C, by Sun et al.[194]

- Between 700°C and 500°C, more NiAl particles precipitated in the α phase (see Fig. 12.13e).
- Quenched at 520°C, the specimen had a microstructure similar to the as-cast condition (see Fig. 12.11c).

Alloy 5
This alloy has a significantly higher aluminium content (9.7%) than the above alloys and this has a marked impact on its microstructural development.

- Quenched at 990°C, the alloy was in the β phase (see Fig. 12.14a).
- Quenched at 830°C, α nucleated at the β boundaries (see Fig. 12.14b). The α phase went on growing as the temperature fell and there were no appreciable changes in microstructure until the temperature reached 640°C.
- Quenched at about 640°C, the eutectoid decomposition of β into α+NiAl began at the α/β boundaries (see Fig. 12.14c).

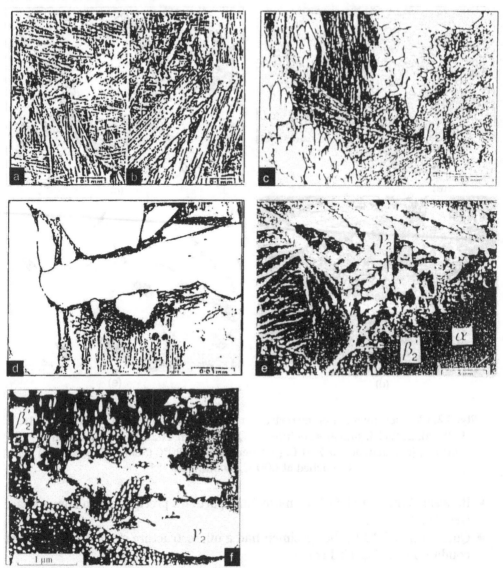

Fig. 12.14   Development of microstructure of Alloy 5: 9.7% Al, 2.7% Ni. (a) Quenched at 990°C; (b) Quenched at 830°C; (c) Quenched at 640°C; (d) Quenched at 540°C; (e) Quenched at 540°C (SEM micrograph); (f) Quenched at 400°C (TEM micrograph), by Sun et al.[194]

- Quenched at 540°C, another eutectoid reaction began to occur as β decomposed into α+γ$_2$ (see Fig. 12.14d and e). The NiAl particles (designated β'$_2$) followed the lines of the previous β boundaries, whereas γ$_2$ was within the β region (see Fig. 12.14e and f).
- Quenched at 400°C, the specimen had a structure similar to the as-cast structure.

Decomposition of the β phase

It was previously thought that nickel tends to stabilise the β phase. This does not seem, however, to apply to the decomposition of the β phase into α+NiAl in the ternary alloy which occurs over a much wider range of temperatures than previously thought.

The temperature at which β decomposes into α+γ$_2$ does not seem to be sensitive to either Ni or Al concentrations, as may be seen in the case of Alloys 1, 2 and 5 where the decomposition starts at similar temperatures.

Composition of phases

The chemical compositions of the various phases of the copper–aluminium–nickel system, carried out by different researchers, are given in Table 12.6. It will be seen that there is similarity in the figures obtained by the various researchers. The composition of the various phases does not seem to be affected by the aluminium and nickel contents of the alloy.

Table 12.6 Chemical composition of equilibrium phases in Cu–Al–Ni system.

Phase	Reference	wt% Al and Ni in alloy content (if known)			%Cu	%Al	%Ni
		Sample	Al	Ni			
α	Brezina[33] Ref 14		10.5	5.0	90	8.5	1.5
	Ref 33				88	10	2
	Ref 34				89.5	8	2.5
Matrix α	Sun et al.[194]	1	9.2	0.1	90.9±0.6	9.1±0.6	–
		4	9.1	4.5	87.8±0.6	8.4±0.6	3.8±0.3
		5	9.7	2.7	88.1±0.7	9.2±0.4	2.7±0.3
Eutectoidal α		1	9.2	0.1	90.4±0.3	9.6±0.3	–
		4	9.1	4.5	91.2±0.4	7.0±0.4	1.8±0.2
		5	9.7	2.7	90.9±0.5	7.2±0.4	1.9±0.4
Martensitic β	Sun et al[194]	1	9.2	0.1	89.2±0.4	10.8±0.4	–
		5	9.7	2.7	85.6±0.6	11.3±0.5	3.1±0.4
γ$_2$	Brezina[33] Ref 14		10.5	5.0	79	14.5	3.5
	Ref 33				72	18.8	9.2
	Ref 34				76	18	6
	Sun et al.[194]	1	9.2	0.1	83.5±0.5	16.5±0.5	–
		5	9.7	2.7	82.0±0.4	14.6±0.5	3.4±0.7
NiAl (designated β'$_2$ by Sun et al.[194])	Brezina[33] Ref 14		10.5	5.0	18	27	55
	Ref 33				21.3	28.5	50.2
	Ref 34				25	29	46
	Sun et al.[194]	4	9.1	4.5	27.2±0.8	26.3±1.0	46.5±0.6
		5	9.7	2.7	22.2±0.6	26.7±0.6	51.1±0.6

Effects of tempering

Sun and al[194] observed the following effects of tempering the above alloys:

Tempering as-cast Alloy 1 at 540°C for 7 hours, resulted in the elimination of martensitic β and in the absorption of $\alpha+\gamma_2$ eutectoid into the α matrix. It also resulted in a coarsening of the structure with tempering time.

Tempering as-cast Alloy 2 at 540°C for 24 hours resulted in the elimination of martensitic β but did not otherwise change the structure significantly. There was little evidence of absorption of the $\alpha+\gamma_2$ eutectoid and only minor coarsening of the structure.

Tempering as-cast Alloys 3 and 4 at 680°C and at 750°C for between 1 and 48 hours resulted mainly in an increase of NiAl precipitates in the α matrix with a higher density of these precipitates in Alloy 4 than in Alloy 3. There was no obvious increase in coarsening of the structure with tempering time. The density of precipitates in the α matrix was higher after tempering at 680°C than after tempering at 750°C.

Tempering the as-cast Alloy 5 at 540°C resulted in the elimination of martensitic β and in an increase in the density of the NiAl precipitates in the prior eutectoid α. The distribution of NiAl and γ_2 phases was similar to that of the as-cast condition, with NiAl mainly at the α/β boundaries and γ_2 inside the prior β region. The density of NiAl precipitates in the α phase increased with tempering time. No overall coarsening of the structure was observed.

Effect of nickel on corrosion resistance

Brezina[33] states that nickel shifts the region of transformation of β to $\alpha+\gamma_2$ to higher aluminium contents. This does not appear to be the case in ternary alloys if one compares the Cu–Al–Ni diagrams (Figs 12.10 a and b) with the binary Cu–Al diagram (Fig. 11.4) or with the Cu–Al–Fe diagrams (Fig. 12.1 a and b). It is only the combination of nickel and iron that significantly moves the region of transformation of β to $\alpha+\gamma_2$ to higher aluminium contents, as will be seen in Chapter 13 (see Fig. 13.1a). The effect of nickel in reducing the likelihood of this transformation in the ternary Cu–Al–Ni alloys would therefore appear to be due to some other reason. One possible explanation is that nickel stabilises the β phase against transforming to $\alpha+\gamma_2$ even though, as mentioned above, it does not seem to stabilise β against transforming to $\alpha+$NiAl.

In the case of the less highly alloyed copper–aluminium alloys, nickel significantly reduces the risk of decomposition into the corrosive prone $\alpha+\gamma_2$ eutectoid provided, as explained in Chapter 13, the relationship of aluminium to nickel content is in accordance with the following formula by Weill-Couly and Arnaud:[183]

$$Al \leq 8.2 + Ni/2$$

The effect of aluminium content and cooling rate on the formation of a semi-continuous γ_2 phase for different percentage nickel additions, is shown on Fig.

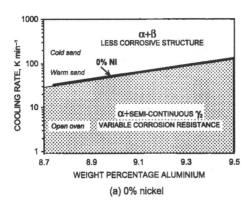

(a) 0% nickel

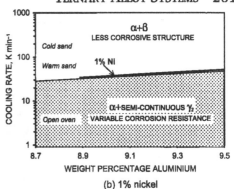

(b) 1% nickel

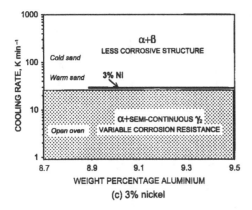

(c) 3% nickel

Fig. 12.15 Effect of aluminium content and cooling rate on the formation of the corrosion-prone γ_2 phase at various nickel additions.[127]. (a) 0% Nickel; (b) 1% Nickel; (c) 3% Nickel.

12.15. It will be noted however that an $\alpha+\beta$ structure is retained, even at very low cooling rates, when 2% nickel is present. As previously mentioned in Chapter 11, the β phase is anodic to the α phase and is therefore vulnerable to de-aluminification in a duplex ($\alpha+\beta$) binary alloy. As explained in Chapter 13, however, the presence of nickel provides protection from this form of corrosion, provided the rate of cooling from the β field is not too great and that the above relationship of aluminium to nickel applies (it would seem that the NiAl precipitate of a copper-aluminium-nickel alloy fulfils the same protective function as the κ_3 precipitate of complex alloys).

It should be noted that, at the minimum nickel content allowed by some standard specifications, the maximum aluminium content allowed by the specification may be higher than the maximum corrosion-safe aluminium content given by the above formula.

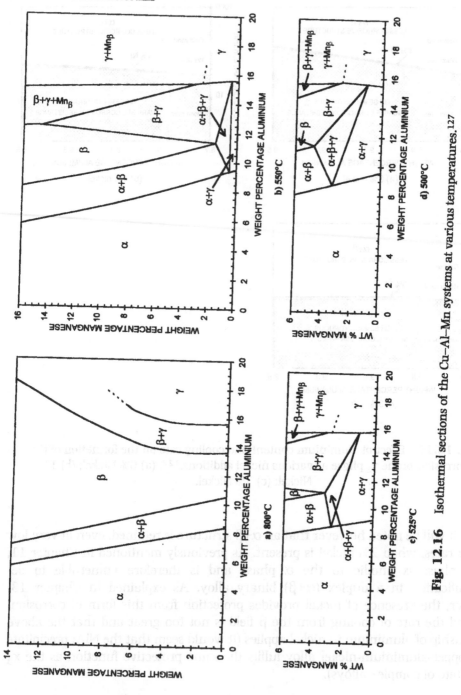

Fig. 12.16 Isothermal sections of the Cu–Al–Mn systems at various temperatures.[127]

The copper-aluminium-manganese system

Effects of manganese

Small addition of manganese are made to a number of aluminium bronzes to improve fluidity in castings. Some consider that the mechanical properties are improved by additions of up to 2% and that the proof strength or general toughness of the alloy is improved. Higher manganese contents of 11–14% are used in complex alloys in association with iron and nickel additions (see Chapter 14). Manganese is seldom used without other alloying elements.

The influence of small additions of manganese on the structure is not marked. With larger additions, however, Fig. 12.16 reveals that there is a significant increase in the proportion of β for any alloy of given aluminium content, 1% manganese being, according to Edwards and Whitaker,[69] equivalent to about 0.25% aluminium. It will be noted that manganese is soluble in all phases except at aluminium contents approaching 14%. It does not therefore appear as such in the microstructure of common alloys. At 650°C, the solubility of manganese is around 8% in the α phase and 26% in the β phase, the latter increasing sharply with temperature[112].

Manganese stabilises the β phase which means that it reduces the risk of its decomposition to the harmful $\alpha+\gamma_2$ eutectoid, but, by the same token, it retards the decomposition of the corrosion-prone β phase into α. The effect of even low manganese contents in reducing the risk of the $\alpha+\gamma_2$ eutectoid being formed is shown in Fig. 12.17.

Alloys containing even small additions of manganese are however more susceptible to corrosion in sea water under conditions of limited oxygen availability, such as at crevices or under deposits. Fig. 12.18 shows that the rate of penetration at these shielded areas is directly proportional to the manganese content (up to 2%). It follows therefore that manganese additions should be kept to the absolute minimum required to ensure the fluidity necessary to produce sound castings. The limits imposed on manganese content in official specifications is aimed at striking a compromise between the requirement of the foundry and reducing susceptibility to crevice corrosion.

Standard copper-aluminium-manganese alloys

There is only one commonly known ternary copper-aluminium-manganese alloy: it is the wrought duplex α/β structured German alloy CuAl9Mn2.

The copper-aluminium-silicon system

The copper-aluminium-silicon equilibrium diagram

Fig. 12.19 shows the 2% silicon section of the ternary copper–aluminium–silicon diagram. The phase boundaries are similar to those in the binary copper–

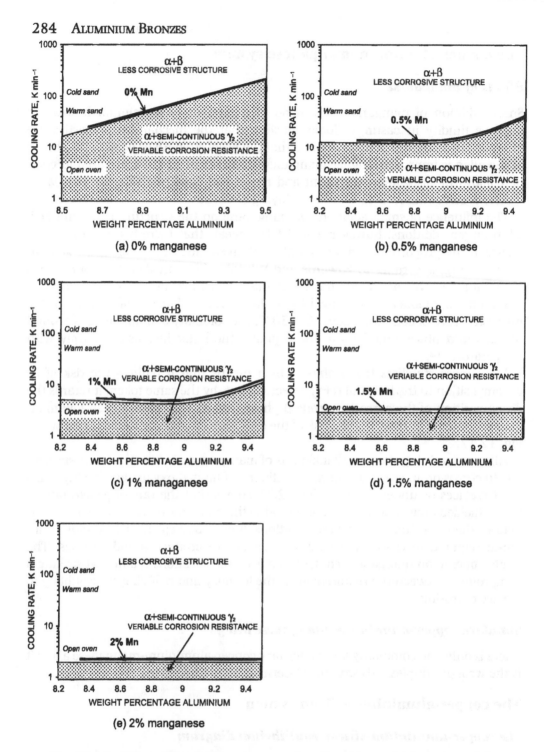

Fig. 12.17 Effect of aluminium content and cooling rate on the formation of the corrosion-prone γ_2 phase at various manganese additions.[127] (a) 0% Manganese, (b) 0.5% Manganese; (c) 1% Manganese; (d) 1.5% Manganese; (e) 2% Manganese.

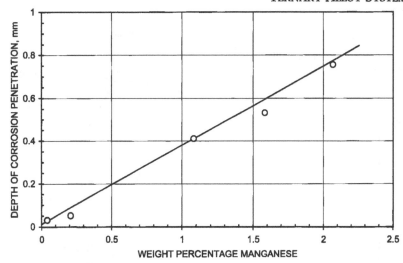

Fig. 12.18   Effect of manganese on crevice corrosion of shielded areas.[127]

aluminium system with the exception that they have been moved a distance equivalent to about 3% lower aluminium. This is because, due to its tendency to β formation, silicon has similar effects to aluminium, 1% silicon being equivalent to about 1.6% aluminium. Hence when silicon is added to an alloy of given aluminium content, the tensile strength and proof strength are raised with a marked drop in elongation. Hardness also increases. If it is desired to add silicon intentionally, the aluminium content should therefore be lowered at the same time.

The standard American copper-aluminium-silicon alloy ASTM 956, has a range of aluminium content of 6% to 8%, whereas the British Standard alloy AB3 has a narrower band of 6% to 6.4% aluminium. ASTM 956 makes no mention of iron, whereas AB3 specifies and iron content of 0.5% to 0.7%. A small quantity of iron refines the grain and the top limit ensures the good magnetic permeability for which this alloy is mostly used.

As may be seen from Fig. 12.19, these alloys have a short freezing range of around 1010–980°C and solidify into an α+β binary structure.

Iqbal, Hasan and Lorimer[100] have investigated the slow cooling (5°K min⁻¹) of a BS1400 AB3 alloy from the molten state by quenching a number of specimens of this alloy in succession from various temperatures with the following findings (see Fig. 12.20a-f). The nature and composition of the various phases are given below:

- Quenched at 965°C, the alloy has a structure as shown in Fig. 12.20a–b consisting of large β grains of an 'acicular' or needle-like structure known as a martensitic structure. There is a small amount of α phase at the grain boundary. The transformation of the β phase to a martensitic structure is brought about by quenching.
- Between 965°C and 940°C the α grains grow progressively in size and small irregularly shaped sparsely distributed $\kappa(Si)_I$ particles, based on Fe_5Si_3, begin

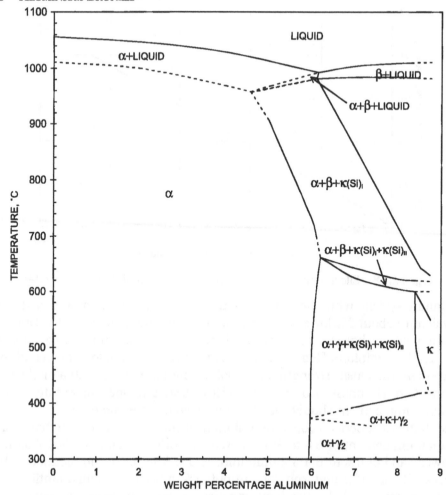

Fig. 12.19 Vertical section of the Cu–Al–Si system at 2% Si.[127]

to precipitate in both the α and β grains as the temperature approaches 940°C. It is thought that they may be present in the melt. Fig. 12.20c shows the structure on quenching at 940°C. The proportion of martensitic β is much reduced whereas that of the α phase is significantly increased. The $\kappa(Si)_I$ particles are sparsely dispersed in the α and β phases, although difficult to discern in the latter.

- Between 940°C and 650°C, the α grains continue to grow, more $\kappa(Si)_I$ particles precipitate in the α and β grains and, as the temperature nears 790°C, a high density mass of fine $\kappa(Si)_{II}$ precipitate, based on Fe_3Si_2, appears at the centre of the α grains, leaving a precipitate-free zone near the grain boundary. Fig. 12.20d shows the structure on quenching at 790°C.
- Between 650°C and 545°C, depending on aluminium content, two new phases appear:

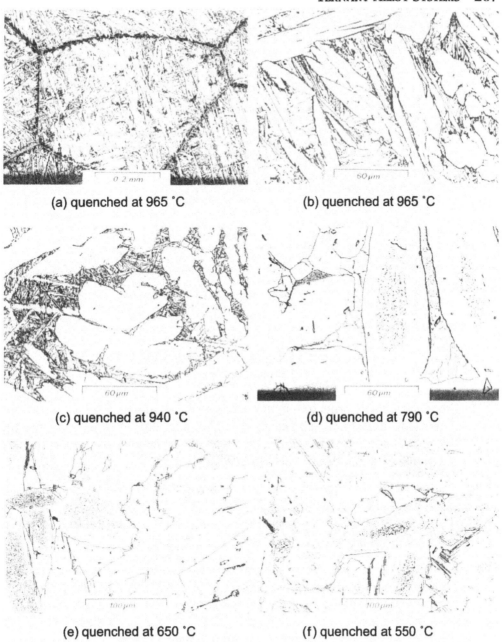

(a) quenched at 965 ˚C

(b) quenched at 965 ˚C

(c) quenched at 940 ˚C

(d) quenched at 790 ˚C

(e) quenched at 650 ˚C

(f) quenched at 550 ˚C

Fig. 12.20   Microstructure of Silicon-Aluminium Bronze quenched at various temperature as it cools slowly from 965°C. (a) Quenched at 965°C; (b) Quenched at 965°C; (c) Quenched at 940°C; (d) Quenched at 790°C; (e) Quenched at 650°C; (f) Quenched at 550°C, by Iqbal et al.[100]

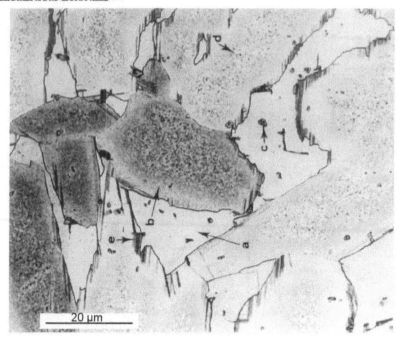

Fig. 12.21   As-cast silicon-aluminium bronze with 6.1% Al and 2.3% Si, by
Lorimer et al.[122]

(1) the β phase transforms to a light etching γ phase and
(2) a new twin-like structure appears at the α/γ boundary which grows into the
α grains. It consists of parallel sided plates of α and γ and goes on increasing
in volume as temperature falls below 650°C.

Fig. 12.20e shows the structure on quenching at 650°C. The γ phase does not
undergo a martensitic transformation on quenching as did the β phase. This may
be due to the fact that the cooling rate down to 650°C was slow enough for the α
phase to separate out from the β phase prior to quenching.

- At 550°C (Fig. 12.20f), the structure is similar to the as-cast structure (Fig.
  12.21) and consists of the light etching γ phase and the darker α phase. Some
  α grains appear light grey and some dark grey due to the different orientations
  of the grains reflecting light differently. Both types of precipitates, $\kappa(Si)_I$ and
  $\kappa(Si)_{II}$, are visible. The lamellar α/γ structure is also visible.
- If the rate of cooling below 400°C is much slower than that experienced in
  castings, the γ phase might transform to $\alpha+\gamma_2$, although it seems very reluc-
  tant to do so.

Nature of phases
The chemical composition of the following phases is given in Table 12.7.

The α phase
The α phase is a copper-rich solid solution which has a face-centred-cubic (fcc) space lattice arrangement.

The β phase
The β phase is copper-rich solid solution which transforms to a martensitic structure on quenching at high temperature (above 650°C). This martensitic structure is an unstable form of α in β.

The γ phase
The γ phase, which forms from β at about 650°C, is a copper-rich phase with a higher silicon content than α. It has a hexagonal close-packed (hcp) space lattice arrangement, as revealed by electron diffraction.[100]

The α/γ lamellar structure
The α/γ lamellar structure consists of plates of fcc α and hcp γ and it forms at about 650°C at the boundary of the α phase. It is thought to be the last to form from β and the γ plates within it are consequently richer in aluminium and silicon (it may be seen from Fig. 12.19, that the solubility of aluminium and silicon in α increases as the temperature falls to 650°C).

The γ_2 phase
Little is known about the γ_2 phase because it is seldom, if ever, found in this alloy as it is formed at much slower cooling rates than those experienced in castings. It forms, if at all, below 400°C and is part of the brittle α+γ_2 eutectoid.

Second phase κ(Si) particles
There are two types of intermetallic κ particles, designated here as κ(Si)_I and κ(Si)_II, to avoid confusion with the totally different κ phases of the Cu–Al–Ni–Fe system

Table 12.7   Composition of a Silicon-Aluminium Bronze alloy and of its phases.[100]

Phase	Technique	Composition wt%				
		Al	Si	Mn	Fe	Cu
Alloy Composition		6.04	2.32	0.06	0.6	90.98
α	Bulk microprobe	5.6±0.9	2.2±0.4	–	–	92.2±1.1
γ	Thin foil	5.7±0.1	2.7±0.1	–	0.9±0.6	90.7±0.5
α/γ lamellar structure	Thin foil	7.1±0.1	3.8±0.2	–	0.9±0.2	88.2±0.7
κ(Si)_I (Fe_5Si_3)	Thin foil	1.5±1.3	17.4±1.6	1.1±1.6	58.8±12.4	21.2±13.2
κ(Si)_II (Fe_3Si_2)	Extraction replica	1.2±0.3	23.1±2.9	2.1±0.4	72±3.4	1.5±1

(see Chapter 13). Although the silicon content of silicon–aluminium bronze is only a third of the aluminium content, it is sufficient to replace the Fe–Al based inter-metallic precipitate of the copper–aluminium–iron system by Fe–Si based precipitates. This shows that there is a stronger affinity between iron and silicon than between iron and aluminium. Since the iron content is less than 1% (0.6% in the above experiment), the occurrence of these precipitates indicate that silicon reduces the solubility of iron in both α and β by comparison with copper–aluminium–iron and copper–aluminium–nickel–iron alloys.

Bradley[29] reports that the Fe_5Si_3 precipitate lowers the magnetic permeability of the alloy but makes no reference to Fe_3Si_2. The effect of both forms of Fe–Si precipitates on magnetic properties needs to be investigated.

The presence of these intermetallic precipitates is thought to be responsible for the good machining properties of silicon-aluminium bronze.

- The $\kappa(Si)_I$ particles precipitate at very high temperatures as relatively large, irregular shaped inter-metallic particles in both the α and β grains. Thin specimen microprobe analyses showed that they were based on Fe_5Si_3. It is thought that they may be formed in the melt although they do not act as nuclei in the way that some Fe_3Al precipitates of the Cu—Al–Fe system may do (see above). Transmission electron microscopy revealed that these particles formed groups of lathes and that they are approximately 40–50μm in length. They have been found to have a hexagonal closed-packed (hcp) space lattice arrangement. [100]
- The $\kappa(Si)_{II}$ particles precipitate as a high density mass of fine particles at the centre of the α phase as the temperature nears 790°C on cooling. Transmission electron microscopy indicates that these fine particles are lath-like in shape and of an average length of approximately 5μm. Microprobe analysis revealed that they are based on Fe_3Si_2. They are reported to have a bcc-type (B2) structure.[100]

In the case of silicon–aluminium bronze, iron is not effective as a grain refiner.

Resistance to corrosion

The good resistance to corrosion of cast silicon-aluminium bronze is thought to be due the absence of the martensitic β phase[100] which, as in the case of the nickel–aluminium bronze, is more susceptible to corrosion (see Chapter 13). The γ phase, to which the β phase transforms at 650°C, has good corrosion resisting properties and undergoes only limited preferential corrosive attack, although it is slightly more vulnerable in the lamellar α/γ structure.[122] This is understandable since, as explained above, the γ constituent of this structure is thought to have a higher aluminium/silicon content than the γ constituent of the $\alpha+\gamma$ phase, which makes it more anodic. It also shows that corrosion rates are highest when the anodic areas are small relative to the cathodic areas. The very corrodible γ_2 phase is, as mentioned above, extremely unlikely to appear in a casting due to the very slow cooling rate required for it to form.

Summary of the characteristics of a Cu–Al–Si alloy

1. The main advantages of this alloy are its low magnetic permeability, its good machining properties and good corrosion resistance.
2. 1% silicon is equivalent to about 1.6% aluminium in its effect on the microstructure and mechanical properties.
3. In the as-cast condition, the microstructure consists of an $\alpha+\gamma$ phase with two types of precipitates: $\kappa(Si)_I$ and $\kappa(Si)_{II}$. A lamellar α/γ structure also appears at the boundary of the α phase.
4. The γ phase is the result of the decomposition of a high temperature β phase. Its structure is not however martensitic. It is unlikely to transform to the corrodible $\alpha+\gamma_2$ eutectoid, even at very slow cooling rate. It has good corrosion–resisting properties and undergoes only limited preferential corrosive attack, although it is slightly more vulnerable in the lamellar α/γ structure.
5. The two types of intermetallic iron-rich κ particles, designated as $\kappa(Si)_I$ and $\kappa(Si)_{II}$, are thought be responsible for the good machining properties of this alloy. They do not appear to have an effect on corrosion resistance.
6. A minimum of 0.5% of iron is required for its grain refining effect but must not exceed 0.7% for good magnetic permeability properties.
7. The $\kappa(Si)_I$ precipitate, which is richer in iron than $\kappa(Si)_{II}$, lowers the magnetic permeability of the alloy.

The copper–aluminium–beryllium system

Copper–aluminium–beryllium alloys were developed for cavitation-resistant weld claddings.[33] No sections of the equilibrium diagram are available for this system.

The solubility of beryllium at 600°C is as follows.[186]
- 1% in the α phase (very dependent on aluminium content)
- 2.5% in the β phase.

At aluminium content of less than approximately 7%, the γ-Be phase is formed which is known from the binary Cu–Be system. At higher aluminium content, the β phase is stabilised. The chemical composition of the κ precipitates is not known.

The copper-aluminium-tin system

Copper–aluminium–tin alloys were developed as tarnishing resistant alloys for architectural purposes. They have been found however to have good corrosion resistance in various media, coupled with good mechanical properties.[33]

As previously mentioned, small additions of tin to copper–aluminium–iron alloys can improve the corrosion resistance of the binary alloys and eliminate intergranular stress corrosion cracking in the single phase alloys.

Table 12.8   Structures at 700°C of a Cu–Al–Sn alloy.[84]

Structure at 700°C	Composition wt%	
	Al	Sn
α	5	5
α+β	7	5
α+β	7	7
β	9	7

Tin is as soluble as aluminium in the α phase and tends to stabilise the β phase. According to Habraken et al.,[84] the structures given in Table 12.8 occur at 700°C:

In a commercial alloy containing 7%Al and 5%Sn, the following concentrations were measured which show tin slightly enriched in the β phase:

Structure	Composition wt%	
	Al	Sn
α	7.2	2.4
β	8.2	6.9

The copper-aluminium-cobalt system

Aluminium bronzes with cobalt additions have excellent corrosion resistance in sea water, and are specially wear-resistant.[33] Cobalt has a similar effect to that of iron. The cobalt-rich phase is designated by the letter C. Based on measurements carried out on an alloy containing 13%Al, 2%Co, 3%Fe and 3%Mn, there is evidence that a κ_2 phase is present.

13
COPPER–ALUMINIUM–NICKEL–IRON SYSTEM

Nickel-aluminium bronzes

Alloys containing 9–12% aluminium with additions of up to 6% each of iron and nickel represent a most important group of commercial aluminium bronzes. The common alloys, which normally contain 3–6% each of these two elements, have been fully investigated in view of their excellent combination of mechanical and other properties. They often contain small additions of manganese.

The main alloys of this type are: **CuAl10Fe5Ni5**, **CuAl10Ni5Fe4**, **CuAl11Ni6Fe5** and **CuAl9Ni5Fe4Mn**. Although they contain varying proportions of alloying elements, they have similar structures which would be difficult to distinguish metallographically from one another. Particulars of these alloys and their standard specifications are given in Chapters 3 and 5.

Because of the volume of information on this important alloy system, this chapter has been divided, for clarity, into the following main sections:

Section A: Microstructure of copper–aluminium–nickel–iron alloys
Section B: Resistance to Corrosion
Section C: Effects of welding
Section D: Effects of hot and cold working – Heat treatment
Section E: Wear resistance

A – Microstructure of copper–aluminium–nickel–iron alloys

The copper–aluminium–nickel–iron equilibrium diagrams

The equilibrium diagram of Cu–Al–Ni–Fe alloys containing 5% each of iron and nickel is shown in Fig. 13.1 from the work of Cook et al.,[47] together with a section of the binary copper–aluminium diagram for comparison. The diagram is similar to the binary system but with some significant differences. The nature and appearance of the various phases will be discussed later. The following is the sequence of phase transformation for an alloy of nominally CuAl10Fe5Ni5 composition:

According to Feest and Cook,[70] two types of iron-rich kappa intermetallic particles begin to precipitate at pre-solidification temperatures. The particles designated 'type-1 pre-primary κ_1 phase' precipitate at a higher temperature than those designated 'type-2 pre-primary κ_1 phase'. In each case, this temperature depends on the iron content of the alloy: the higher the iron content, the higher the temperature at

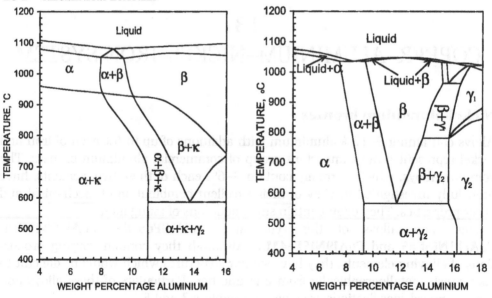

Fig. 13.1 Comparison of Cu–Al–Ni–Fe equilibrium diagram with binary diagram.[33] (a) Vertical section of the Cu–Al–Ni–Fe system at 5% each Ni and Fe; (b) Binary diagram for comparison.

Table 13.1 Compositions of alloys cooled slowly from 1010°C. [102]

Elements	Alloy 1	Alloy 2	Alloy 3	Alloy 4
Cu	80.02	79.6	80.0	80.0
Al	9.37	9.02	8.9	9.1
Fe	4.38	5.09	4.5	4.57
Ni	4.84	4.35	5.1	5.1
Mn	1.18	1.37	1.2	1.13
Si	0.07	0.08	0.26	0.14
Ni/Fe	1.105	0.855	1.133	1.116

Table 13.2 Temperature (°C) at which phases first appeared for each alloy. [102]

Alloy	Phase				
	κ_I	α	κ_{II}	κ_{IV}	κ_{III}
1	not observed	900	900	850	800
2	1010	950	950	900	800
3	not observed	1000	1000	940	840
4	not observed	940	890	840	810

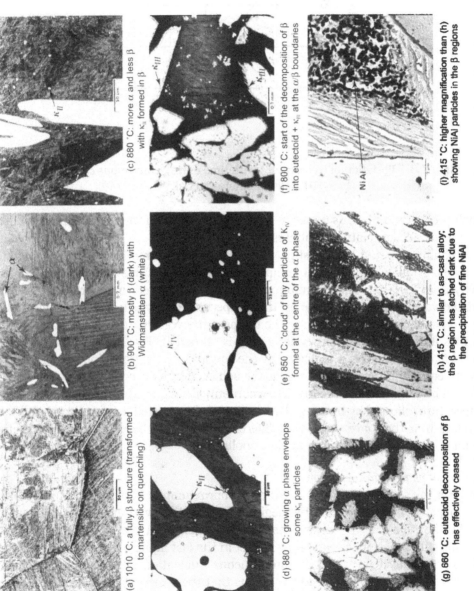

Fig. 13.2 Development of microstructure of CuAl10Fe5Ni5 type alloy slow cooled from 1010°C, by Jahanafrooz et al.[12]

(a) 1010 °C: a fully β structure (transformed to martensitic on quenching)

(b) 900 °C: mostly β (dark) with Widmanstätten α (white)

(c) 880 °C: more α and less β with κ$_{II}$ formed in β

(d) 880 °C: growing α phase envelops some κ$_{II}$ particles

(e) 850 °C: 'cloud' of tiny particles of κ$_{IV}$ formed at the centre of the α phase

(f) 800 °C: start of the decomposition of β into eutectoid + κ$_{III}$ at the α/β boundaries

(g) 660 °C: eutectoid decomposition of β has effectively ceased

(h) 415 °C: similar to as-cast alloy; the β region has etched dark due to the precipitation of fine NiAl

(i) 415 °C: higher magnification than (h) showing NiAl particles in the β regions

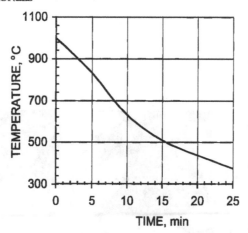

Fig. 13.3 Variation of temperature with time during continuous cooling from 1010°C, by Jahanafrooz et al.[102]

which precipitation begins to occur. More information on these precipitates will be given below under 'Nature of phases'.

A. Jahanafrooz et al.[102] investigated the phase transformations of a group of four Cu–Al–Fe–Ni alloys, of different compositions shown in Table 13.1, as they cooled slowly from 1010°C. The variation in compositions served to illustrate the effects that the various elements have on the temperature at which phase changes take place on cooling (see Table 13.2). The temperature was first held at 1010°C for 30 minutes and the subsequent cooling rate is shown in Fig. 13.3. Phase changes tended to occur later than indicated by the equilibrium diagram since, as Brezina[33] reports, equilibrium conditions can not be achieved by slow cooling, even when it is followed by long-term annealing. They can only be achieved by quenching at high temperature followed by prolonged annealing at the selected temperature. It follows therefore that there are significant differences between the equilibrium state and the microstructure resulting from slow cooling.

Sequence of phase transformations
- Over the solidification range, which is approximately from 1080°C to 1050°C, copper-rich β phase dendrites begin to form and grow, some of them nucleated by the type-2 κ_1 particles. The type-1 particles, on the other hand, do not nucleate other phases and they collect in the last liquid to solidify between the arms of the dendrites. The simultaneous nucleation of copper-rich dendrite crystals by the type-2 particles means that these crystals hinder each other's growth and the result is a fine grain structure in the alloy. Some of these precipitates are thought to re-dissolve in the solid state at high temperatures provided the rate of cooling is sufficiently slow.
- At 1010°C, the microstructure of alloys with the lower iron contents (Fig. 13.2a) consists entirely of a β-phase that has been transformed to a marten-

Fig. 13.4   Microstructure of alloy containing ~5% Fe, quenched from 1010°C showing κ_I particles in martensitic β-phase, by Jahanafrooz et al.[102]

sitic structure by quenching, whereas, with the higher iron contents of ~5% (Fig. 13.4), the microstructure consists of β+κ_1. The κ_1 particles are presumably type-2 pre-primary κ_1 particles which nucleated the β-grains and have not re-dissolved.

- Below ~1010°C, the β phase breaks down progressively during cooling into an intermediate α+β structure. The α-phase grows initially at the β grain boundaries and along crystallographic planes in a typical needle-like form known as a Widmanstätten structure (see Fig. 13.2b). The lower the aluminium content of the alloy, the higher the temperature at which the α-phase begins to nucleate (see Fig. 13.1). Also the higher the temperature of nucleation of the α-phase, the higher its iron content is likely to be. In the case of alloys with a higher iron content of ~5%, the κ_1 particles, which were present in the β matrix at 1010°C, are thought to act as nucleation site for the α-phase. They are few in numbers and they grow, on cooling over a wide range of temperatures, into large dendritic shaped 'rosettes' at the centre of the α-phase. This is why they only appear in the α-phase in as-cast microstructures (see Fig. 13.5b).[102]
- Between ~1000°C and ~900°C, the iron-rich inter-metallic κ_{II} particles begin to nucleate in the β-phase. It will be seen from Table 13.2, that the higher the aluminium content of the alloy, the lower the temperature at which these particles appear. It is also thought that nickel enhances the solubility of iron and hence, the higher the Ni/Fe ratio, the lower the temperature of nucleation of these particles.[102] These particles are small and dendritic in shape (see Fig. 13.2c) and some, which nucleated initially at the α/β boundary, become enveloped by the growing α-phase (see Fig. 13.2d).

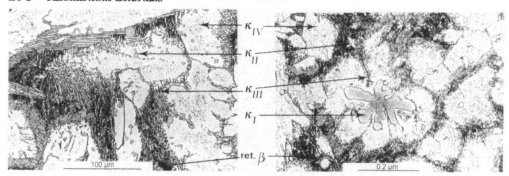

Fig. 13.5 Effect of iron content on microstructure of as-cast nickel aluminium bronze alloys, by A. Jahanafrooz et al.[102] (a) Alloy containing 9.37% Al, 4.38% Fe, 4.84% Ni, (b) Alloy containing 9.02% Al, 5.09% Fe, 4.35% Ni.

- Between 940°C and 840°C, the solubility of iron in the α-phase is exceeded and a 'peppering' of tiny κ_{IV} particles starts to appear at the centre of the α-phase (see Fig. 13.2e). As in the case of κ_{II} above, it is thought that nickel enhances the solubility of iron and hence, the higher the Ni/Fe ratio, the lower the temperature of nucleation of these particles. Also the higher the iron content of the α-phase, the higher the temperature at which the κ_{IV} particles appear. Iron-rich grains are the first to precipitate from β and contain the highest amount of iron in solution. They are therefore the first to reach the solubility limit of the α-phase which explains the concentration of κ_{IV} particles at the centre of the α-grains and the particle-free zone at the periphery of the grain.[102]
- Between 840°C and 800°C, depending on alloy composition and cooling rate, the remaining β begins to transform to a finely divided eutectoid designated $\alpha+\kappa_{III}$. A few initially formed κ_{III} particles are globular in shape but subsequent particles are mostly of lamellar or pearlitic appearance and form at the α/β grain boundaries (see Fig. 13.2f). Unlike the preceding κ phases that are all iron-rich and based on Fe_3Al, κ_{III} is nickel-rich and based on NiAl. Other nickel-rich particles and more κ_{II} particles also precipitate at about the same time in the martensitic β-phase.
- The eutectoid decomposition of β into $\alpha+\kappa_{III}$ becomes progressively slower as the temperature falls and, at 660°C, has effectively ceased (see Fig. 13.2g) – it does not reach completion at normal cooling rates. The remaining β consequently transforms to a martensitic structure on cooling below its metastable temperature. Meanwhile, the tiny κ_{IV} precipitates which had begun to form in the α areas at 850°C, have grown and more have nucleated on cooling.
- At 415°C, the microstructure is similar to that of the as-cast alloy (Fig. 13.5a) and the structure does not change appreciably below this temperature. The β regions have etched dark due to the precipitation of fine NiAl particles between 660°C and 415°C (see Fig. 13.2h). These particles may be seen more clearly at a higher magnification in Fig. 13.2i. They are perhaps the very fine globular

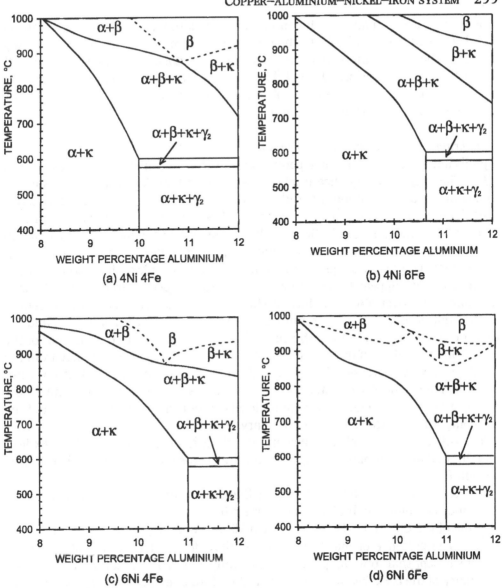

Fig. 13.6 Vertical sections of the Cu–Al–Ni–Fe system with various amounts of iron and nickel, by Cook et al.[47].

particles, which may also take the form of isolated laths, reported by Brezina[33] and Culpan and Rose[62] and which they designate κ_V. The as-cast structure of the alloy with the highest iron content of 5.09% is shown in Fig. 13.5b for comparison. It contains a large iron-rich dendritic κ_I precipitate at the centre of an α-grain, previously mentioned.

- In the case of an alloy having more than 11% aluminium, and at temperatures between 600°C and 575°C, the $\alpha+\beta+\kappa$ structure transforms on slow cooling into the brittle and corrodible $\alpha+\kappa_{III}+\gamma_2$ eutectoid. According to Brezina,[30] whereas the transformation of β into $\alpha+\kappa_{III}$ takes place even at cooling rates as fast as 5°K min^{-1}, for a complete transformation of β into $\alpha+\gamma_2$ the cooling rate must be less than 0.5°K min^{-1}.

Equilibrium diagrams for other Ni/Fe combinations

Equilibrium diagrams for other percentages of iron and nickel are shown in Fig. 13.6. The only really significant difference between these diagrams is the aluminium content that determines the right hand boundary of the $\alpha+\kappa$ field. Beyond this right hand boundary, the β phase in the $\alpha+\beta+\kappa$ structure will transform on slow cooling between 575°C and 600°C to the brittle and corrodible $\alpha+\gamma_2$ eutectoid. In alloys containing 4% each of nickel and iron, the $\alpha+\kappa$ field extends to about 10% aluminium, whereas with greater amounts of iron and nickel it extends to 11%. Nickel additions appear to have more effect in this respect than iron. Additions of iron and nickel over 5% would have no further influence on this particular feature, as the limit of solubility of aluminium is identical for alloys containing 5% and 6% each of iron and nickel. As in the case of the binary diagram, the location of the right hand boundary of the $\alpha+\kappa$ field is therefore of particular importance from the point of view of corrosion resistance. If, due to fast cooling, $\alpha+\beta+\kappa$ does not transform into the $\alpha+\gamma_2$ eutectoid, it survives as an unstable structure below 600°C. In practice the aluminium and nickel contents are normally chosen to keep the composition to the left of this boundary.

The role played by nickel in affecting the right hand boundary of the $\alpha+\kappa$ field and in making complex alloys resistant to corrosion is explained below (see Section B).

The $\alpha+\kappa$ alloys generally contain little β and no γ_2 and have excellent corrosion resistance combined with high tensile strength.

Some of the Cu–Al–Ni–Fe alloys have additions of manganese which may be up to 3%. The presence of manganese in the alloy does not affect the above observations except that, 6% manganese being equivalent to 1% aluminium, moves the boundary of the $\alpha+\kappa$ field further to the right. Manganese, being fully soluble, does not otherwise affect the equilibrium diagram and microstructure. It has however implications for corrosion resistance as we shall see.

Microstructure and nature of the various phases

Microstructure of type 80–10–5–5 alloys

Microstructures of the various phases of two type 80–10–5–5 alloys are shown on Figs. 13.7a and 13.7b by F. Hasan et al.[87] It will be seen that they are very similar to the microstructures of similar alloys shown on Figs. 13.5a and 13.5b, with the

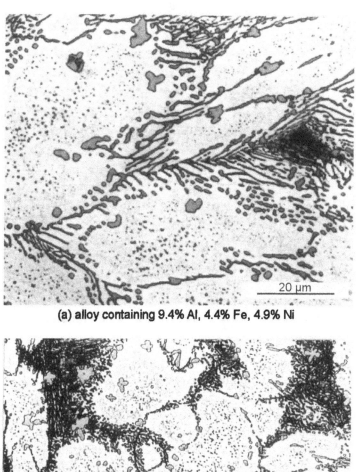

(a) alloy containing 9.4% Al, 4.4% Fe, 4.9% Ni

(b) alloy containing 9.0% Al, 5.1% Fe, 4.4% Ni

Fig. 13.7   Various phases of a type 80–10–5–5 aluminium bronze, by F. Hasan et al.[87]

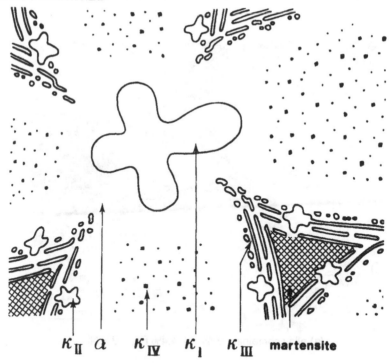

κ_{II} α κ_{IV} κ_{I} κ_{III} martensite

Fig. 13.8 Diagrammatic representations of the various phases in a type 80–10–5–5 cast aluminium bronze, by F. Hasan et al.[87]

large κ_I particle appearing only in the structure of the alloy with the higher iron content. A diagrammatic representation of these phases by Hasan et al.[87] is given in Figs. 13.8. A summary description of the microstructure and crystallography of the various phases is given in Table 13.3 and the corresponding chemical analyses in Table 13.4. It will be seen that, judging by their composition, some particles designated κ_I by Culpan and Rose,[62] should in fact be designated κ_{II} and some designated κ_{II} should in fact be designated κ_{III}. It does seem, however, that the chemical composition of the various phases in nickel-aluminium bronzes can vary markedly between specimens and even within the same specimen. This is particularly so in the case of the κ phases as is evident from the work of Brezina[33] and others.

Alloys with low nickel and iron

The chemical analyses of the phases (other than κ) of a low nickel and iron alloy are given in Table 13.5. The best known alloy of this kind is the CuAl9Ni3Fe2 wrought alloy (see Chapter 5, 'Duplex (twin-phase) alloys'. It is of technical interest as it represents a compromise, both in structure and mechanical properties, between the above alloys and the binary or ternary alloys (Chapters 11

Table 13.3   Morphology and crystallography of phases in as-cast 80/10/5/5 type aluminium bronzes.

Phase	Temperat. at which phase appears	Morphology		
		Distinctive appearance in microstructure	Size of particles	Crystalline structure
[1] α	~ 1010°C	Light etching grains		fcc – solid solution
[1] β'	< ~ 840°C	Dark etching needle-like phase known as 'martensite' by analogy with steel		Super-saturated solid solution with distorted lattice structure known as martensitic
γ₂	< 600°C	Forms eutectoid with α and κ on very slow cooling		Stable intermediate solid solution based on Cu_9Al_4.
κ phases based on Fe_3Al (but including some κ$_I$ that are based on FeAl)				
[3] Pre-primary κ$_I$ Type-1		Globular particles between arms of dendrites.	not stated	disordered solid solution
Type-2		Dendritic shaped particle at nucleus of β grain.	"	ordered solid solution
[1] κ$_I$	~920–900°C	Large dendritic light-etched 'rosette' at centre of α phase, or of β at higher temperatures	20–50μm	Some disordered bcc – some ordered bcc based on Fe_3Al – some ordered bcc based on FeAl
[1] κ$_{II}$	~920–900°C	Unevenly distributed small dentritic 'rosette' at α/β boundary	5–10μm	Ordered bcc (DO_3) based on Fe_3Al
[1] κ$_{IV}$	< 840°C	Dense mass of small equi-axed particles in α phase leaving particle-free zone near boundary	< 2μm	Ordered bcc (DO_3) based on Fe_3Al
κ phases based on NiAl				
[2] κ$_{II}$ (κ$_{III}$?)	~920–900°C	Unevenly distributed particles at α/β boundary – some lamellar, some globular.	5–10μm	Ordered bcc (B2) based on NiAl
[1] κ$_{III}$	870–840°C	Fine precipitates – some lamellar, some globular – forming eutectoid with α, normal to α/β boundary.		Ordered bcc (B2) based on NiAl
(1) Particles in β (probably related to κ$_{III}$)		Dense mass of small spherical or cubic particles in martensitic β phase. Size depends on cooling rate.		Ordered bcc (B2)
[2] κ$_V$	< 840°C	Unevenly distributed lath-like particles at α/β boundary. Grow in size with HT	1×0.1μm to 10×2μm	Ordered bcc (B2) based on NiAl

[1] By Hasan et al. [87–88]   [2] By Culpan and Rose[62]   [3] By Feest and Cook[70]

Table 13.4   Chemical composition of phases in as-cast 80/10/5/5 type aluminium bronzes.

Phases	No of Analyses	Technique	% Composition (wt)					
			Al	Si	Mn	Fe	Ni	Cu
α phase								
[1] α	8	Thin foil°	7.2±0.4	<0.1	1.1±0.1	2.8±0.3	3.0±0.2	85.8±0.4
[3] α	20–30	Thin foil*	8±2	–	0.8±0.3	2.4±1	3.0±2	86±4
"		Bulk+	8.3±1.7	–	1.4±0.1	2.7±2	2.5±1.4	85.4±4
[6] α	20	Bulk	6.8±0.5	–	1.3±0.1	2.2±0.4	2.9±0.7	87.0±1.4
β' phase								
[3] β'	<10	Bulk+	8.7	–	1.0	1.6	3.5	85.2
κ phases based on Fe₃Al (DO₃ structure)								
[4]–[5] Pre-primary κ$_I$								
Type-1	Not stated	Electron-probe	7.6	0.6	1.5	70.6	4.8	15.3
Type-2	"		9.8	0.7	1.2	64.0	6.6	20.1
[2] κ$_I$	12	Bulk+	9.3±0.5	1.6±0.4	2.9±0.5	72.2±1.4	3.5±0.4	10.5±1.0
[3] κ$_I$(κ$_{II}$?)	20–30	Bulk+	13±5	–	2±0.4	55±7	15±3	15±5
[1] κ$_{II}$	10	Thin foil°	12.3±1.3	4.1±0.8	2.2±0.2	61.3±4.9	8.0±1.8	12.1±3.1
[1] κ$_{IV}$	12	Extr replica°	10.5±1.7	4.0±0.5	2.4±0.2	73.4±2.3	7.3±1.5	2.6±0.7
[3] κ$_{IV}$	20–30	Extr replica*	14±2	–	1.1±0.4	63±6	14±4	8±3
"		Bulk+	20±3	–	1.5±0.3	62±4	4±1	13±1
"		Thin foil*	9±4	–	1.6±0.4	60±8	6±4	23±6
κ phases based on NiAl (B2 structure)								
[3] κ$_{II}$(κ$_{III}$?)	20–30	Thin foil*	18±4	–	1.6±0.3	34±5	24±5	23±4
"		Bulk+	19±3	–	2.2±0.6	32±3	27±4	21±5
"		Extr replica*	19±5	–	1.3±0.1	34±5	30±3	15±5
[1] κ$_{III}$	10	Extr replica°	26.7±1.0	<0.1	2.0±0.4	12.8±1.6	41.3±6.0	17.0±4.6
[3] κ$_{III}$	20–30	Bulk+	18±6	–	2±0.3	22±0.7	32±2	26±4
"		Thin foil*	22±4	–	1.6±0.4	22±5	28±5	26±4
[6] κ$_{III}$	21	Bulk	18.5±1.9	–	2.1±0.3	28.9±6.4	30.3±3.9	20.3±2.9
Particles in β[1]	10	Extr replica°	28.1±0.8	0.4±0.3	2.2±0.3	14.0±6.0	35.1±8.6	20.2±3.7
[3] κ$_V$	<10	Bulk+	26	–	1.1	26	21	26
	20–30	Extr replica*	27±4	–	1.5±0.3	27±4	35±3	10±2
Composition of alloys								
[1] Alloy I by Hasan et al.[87]			9.4	0.07	1.2	4.4	4.9	80.0
[2] Alloy II by Hasan et al.[87]			9.0	0.07	1.4	5.1	4.4	80.1
[3] Alloy III by Culpan and Rose[62]			9.42	–	1.09	4.24	4.70	80.55
[4] Alloy IV by Feest and Cook[70]			9.02	Zn: 0.46	1.37	5.09	4.35	Bal
[5] Alloy V by Feest and Cook[70]			9.32	Zn: 0.04	0.48	4.93	5.11	Bal
[6] Alloy VI by Jones and Rowlands[104]			9.04	–	1.1	4.65	5.20	Bal

+ By scanning electron microscope (SEM)   * by scanning transmission electron microscope (STEM)   °By analytical electron microscope

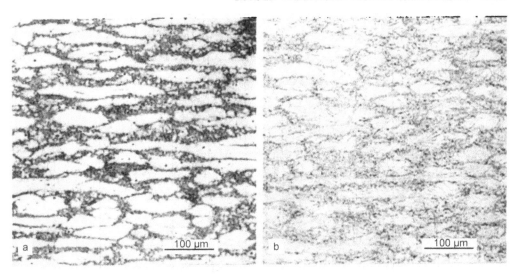

Fig. 13.9 Hot worked 9.5% Al – 2.5% Fe – 2.5% Ni, alloy;[127] (a) as-rolled; (b) soaked at 600°C for 18 hours.

and 12 respectively). The impact and elongation values of binary and ternary alloys are often superior to those of complex alloys, while the latter offer a higher tensile strength and proof strength. This alloy offers a good in-between combination of ductility and toughness with moderate levels of proof strength. It is in considerable demand in France.

Fig. 13.9 shows the structure of a similar alloy, containing 2.5% each of nickel and iron, in the as-rolled condition and after prolonged tempering at 600°C. It contains the κ precipitate distributed throughout both the α and β phases in a rather more finely divided form. It will normally contain a proportion of partly dissociated β, but with an aluminium content around 9%. Tempering at 600°C can result in complete removal of the martensitic β phase.

Table 13.5 Analysis of phases of a low nickel and iron aluminium bronze by Weill-Couly and Arnauld.[183]

Element	Composition, wt %			
	α phase	α+γ₂	γ₂ phase	β' phase
Iron	1.55	0.65	0.70	0.85
Aluminium	6.94	11.69	16.00	12.30
Nickel	1.24	2.85	3.77	3.53
Manganese	1.70	1.70	0.95	2.15
Balance (mostly copper)	88.6	83.1	78.6	81.2
Alloy Composition % (wt)				
Al	Ni	Fe	Mn	Cu
10.3	1.5	1.4	1.4	85.4

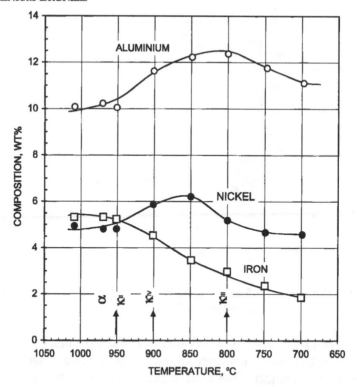

Fig. 13.10 Variations of the iron, nickel and aluminium contents of the intermediate β phase with temperature, by Jahanafrooz et al.[102]

The α phase

The white areas on Figs. 13.7a and 13.7b are of the α phase which is a copper-rich stable solid solution with a face-centred cubic (fcc) structure. The composition of the α phase of a given specimen remains constant except for the iron content which reduces with temperature as the κ phases precipitate. It also contains some dissolved nickel and manganese. The percentage of these elements in the α phase is influenced by the alloy composition, as may be seen by comparing the analysis of α in Table 13.4.

The α phase provides ductility to the alloy whereas other phases increase tensile strength, proof strength and hardness.

The β phase

The high temperature β phase is an intermediate solid solution with a random body-centred cubic (bcc) structure and has a higher percentage of aluminium than the alloy as a whole. It contains nickel, iron and, in some cases, manganese in solution. It may also contain some κ_I precipitates as previously mentioned.

Fig. 13.10 by Jahanafrooz et al.,[102] shows the variations in the iron, nickel and aluminium contents of the β-phase, as temperature falls and other phases precipitate from it. This information was obtained by quenching from decreasing temperatures from 1010°C to 700°C. The iron content of the β-phase steadily falls over this range as the iron-rich κ particles precipitate from it. The nickel and aluminium contents of the β-phase initially increase but then fall as the $κ_{III}$ (NiAl) particles precipitate. It should also be noted that the 11% aluminium content of the β phase at 700°C is higher than that of the alloy, whereas the nickel content has dropped to 4%.

The 'retained β' or martensitic β phase

The very dark-etched areas in Figs. 13.7a and 13.7b are the 'retained β' phase, otherwise called the martensitic β or β' phase[127] to distinguish it from the high temperature β phase above. It is the result of the transformation of this high temperature β phase as it cools to room temperature. It is needle-like or martensitic and has an approximately closed-packed hexagonal (cph) structure. Hasan et al. report[87] that there are two forms of the martensitic β phase, one of which, referred to as 3R, has a high density of precipitates that are similar to $κ_{III}$.

Hasan et al.[89] point out that the proportion of β in the Cu–Al–Fe–Ni alloy is smaller than in the ternary alloy (see Chapter 12) due to the substantial amount of aluminium required to form the high nickel precipitates.

The $γ_2$ phase

The $γ_2$ phase (not shown on Fig. 13.7), which forms an eutectoid with $α+κ_{III}$, is a stable intermediate solid solution. According to Toner[175] and Jellison and Klier,[103] it contains 15.6% aluminium. Bradley et al. report that it is based on Cu_9Al_4. It is a corrosion-prone phase. Where corrosion is not a consideration however, alloys containing this phase offer excellent wear properties although their brittleness restricts them to compressive loads.

Forms of the inter-metallic kappa phase

The κ particles are intermetallic compounds that differ from solid solution in that the constituent metals have reacted chemically to form a definite combination. With one exception, the κ particles fall into one of two combinations: Fe_3Al and NiAl. The only exception is $κ_I$ which may be based on Fe_3Al or FeAl or may have a disordered structure. In the case of each combination, one of the constituent metals can be partially replaced by some other metal present in the alloy; this is known as 'substitution'. Thus, in the Fe_3Al combination, copper, manganese and nickel may partly substitute for iron, and silicon may partly substitute for aluminium, creating in effect different compounds that have the same basic structure as the Fe_3Al

combination. This structure is designated 'DO₃'. Likewise substitution can occur in the case of the basic NiAl combination, the structure of which is designated 'B2'. Both the Fe_3Al and the NiAl combinations have an ordered body-centred cubic (bcc) space lattice structure. The resultant variety of compounds of these two basic combinations explain the significant variations in chemical compositions of any one type of κ phase.

In practice the κ phases have been designated κ_I, κ_{II}, κ_{III} etc., according to the order in which they appear in the microstructure, as the temperature of the alloy falls on cooling. Although there are similarities between κ phases, they are distinguishable by the combination of their morphology, their location and their distribution in the microstructure.

Brezina[33] and Culpan and Rose,[62] identify five different forms of the κ phase, designated κ_I, κ_{II}, κ_{III}, κ_{IV} and κ_V. Hasan et al.[87] report the presence of nickel-rich particles in β which are similar to κ_{III}. Perhaps they are related to the κ_V particles which are also reported to form in the β-phase. As reported above, Feest and Cook[70] have identified two types of κ phases which precipitate in the melt and which they have designated type-1 and type-2 pre-primary κ_I phases. As will be seen below, further changes in the κ phases occur, following heat treatment, resulting in the formation of a composite κ phase.

Various researchers have designated the various κ phases in different ways. Hasan et al.[87] and Weill-Couly et al.[183] do not refer to κ_V as a separate phase. Weill-Couly et al.[183] have used the designation κ_I for both κ_I and κ_{II}. The κ phases shown diagrammatically on Fig. 13.8 by Hasan et al.[87] do not include κ_V which Brezina and Culpan and Rose[62] include in their range of κ phases.

The morphology and crystalline structure of the various κ phases are summarised in Table 13.3 and their chemical composition in Table 13.4. The κ phases have been grouped in two categories: the iron-rich κ phases based on Fe_3Al, and the nickel-rich κ phases based on NiAl.

As in the case of the Fe(δ) and NiAl precipitates in the ternary alloys, κ precipitates have a tendency to return into solution at higher temperatures (see below 'Effects of heat treatment on microstructure').

The κ phases absorb aluminium from the matrix and hence extend the apparent range of the α field. They have a pronounced effect on properties and considerably increase the mechanical strength. At the same time the reduction in ductility is not as marked as in the case of a β-containing binary alloy of equivalent strength. This is the main advantage of the alloys over other aluminium bronzes.[62]

The form and chemical composition of the κ particles is affected by the rate of cooling (see below: 'As-cast microstructure and effects of cooling rate').

Pre-primary κ_I phases
According to Feest and Cook,[70] there are two types of κ_I particles that precipitate in the melt prior to solidification which they have designated type-1 and type-2. They have established by test that these particles were precipitates and not due to

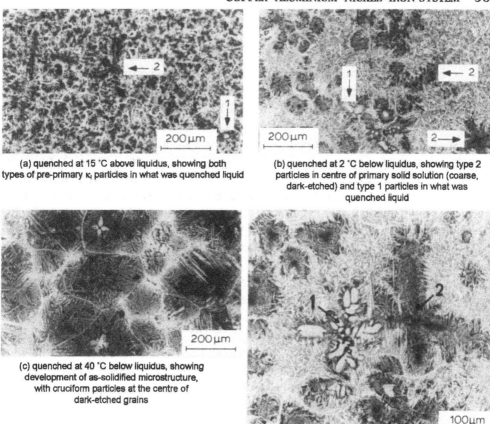

(a) quenched at 15 °C above liquidus, showing both types of pre-primary κ_I particles in what was quenched liquid

(b) quenched at 2 °C below liquidus, showing type 2 particles in centre of primary solid solution (coarse, dark-etched) and type 1 particles in what was quenched liquid

(c) quenched at 40 °C below liquidus, showing development of as-solidified microstructure, with cruciform particles at the centre of dark-etched grains

(d) quenched at 11 °C below liquidus, showing type 1 and type 2 pre-primary κ particles as in (b) but larger

Fig. 13.11 Microstructure of samples quenched at different liquid and solid temperatures by Feest and Cook.[70]

incomplete dissolution during alloying. The higher the iron content of the alloy, the greater the number of both types of particles. Figs. 13.11a to d show the microstructure of samples quenched at various liquid and solid temperatures with the two types of κ_I particles indicated.

The type-1 particles appear white on the micrographs. They have a coarse globular morphology and a higher iron content (71%) than type-2 (64%). The temperature at which they precipitate is dependent on the alloy composition: the higher the iron content, the higher the initial temperature of precipitation. Thus, in an alloy containing 5.09% Fe and 4.35% Ni, they begin to precipitate at 1133°C–1104°C, whereas with an alloy containing 4.93% Fe and 5.11% Ni, precipitation occurs at 1075°C–1065°C. They do not nucleate copper-rich dendrites but congregate between the arms of the dendrites in the last liquid to solidify. They are a solid solution with a non-ordered, non-faceted structure.

They are likely to form local iron-rich segregation in the cast material which may account for the 'rust' staining of castings exposed to a saline atmosphere. These particles act as impurities, reduce ductility and may have an adverse effect on impact value which could be significant in the case of applications involving shock conditions. Although they do not appear to lead to corrosion of the casting, the rust staining spoils the appearance of a casting and undermines confidence. Ensuring that the Ni–Fe ratio in the alloy composition is greater than 1, has been shown to minimize the occurrence of these precipitates. It would appear from the work of Hasan et al.[87] (see below), that keeping the iron content below 4.5% might prevent Type-1 particles arising. Consideration might be given to reducing the maximum allowable iron content of BS1400 AB2 (5.5%) and of ASTM 955 (5%) and 958 (5.5%) for applications where ductility and toughness are important.

The type-2 particles appear dark and 'slaty' on the micrograph. They a have a more pronounced dendritic appearance than the type-1 particles and a faceted ordered structure. They have a lower iron content (64%) and their solubility changes more rapidly with temperature than that of the type-1 particles. They precipitate at lower temperatures than the type-1 particles, for the same alloy composition, and nucleate the β-phase crystals. Type-2 particles are clearly beneficial because of their grain refining function.

The κ_I phase

The post-solidification κ_I phase consists of iron-rich intermetallic particles which form initially in the β-phase of alloys of relatively high iron content. Thus Hasan et al.[87] report that they were unable to find it in Alloy I (Table 13.4) which has a 4.4% iron content but only in Alloy II which has a 5.1% iron content. On cooling, they nucleate some α-grains and it is only in the α-phase, in the form of large dendritic κ_I 'rosettes', that they are found at room temperature, as shown in Fig. 13.7b and diagrammatically in Fig. 13.8.

The iron-rich κ_I precipitates do not all have the same composition and crystalline structure. Some have an ordered bcc structure based on Fe_3Al, some have an ordered bcc structure based on FeAl[87] and some have a disordered structure. FeAl is related to the δ (Fe) particles of the Cu–Al–Fe ternary system (Figs 12.1 and 12.2). Iron is partly substituted in the Fe_3Al lattice structure by copper, nickel and manganese. The κ_I particles are typically 20 to 50 μm in diameter and are 'cored', that is to say they are copper-rich at their centre.[87]

The κ_{II} phase

The κ_{II} particles are thought to precipitate initially in the β phase at high temperature and to be enveloped in the α phase as the β phase breaks down into the α+β+κ structure at about 900°C during cooling. They go on forming as the temperature falls to 840°C and tend to occur near the α/β boundary in company with κ_{III} below (see Fig. 13.8).

The κ_{II} particles are coarse and rounded, and take the form of dendritic 'rosettes' which are smaller than the κ_I rosettes (5 to 10 μm).[87] Brezina[33] reports some κ_{II} particles were located in the martensitic β phase. These may not be iron-rich κ_{II} particles but the nickel-rich particles in β reported by Hasan et al.[87]

This iron-rich κ_{II} phase has an ordered bcc structure based on Fe_3Al with nickel, copper and manganese substituting for iron, and silicon substituting partly for aluminium. It is closely related to κ_I above and to the δ(Fe) particles previously mentioned in Chapter 12 which precipitate in the β phase. Particles of κ_{II} are less than 10 μm in diameter.[120] Weill-Couly and Arnaud[183] report that the iron content of κ_{II} reduces as the aluminium content increases. The κ_{II} phase has no significant effect on corrosion beyond a superficial rusting.

It is conceivable that some of the κ_{II} particles may originate from the pre-primary (type-1) κ_I particles which congregated between the arms of the β dendrites during solidification and re-dissolved in the solid state, only to re-precipitate at about 900°C as κ_{II} particles. The fact that they are iron-rich would explain the superficial rusting found on some castings exposed to marine atmosphere.

The κ_{III} phase
The κ_{III} particles precipitate between approximately 840°C and 600°C, when the remaining β transforms to a finely divided eutectoid designated $\alpha + \kappa_{III}$. This phase has a lamellar or pearlitic form (visible at the higher magnification on Fig. 13.2i) and sometimes a coagulated or globular (degenerate lamellar) form. It grows at right angles to the α/β boundary and also forms at the boundary of the large κ_I rosettes. The κ_{III} precipitate is a nickel-rich inter-metallic compound with an ordered bcc structure based on NiAl in which iron, copper and manganese substitute for nickel.[87] It is similar to the precipitate of the ternary copper–aluminium–nickel system (Figs. 12.10 Chapter 12).

Hasan et al.[87] report the presence of particles in the martensitic β which have a very similar composition to the κ_{III} particles (see Table 13.4) being also based on NiAl with an ordered bcc structure. They have a spherical or cubic morphology and their size depends on the cooling rate.

Sarker and Bates[158] report that the amount of κ_{III} precipitate is increased by an increase of aluminium and nickel or by a reduction in iron. Crofts et al.[58] report that lamellar κ_{III} increases with high nickel content or with high nickel/iron ratios associated with low aluminium contents. They state that at 8.6% Al, this lamellar structure gives a higher proof strength and lower elongation than the globular κ_{III} which lowers proof strength, although it increases ductility and tensile strength.

The κ_{IV} phase

If the rate of cooling below 850°C is sufficiently slow, κ_{IV} precipitates in the α grains in the form of finely divided iron-rich particles, leaving a precipitate free zone near the grain boundary. Their appearance is due to the solubility of iron falling to 0.03% as a result of very slow cooling. Hasan et al.[87] report that the κ_{IV} particles always appear as small (< 2μm dia) equi-axed particles, similar to, but smaller than, κ_{II}.

The κ_{IV} particles have a composition and crystal structure which are also similar to those of κ_{II} and have likewise an ordered bcc structure based on Fe_3Al. The κ_{IV} precipitates can be seen in the (white) α grains in Figs 13.2e, 13.7a and 13.7b and diagrammatically in Fig. 13.8.

It will be seen below that, as a result of heat treatment, these particles act as nucleants for 'composite' particles whose extremities have a NiAl structure and composition.

The κ_V phase

Together with the globular κ_{IV} particles, a small number of lath-like particles may appear which Brezina[33] and Culpan and Rose[62] designate as κ_V. This phase may not appear in the as-cast structure but become prevalent as a result of heat treatment (see below). It has an ordered bcc structure based on NiAl. Most researchers do not, however, refer to this phase by a separate designation and consider it as a modified form of κ_{III}.

Summary of effects of alloying elements on the structure

1. Aluminium is primarily responsible for the excellent tensile properties of aluminium bronzes. In conjunction with nickel, it determines the boundary line between an $\alpha+\kappa$ alloy and an $\alpha+\kappa+\beta$ alloy. An $\alpha+\kappa$ alloy has excellent corrosion resisting properties and the best combination of tensile strength, proof strength and elongation. An $\alpha+\kappa+\beta$ alloy has higher tensile but lower proof and elongation properties and it also contains the more corrodible martensitic β-phase. As will be seen in Section B, the cooling rate significantly affects the degree of retention of the martensitic β-phase at room temperature and determines therefore the aluminium content marking the transition between these two types of alloys. For most applications a 9.5% aluminium content gives the best combination of properties. An aluminium content of 10% or more adversely affects mechanical and corrosion resisting properties.
2. Nickel also gives strength and toughness to aluminium bronzes and improves corrosion resistance. It gives rise to nickel-rich κ precipitates which contributes to the good mechanical properties of the $\alpha+\kappa$ alloy. No advantage accrues from increasing the nickel content above 5%. As will be seen in Section B, cooling rate and the relationship of nickel content to aluminium content determines whether the very corrodible γ_2 phase can be avoided.

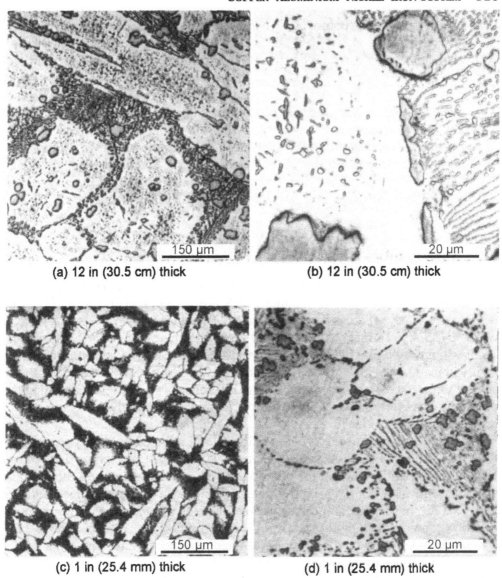

(a) 12 in (30.5 cm) thick (b) 12 in (30.5 cm) thick

(c) 1 in (25.4 mm) thick (d) 1 in (25.4 mm) thick

Fig. 13.12 Effect of section thickness on the sand cast structure of nickel-aluminium bronze with 10% Al, 5% Fe and 5% Ni.[127]

Specimen	Proof Strength (N mm^{-2})	Tensile Strength (N mm^{-2})	Elongation %
a and b	186	603	27
c and d	232	680	25

3. Iron refines the structure and thereby gives toughness to the alloy. It also gives rise to iron-rich κ precipitates that contribute to the strength of the α+κ alloy and which have no detrimental effects on corrosion resistance. No advantage accrues from increasing the iron content above 4.5% and too high an iron content can give rise to cavitation problems (see Section B). It is recommended that the iron content should be less than the nickel content.

4. Manganese improves the fluidity of the alloy and therefore facilitates the casting of thin sections. Small additions of manganese remain in solution and therefore are not noticeable in the microstructure. Manganese stabilises the β phase and hence hinders its decomposition to α+κ, with detrimental consequence for corrosion resistance. It should therefore not exceed 2% (see Section B).

Effects of cooling rate on microstructure

Fig. 13.12 shows the as-cast structures of two samples of a CuAl10Fe5Ni5 alloy. One sample ('a' and 'b') was 12 inch (305 mm) thick and would have cooled particularly slowly whereas the cooling rate of the other sample ('c' and 'd') which was 1 in (25.4 mm) thick, would be more representative of the cooling rate of most sand castings. Two magnifications are given in each case. The structure of the 12 inch sample shown in Figs. 13.12a and 13.12b, being slowly cooled, is much coarser than that of the 1 inch sample. This difference of structure is reflected in the mechanical properties, the 1 inch sample having significantly better proof and tensile strength than the 12 inch sample, but slightly lower elongation.

It should be noted that the α phase of the 1 inch sample has no κ_{IV} precipitates (Figs 13.12c and 13.12d). As explained above, κ_{IV} will only precipitate if the rate of cooling is sufficiently slow. On the other hand, the cooling rate of the 1 inch sample was sufficiently slow for κ_{III} particles to have precipitated at the α grain boundaries (Fig. 13.12d). In the case of the slowly cooled 12 inch sample, there is a concentration of κ_{IV} precipitates at the centre of the α grains (Figs. 13.12a and 13.12b) which are clearly of two types: some with a globular appearance and some with a thin sliver-like appearance which Brezina designates κ_V.

Weill-Couly and Arnaud[193] report that the transformation of β into α+κ_{III} is closely related to the cooling rate as illustrated by Figs. 13.13a to f. It is completely suppressed at high cooling rates (i.e. oil or water quenched – Figs. 13.13a and b). At moderate cooling rate (i.e. in sand – Figs 13.13c and d), α+κ_{III} forms a border around isolated areas of non-decomposed β, as illustrated diagrammatically on Fig. 13.8. For β to transform completely, the cooling rate between 900°C and 600°C must be less than 0.80°K min⁻¹ (50°K h⁻¹) which can only be achieved in a controlled oven (Fig.13.13f). In practice this means that martensitic β will inevitably be present in a sand casting where the cooling rate is 200°K min⁻¹ in cold sand and 65°K min⁻¹ in warm sand. In the case of a thick casting such as the 12 inch sample in Figs 13.12a and b, the temperature of the sand mould would rise, the metal initially chilled by the cold mould would be re-heated and the subsequently slow cooling mould would allow more time for β transformation into α+κ$_{\text{III}}$. It will be seen below that the martensitic β phase can, however, be eliminated by heat treatment.

Wenschot[187] has investigated the effects on structure and properties of the section thickness of castings ranging from 25 mm to 450 mm. The effect on mechanical properties and on fatigue strength has been dealt with in Chapter 3. The following are the effects on the structure:

- As mentioned above, the thicker the section, the slower the rate of cooling and the larger the grain size.
- At heavy sections, there is a likelihood of segregation occurring. This may take two forms:
  (a) local differences in concentration of alloying elements and
  (b) local concentration of impurities giving the appearance of casting defects. It should be said that low concentration of hydrogen gas which may remain in solution over a range of section thicknesses may come out of solution at the heaviest sections and encourage shrinkage defects.
- Both these forms of segregation will locally weaken the casting.
- Below approximately 100 mm, the smaller the section the greater the likelihood of some of the martensitic β phase not transforming to α+κ$_{\text{III}}$, making the material less ductile and corrosion resistant (see below).
- Above 100 mm, there is a growing danger of the γ$_2$ phase appearing making the material even less ductile and corrosion resistant.

The combination of grain size and distribution of the κ precipitates largely determine the strength and fatigue properties of the alloy. The κ precipitates harden the alloy and a fine grain size favours a fine, regular precipitation of these precipitates.

Summary of effects of cooling rate

1. A fast cooling rate, as in die-casting, produces a fine structure with an even distribution of fine precipitates resulting in significantly better tensile and proof strength but lower elongation. It will however result in a high volume of the corrodible martensitic β-phase in the structure.
2. A relatively slow cooling rate, as in sand castings, reduces the volume of the corrodible martensitic β-phase in the structure (see below). This phase can be further reduced or even eliminated by heat treatment (see below).
3. Very slow cooling, as in sand castings with very large section thicknesses, may give rise to the highly corrodible and brittle γ$_2$ phase. It may also lead to segregation of impurities and to the release of residual dissolved hydrogen, with detrimental effect on properties. Finding ways of increasing the cooling rate of heavy sections would be beneficial.

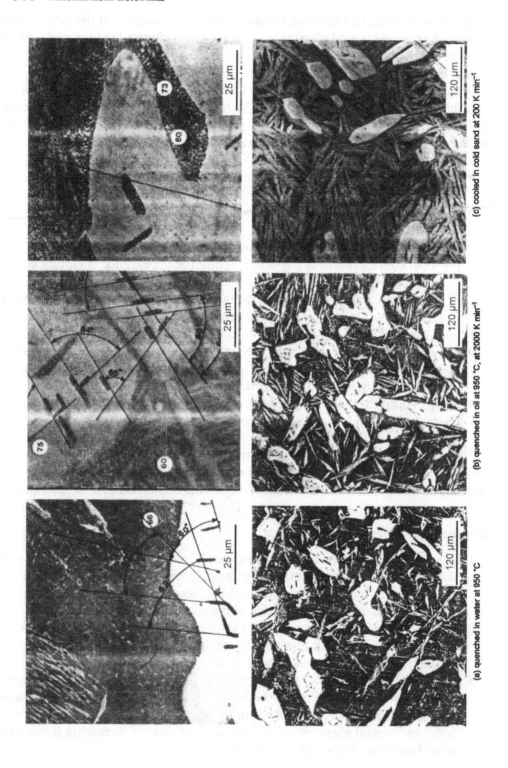

(a) quenched in water at 950 °C

(b) quenched in oil at 950 °C, at 2000 K min⁻¹

(c) cooled in cold sand at 200 K min⁻¹

Fig. 13.13 Effect of cooling rate on the transformation of the martinsitic ß phase into α+κ$_m$. by P. Weill-Couly.[185]

(f) oven cooled at 0.8 K min^{-1} (50 K h^{-1})

(e) cooled in open oven at 3 K min^{-1}

(d) cooled in warm water at 65 K min^{-1}

B – Resistance to corrosion

Microstructure and resistance to corrosion

Nickel–aluminium bronze has a high resistance to sea water corrosion thanks to its protective oxide film which is only slightly permeable to liquids. Its more corrodible phases can be prevented from arising provided, as will be explained, that certain limits are set on its aluminium, nickel and manganese contents and that its cooling rate from high temperature is within certain limits. Only one phase, the γ_2 phase, is prone to severe corrosion, due to its high anodic value, but the composition of this alloy is usually selected so as to avoid its occurrence. Weill-Couly and Arnaud[183] have observed that, in alloys that contained elements of γ_2 within the martensitic β phase, no severe corrosion had occurred.

The martensitic β phase may nevertheless experience limited attack. Lorimer et al.[122] carried out corrosion tests on a nickel aluminium bronze containing 9.4%Al, 4.4%Fe, 4.9%Ni and 1.2%Mn which they immersed in artificial sea water for 48 hours. They reported that the alloy underwent limited corrosive attack on two phases: the martensitic β phase, which is anodic to the α phase, and the $\alpha+\kappa_{III}$ eutectoid in which α is anodic to κ_{III}. Hasan et al.[87] report that the high chemical reactivity of the metastable martensitic β phase may be responsible for its accelerated corrosion.

Although the α constituent of the eutectoid was preferentially attacked, the α grains were unaffected. Al-Hashem et al.[6–7] reported that the α phase corroded at the interface with the κ_{III} precipitate at a rate of 0.1 mm per year. The κ_{II} particles, present in these phases, showed no sign of attack. J. C. Rowlands[155] reports however that whereas the κ_{III} phase was cathodic to the α phase under ordinary sea water conditions, it became anodic to it under crevice conditions and corroded at 0.7–1.1 mm/year. He also showed that the pH value would seem to account for this reversal of galvanic effect. Thus, in slightly alkaline (pH ~8.2) ordinary sea water, the α-phase is anodic to the κ_{III} phase, leading to the α phase corroding preferentially. The corrosion products of the α-phase will be cuprous oxide and aluminium chloride[42] which both experience hydrolysis in corrosion pits in copper or in aluminium, giving rise to cuprous oxide and hydrochloric acid, in the case of cuprous oxide, and to aluminium hydroxide and hydrochloric acid, in the case of aluminium. This explains the reduction in the pH value within the crevice and, if it falls to a value of 3, it is the κ_{III} phase that becomes anodic and corrodes. Jones and Rowlands,[104] who came to the same conclusion, report that this crevice effect occurred regardless of whether the seawater was aerated or not. They suggest that the effect of the crevice is to prevent the diffusion of hydrogen ions out of the crevice resulting in a build up of hydrogen ions and therefore of acidity in the crevice. In confirmation of this, they demonstrated that if chalk was cathodically deposited on the surface of the nickel–aluminium bronze, preferential phase corrosion was prevented.

The incidence of crevice corrosion in nickel–aluminium bronze is rare. If the alloy is cathodically protected by the vicinity of a steel structure or by a sacrificial anode, crevice corrosion is unlikely to occur.

Culpan and Rose[62] report that in crevice corrosion tests which they carried out on nickel-aluminium bronze castings, corrosion occurred *around* the crevice and was very similar to that seen at the heat-affected zone of a welded specimen.

Apart from the above case of crevice corrosion in the absence of cathodic protection, the $\alpha+\kappa_{III}$ eutectoid is less vulnerable to corrosion than the martensitic β phase and, as the latter transforms on cooling to $\alpha+\kappa_{III}$, this transformation, once begun, reduces the alloy's vulnerability to corrosion. Weill-Couly and Arnaud[183] report that, even a partial transformation of the β phase into $\alpha+\beta+\kappa_{III}$, is sufficient to protect the alloy against severe corrosion in sea water. This, they explain, is because $\alpha+\kappa_{III}$ forms a protective envelope around the β grains, thus isolating the more anodic elements. This is confirmed by accelerated corrosion tests carried out by Soubrier and Richard[165] who found, however, that the presence of exposed porosity or oxide inclusions can give rise to corrosion due to differential aeration (see effect of welding below).

Observations of corroded samples indicate that coarse κ precipitation due to excessively slow cooling may adversely affect corrosion resistance. As the composition of the κ phase varies, it may act anodically or cathodically to the matrix, but within the normal alloy range (4–5% each of iron and nickel), this effect is rarely, if ever, significant.

To improve the resistance to corrosion, the cast alloy may be given, as explained below, an annealing treatment at 675°C for two to six hours. The treatment results in the elimination of the more corrosion vulnerable martensitic β phase and in an increased density of κ precipitates in the α grains.

Role of nickel in resisting corrosion

As previously mentioned, one of the principal reasons for the addition of nickel as an alloying element, is to improve the corrosion resistance of aluminium bronzes. It does so in a number of complementary ways:

- it dissolves preferentially in the aluminium-rich (and therefore anodic) β phase and, being cathodic to it, reduces its potential difference with other phases,
- it creates, as explained above, a nickel-rich, relatively protective, envelope of $\alpha+\kappa_{III}$ around the β phase, provided the rate of cooling from the β range temperatures is not too high,
- it extends the boundary of the $\alpha+\kappa$ field to a higher aluminium content and therefore permits higher strength alloys without the risk of vulnerability to γ_2,
- even if this higher aluminium content is exceeded, it reduces, together with iron, the rate of transformation of β into γ_2.

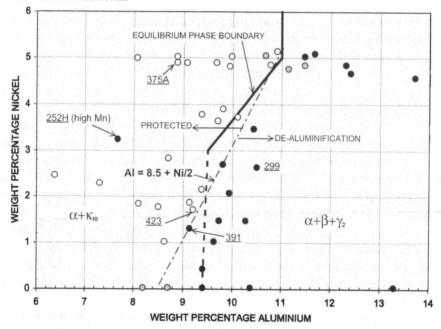

Fig. 13.14   Relationship of nickel to aluminium to ensure resistance to corrosion of sand casting cooled in sand at ~220°K min⁻¹, by Weill-Couly and Arnaud.[183] (See Figs 13.15 and 13.16 for photomicrographs of numbered samples).

Since κ_{III} absorbs aluminium from the α matrix and hence extends the apparent range of the α field, the location of the $\alpha+\kappa$ / $\alpha+\kappa+\gamma_2$ boundary is influenced by the nickel content. There is necessarily therefore a relationship between the aluminium and nickel contents which determines, for slowly cooled alloys, the limit of aluminium content in relation to nickel content within which the occurrence of the corrodible γ_2 is safely avoided.

Weill-Couly and Arnaud[183] have established by means of a series of corrosion tests in warm sea water on fifty samples of varying aluminium and nickel contents that this relationship is given by the following formula previously quoted in Chapters 1, 7 and 12:

$$Al \leq 8.2 + Ni/2$$

The results of their work are shown on Fig. 13.14 and the corresponding micrographs on Fig. 13.15. To the left of the line given by Al = 8.5 + Ni/2, the transformation of β into $\alpha+\kappa_{III}$ has taken place, at least partially, and the alloys are relatively protected against de-aluminification. It should be pointed out that this line indicates a 'zone' rather than a clear-cut border line and, for this reason, the figure 8.5 is reduced to 8.2 in the recommended working formula in order to allow for a margin of safety. To the right of that line this transformation has not taken place, the structure contains α, non-decomposed β and γ_2 and both the latter phases are corrosion-prone. The alloys are therefore not protected. It must be noted

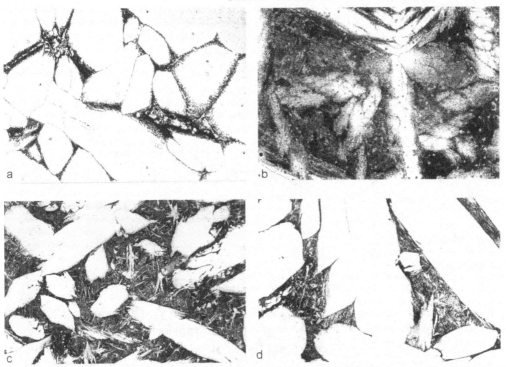

Fig. 13.15 Photomicrographs of samples 423 and 375A (protected from corrosion) and samples 299 and 391 (not protected from corrosion) – see Fig. 13.14.[183] (a) Sample 423: protected (higher magnification); (b) Sample 375A: protected, (c) Sample 299: non-protected, (d) Sample 391: non-protected.

however that these experiments were carried out on cast alloys cooled slowly in sand (at 220°K min⁻¹). As mentioned above, this transformation will not take place at high cooling rates (water or oil quenched). Weill-Couly and Arnaud report however that experiments were carried out with similar results with alloys subjected to different cooling rates and that the formula applies even in the case of alloys air-cooled at 250°K min⁻¹. Furthermore, the experiments were carried out with samples that had been annealed and were duplicated with non-annealed samples with the same results.

It should be noted that, at the minimum nickel content allowed by some standard specifications, the maximum aluminium content allowed by the specification may be higher than the maximum corrosion-safe aluminium content given by the above formula.

It will be seen that the line given by this formula diverges slightly from the line of the aluminium/nickel relationship which corresponds to the $\alpha+\kappa$ / $\alpha+\kappa+\gamma_2$ boundary in the constitutional diagrams (see Figs. 13.1 & 13.6). This indicates that, if the nickel content is less than 5%, protection from de-aluminification is obtained at higher aluminium values than would ensure, according to these diagrams, the

Table 13.6 Conditions required for protection from de-aluminification of complex alloys, with or without Fe, containing 1.5–5.5% Ni and 2% max. Mn. by Weill-Couly and Arnaud.[183]

Aluminium content	< 8.2 +Ni/2			> 8.2 +Ni/2	
Cooling rate from β field	Rapid (quenched)	Medium (cold sand or air cooled)	Slow (oven cooled)	Medium to Rapid	Slow
Structure	$\alpha+\beta'$	$\alpha+\beta+\kappa_{III}$	$\alpha+\kappa_{III}$	$\alpha+\beta'$	$\alpha+\beta+\gamma_2$
Protected against de-aluminification	no	yes	yes	no	no
Reason for vulnerability to de-aluminification	β' not protected by $\alpha+\kappa_{III}$			β' not protected by $\alpha+\kappa_{III}$	Presence of γ_2 phase

complete transformation of β into $\alpha+\kappa_{III}$ and freedom from γ_2. This bears out Weill-Couly and Arnaud's contention that, even a partial transformation of the β phase into $\alpha+\kappa_{III}$, is sufficient to protect the alloy against corrosion in sea water. It may also explain their observation that no corrosion occurred even if γ_2 was present provided it was surrounded by β. It follows that, provided the relationship of nickel to aluminium is according to the above formula, there is every advantage in allowing a casting to cool in its sand mould: not only does it prevent internal stresses arising but it ensures that the β phase has time to transform to $\alpha+\kappa_{III}$. The equilibrium diagrams, (compare Fig. 13.1a with 13.6c and d), show that increasing nickel beyond 5% has no effect on the $\alpha+\kappa_{III}$ / $\alpha+\kappa_{III}+\gamma_2$ boundary.

Table 13.6 summarises the above findings for complex alloys.

Effect of manganese additions on corrosion resistance

It was explained in Chapter 12, that manganese is used to give fluidity to the alloy and thereby improve its castability, but it has the effect of stabilising the β phase and therefore to hinder its decomposition to $\alpha+\kappa_{III}$. Since β is vulnerable to selective phase attack, its retention is undesirable, as was strikingly illustrated in the corrosion tests by Weill-Couly and Arnaud[183] mentioned above (see Fig. 13.14). All the fifty samples tested except one had manganese contents between 0.5% and 1.6%. The exception (sample 252H) had 3.72% manganese and, although the aluminium and nickel contents were well within the safe range given by the formula: Al ≤ 8.2 + Ni/2, this sample had strongly corroded (see Fig. 13.16). Weill-Couly and Arnauld[183] recommend that the manganese content should not exceed 2%. Although some standard specifications set a limit of 1.5%, others allow a manganese content well in excess of 2% which is clearly undesirable.

Fig. 13.16   Photomicrographs of sample 252H with 3.7% Mn, see Fig. 13.14.[183]

Effects of iron addition on corrosion resistance

Weill-Couly and Arnaud[183] report that iron contents between 0 and 5.5% made no difference to the results of the above experiments on corrosion.

As explained in Section A above and in Chapter 12, a light and widespread rust 'staining' occasionally forms on iron-containing aluminium bronze components exposed to a corrosive atmosphere, such as a marine environment. If this rust staining is superficial, it may be due to the presence of precipitates (such as type 1 pre-primary κ_I or κ_{II} precipitates previously mentioned) and is likely to be of no consequence,[183] apart from the unsatisfactory appearance of the component. If, on the other hand, localised 'rust spots' form, which reveal the presence of large iron particles, caused by poor foundry melting techniques, they are likely to be corroded areas and have harmful consequences. They have been found to initiate cavitation on impellers, propellers and other components and to lead to their early failure. As previously recommended, good melting practice combined with ensuring that the ratio of nickel to iron content is greater than one, and keeping the level of iron below 4.5% will reduce the likelihood of this rust staining occurring.

Effect of differential aeration

As explained in Chapter 8, if there is a difference in oxygen concentration at two different points on the surface of a metal object, or on the surfaces of two different components of the same metal in contact with each other and immersed in the same electrolyte, the less aerated surface becomes anodic to the better aerated surface. Weill-Couly and Arnaud[183] found that, whereas differential aeration had only a slight corrosive effect on an alloy conforming with the formula: Al ≤ 8.2 + Ni/2, the attack was extensive in the case of alloys not conforming with it.

Effect of microstructure on resistance to cavitation erosion

J. L. Heuze et al.[91] investigated the resistance to cavitation erosion of binary and complex aluminium bronzes of the following nominal compositions: CuAl2, CuAl6, CuAl9, CuAl9Ni3Fe2 and CuAl9Ni5Fe4. The metal loss against time of exposure of these alloys is given in Fig. 8.4, Chapter 8. It shows that complex alloys are much more resistant to cavitation erosion than binary alloys.

The resistance to cavitation erosion of single phase alloys increases significantly with the aluminium content. It is a function of the degree of work hardening and type of deformation of the α-phase resulting from the hammering effect of cavitation. In the case of the two complex alloys tested, the alloy with the higher nickel and iron content was the most resistant to cavitation erosion. Resistance is due to the presence of the κ intermetallic precipitates. The smaller and more evenly distributed κ_{III} and κ_{IV} precipitates provide the best resistance to cavitation erosion. We have seen above that large iron rich particles, which may be due to bad melting practice, can give rise to cavitation erosion. This is confirmed by J. L. Heuze et al. who found that large κ_I and κ_{II} precipitates are made to sink into the α matrix by the hammering effect of cavitation, breaking the bond between them and the matrix and resulting in the precipitates working loose.

Summary of factors affecting resistance to corrosion

1. The oxide film, which is only slightly permeable to liquids, gives a high degree of protection against corrosion.
2. The following is the order of vulnerability to corrosion of certain phases:
   (a) the γ_2 phase – most corrodible but can be avoided by ensuring that the relationship of aluminium to nickel content is in accordance with the formula: Al ≤ 8.2 + Ni/2; alloys conforming to this formula are not vulnerable to corrosion under differential aeration.
   (b) the martensitic β-phase – it is anodic to the α-phase and may experience limited attack which could penetrate the structure if the β-phase is continuous; cooling too fast retains this phase; even a partial transformation of the β phase is sufficient to protect the alloy against severe corrosion, because $\alpha+\kappa_{III}$ forms a protective envelope around the β grains, thus isolating the more anodic elements; heat treatment will reduce the β-phase (see below);

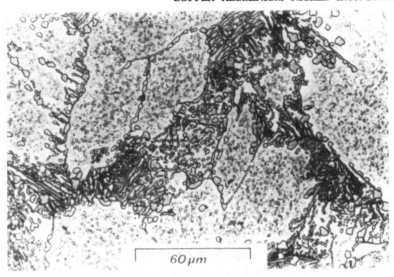

Fig. 13.17 Microstructure of as-cast nickel–aluminium bronze,
by Lorimer et al.[122]

(c) the $\alpha+\kappa_{III}$ phase – the α constituent of this eutectoid is anodic to κ_{III} and may experience some degree of attack, although it is significantly less vulnerable to corrosion than the β-phase; the κ_{III} constituent becomes anodic to α under crevice corrosion, but crevice corrosion is rare.

3. The presence of exposed porosity or oxide inclusions can give rise to corrosion due to differential aeration.

4. Small and evenly distributed κ_{III} and κ_{IV} precipitates provide resistance to cavitation erosion whereas large iron-rich κ_I and κ_{II} precipitates, which may be due to too high an iron content or poor melting practice, can lead to cavitation problems. It would be advisable to restrict the iron content to 4.5% in applications subject to cavitation or requiring good ductility and shock resistance.

5. Iron-rich phases do not normally give rise to ordinary corrosion.

6. Manganese hinders the decomposition of the β-phase and should be limited to 2% for good corrosion resistance.

C – Effects of welding

Effects of welding on cast structure

Lorimer et al.[122] investigated the changes in structure which occur in the heat-affected zone during welding of an 80/10/5/5 nickel aluminium bronze. Fig. 13.17

Fig. 13.18 Microstructure across the heat-affected zone of an electron-beam welded specimen of nickel–aluminium bronze, by Lorimer et al.[122]

shows the pre-weld structure and Fig. 13.18 the changing microstructure across the area affected by the weld. The left hand end of Fig. 13.18 shows the area least affected which resembles the structure shown in Fig. 13.17. The central portion is the most heat-affected area. The fusion line between the heat-affected zone and the weld area to the right is very clearly marked. The rapid cooling of the weld area to the right has resulted in a fine grain structure.

The main effect of the heat generated by the weld is to raise the temperature of the adjoining zone to the point where the α phase and the various κ precipitates reconstitute the high temperature β phase. The subsequent rapid cooling converts the high temperature β phase to large areas of martensitic β in the area adjacent to the weld. The greater the distance from the weld area, the less these transformations occur. The κ phases, especially the iron-rich κ_{II}, only partially re-dissolve. This is because the temperature reached by the heat-affected zone is lower than the temperature at which κ_{II} begins to precipitate after solidification but higher than the temperature at which the $\alpha+\kappa_{III}$ eutectoid begins to form.

The following are some details of the effects of the weld on the microstructure of the heat-affected zone:

(a) The dark-etched martensitic β phase in the heat-affected zone has significantly increased in volume and has replaced the $\alpha+\kappa_{III}$ eutectoid, and a small number of dendritic κ_{II} particles, which had only partially dissolved, have become rounded in appearance.

(b) A thin dark line, located at the boundaries of adjacent α grains, represents a narrow zone of martensitic structure which formed as a result of the partial dissolution of lamellar κ_{III}. It creates a continuous strand of martensitic structure between the large martensitic areas. This continuous martensitic zone aggravates the effects of corrosion.

(c) Another thin dark line, cutting through a light-etched α grain, also represents a narrow strip of martensitic structure which, in this case, has formed around partially dissolved κ_{II} particles that were located in the α grain.

(d) As the distance from the fusion line increases, the proportion of dark-etched areas reduces and the κ_{III}, which was only partially transformed into the β-phase by the heat, has become globular and is surrounded by narrow bands of martensitic structure. This also creates continuous strands of martensitic structure through the area parallel to the line of the weld.

(e) The light coloured α grains in the heat-affected zone have lost their 'peppering' of minute κ_{IV} particles which are a feature of the non heat-affected area to the left and in Fig. 13.17, indicating that these particles re-dissolved in α during heating but did not re-precipitate because of the subsequent rapid cooling.

(f) The α grains nearest the fusion line do however possess particles which have assumed a lath shaped appearance. Hasan et al.[90] examined these particles by transmission electron microscopy and found that they consisted mainly of martensitic structure with a small undissolved portion of κ_{IV} particles at their centre. They concluded that these had been relatively large κ_{IV} particles which did not redissolve completely because of their size whereas smaller particles had done so. Although the temperature would have been at its highest in the proximity of the fusion line, there would not have been sufficient time to dissolve completely the large κ_{IV} particles.

Effect of welding on corrosion resistance

Lorimer et al.[122] had previously carried out corrosion tests on MIG welded samples which also showed that, during welding, the α and κ phases had reconstituted the high temperature β phase in the heat-affected zone, and that, on subsequent cooling, the high temperature β phase transformed to the martensitic β-phase. This phase corroded during the test. The transformations that occur in the heat-affected zone, as the temperature rise and then falls, may result in different amount of martensitic β remaining in its structure, depending on the welding conditions. Interestingly, Lorimer et al.[122] have found that the attack was most severe in the parts of the heat-affected zone where the areas of martensitic β were smallest. This more severe corrosion was thought to be due to the low volume of the anodic martensitic β compared with the larger volume of cathodic κ and α phases.

This test shows the need for the heat treatment recommended below (annealing at 675°C for 2 to 6 hours) to be carried out after welding to restore the heat-affected zone to its pre-welded structure and to relieve thermal stresses which profoundly affect the corrosion behaviour of the alloy.[62] In the case of thin sections (< 6 mm), slow cooling is necessary after annealing to allow time for the decomposition of the β phase. Weill-Couly[185] reports that, in the case of thick sections, multiple passes during welding has the same effect as an annealing treatment and renders the latter superfluous. TIG welding would also have the same effect.

The effectiveness of post-weld heat treatment was confirmed by accelerated corrosion tests (1000 hours in sea water at 80°C) carried out by Soubrier and

Richard.[165] They tested 20 mm thick samples of the following wt % compositions which had been welded by the process indicated below and annealed at various temperatures within the range 600°C to 800°C and air cooled:

	Rolled plate CuAl9Ni5Fe4	Cast plate CuAl9Fe5Ni5	Cast plate CuAl9Ni3Fe2
Cu	81.76	80.00	84.51
Al	9.20	8.90	9.15
Fe	3.72	4.45	3.30
Ni	4.20	5.27	2.30
Welding process	**Carbon arc**	**TIG**	**MIG**

The only corrosion they observed in any of the samples, irrespective of the welding process and annealing temperature, were due to the presence of exposed defects, such as porosity and oxide inclusions, which gave rise to local differential aeration.

If the heat input from welding is limited by allowing time for the metal to cool between passes, the α and κ phases might not reconstitute the high temperature β phase in the heat-affected zone. This would appear to have been the case in experiments carried out by E. A. Culpan and A. G. Foley[63] in which the α constituent was anodic to the κ_{III} constituent of the $\alpha + \kappa_{III}$ eutectoid and corroded preferentially in the heat-affected zone. They suggest that the inability of the oxide film to protect the alloy from corrosion in such a case may be due to residual welding stresses, which accelerate markedly the anodic dissolution by opening up the corrodible areas to ingress by the corrodant, or to the disruptive effect of thermal stresses on the protective oxide film. Had the welded sample been heat treated, the stresses would have been relieved and the corrosion might not have occurred. Extensive corrosion tests carried out by D. Arnaud[185] have shown that annealing at 675°C after welding is totally effective in resisting subsequent corrosion.

Experiments carried out by Jones and Rowlands,[104] showed that preferential phase corrosion in the heat-affected zone only occurred if corrosion products could settle (i.e. at low flows) thereby creating a crevice (see 'Resistance to corrosion' above and Chapter 9).

Summary of effects of welding

1. The heat of welding reconstitutes the high temperature β-phase and the subsequent rapid cooling results in large and continuous areas of the corrodible martensitic β-phase in the heat-affected zone. It creates internal stresses which ruptures the protective oxide film and allow ingress of corrosive liquid. It also decreases ductility and toughness.

2. Annealing at 675°C for 2 to 6 hours after welding restores the pre-weld structure and relieves stresses. It is very effective in resisting subsequent corrosion.

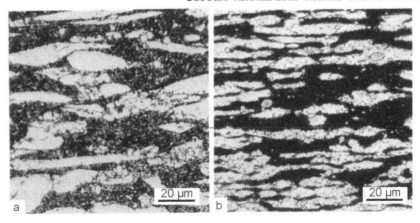

Fig. 13.19   Hot-rolled microstructure of CuAl10Fe5Ni5 alloy.[127] (a) 9/16 in. plate
as hot rolled, commencing at 975°C: α in decomposed β, (b) 3/16 in. plate as hot
rolled, commencing at 925°C: α in partially decomposed β and κ.

3. The presence of exposed porosity or oxide inclusions in the weld or parent
   metal can give rise to corrosion due to differential aeration effect.
4. Multiple passes during welding or the use of TIG welding results in a slower
   rate of cooling, allowing time for the β phase to decompose. This lessens the
   adverse effects of welding on corrosion resistance. The reheating and slow
   cooling of each succeeding pass effectively acts as an annealing heat treatment and may make post weld heat treatment superfluous.

D – Effects of hot and cold working and heat treatment

Effects of hot and cold working on microstructure

Figs. 13.19a–b show the hot-rolled structure of two plates in a CuAl10Fe5Ni5
alloy. At the initial working temperature (925–975°C), the alloy has a β+κ matrix
containing areas of α which become elongated in the direction of working. The
structure of a hot-rolled plate therefore consists of elongated α-phase (white) containing traces of κ_{II} (rosettes) surrounded by dissociated β in the form of α+κ_{III} (see
Fig. 13.1a). The dissociation may well be incomplete depending on the rate of
cooling and the original working temperature.

Effect of grain size on mechanical properties

The relationship between coarseness of the structure and proof strength can be
illustrated by the micro-sections of extruded rod. The back-end of an extruded

Table 13.7   Effect of grain size on properties of extruded rod in CuAl10Fe5Ni5 alloy.[127]

Position	0.1% Proof Strength N mm²	Tensile Strength N mm²	Elongation %	Hardness HV
Front	404	804	16.0	217
Middle	371	792	19.5	223
Back	547	826	14.5	248

product is worked at a lower temperature than the front-end, as a result of chilling in the container, and consequently the structure varies along the length. Fig. 13.20 gives photomicrographs of an extruded rod in a CuAl10Fe5Ni5 alloy taken from the front, middle and back ends.

It will be noted that as the extrusion progresses and the working temperature correspondingly falls, the structure becomes progressively finer. The corresponding mechanical properties are given in Table 13.7 which shows that while the tensile strength and elongation were barely affected, the proof strength increased from 371 N mm⁻² in the middle to 547 N mm⁻² at the back end. In practice, however, little variation is encountered in commercial products, as it can be minimised by careful control of the extrusion process. Any variation, which may occur, can be reduced to a very small amount by subsequent heat treatment.

This effect of working temperature on mechanical properties cannot be explained solely on the basis of strain hardening; the marked increase in proof strength must also be related to the lower working temperature having a refining effect on the microstructure. It will now be shown that a similar combination of properties can be obtained by heat treating the material after it has been submitted to a small amount of cold work.

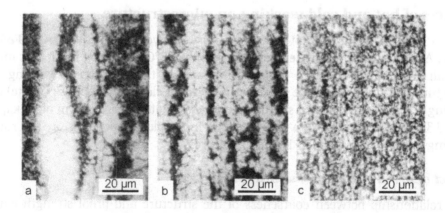

Fig. 13.20   Effect of grain size on microstructure of extruded rod in CuAl10Fe5Ni5 alloy:[127] (a) Front end; (b) Middle; (c) Back end.

Summary of effects of hot and cold working

1. Hot or cold working elongates the grain structure which results mainly in improved proof strength in the direction of the 'flow' of the grain.
2. Cold working additionally refines the grain which improves tensile strength.
3. The hot working temperature and the subsequent rate of cooling determines how much of the β-phase will decompose to α+κ, with important implications for strength and corrosion resistance as previously explained.
4. Hot-working, and especially cold working, introduce internal stresses which need to be relieved by heat treatment.

Effects of heat treatment on microstructure

Heat treatment of castings

Nickel aluminium bronze castings benefit from being annealed at 675°C for to 2–6 hours, depending on section thickness, in order to improve corrosion resistance by the elimination of the martensitic β-phase and modifying the distribution of κ phases. This temperature is recommended by Weill-Couly[185] as a compromise between 650°C, below which there is insufficient dissolution of α+κ$_{III}$, and 700°C above which grain growth becomes significant. It is claimed by some, however, that better results are obtained at 700°C.[178–74] R. Francis[74] reports that heating for 4 to 6 hours at 700°C ± 20°C is more effective in eliminating the β phase in thick wrought sections such as hot rolled plates. Unacceptably long heat treatment times may be necessary to eliminate the β phase in large castings,[71] but the rate of cooling in sand is sufficiently low to have the same effect as the above heat treatment.[185]

Hasan et al.[88] and Culpan and Rose[62] investigated the effect of various heat treatments on the microstructure of cast samples of nickel-aluminium bronze of the following similar (wt %) compositions:

	Al	Ni	Fe	Mn	Si	Cu
Hasan et al.[88]	9.4	4.9	4.4	1.2	0.07	Rom.
Culpan and Rose[62]	9.5	4.7	4.3	1.0	–	Rom.

Both sets of researchers investigated the effects of annealing at 675°C for varying periods of time. Culpan and Rose[62] additionally investigated higher annealing temperatures.

Heat treatment at 675°C

Figs 13.21b to 13.21h – some by Hasan et al.[88] and some by Culpan and Rose[62] – show the effects on the microstructure of annealing at 675°C for various lengths of time, followed by cooling in air. An as-cast structure, by Hasan et al.,[88] is shown on Fig. 13.21a for comparison.

The following are the resulting effects on the microstructure:

- There was coarsening of the structure with increased annealing time (compare Fig. 13.21a with Figs. 13.21e and 13.21h).

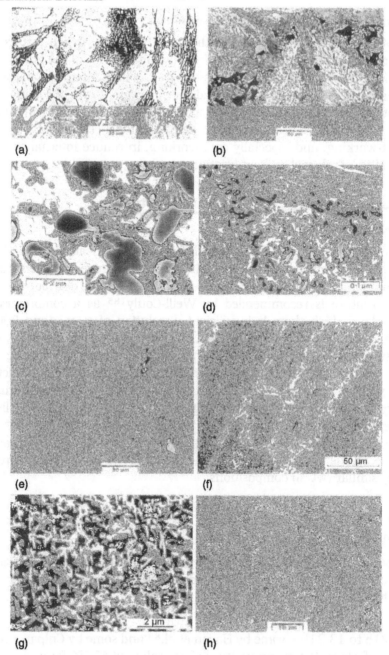

Fig. 13.21 Effect on microstructure of annealing a nickel–aluminium bronze alloy at 675°C for different lengths of time followed by air cooling: (a) *As-cast, (b) *Annealed for 2 hours, (c) Precipitates in α after annealing for 2 hours, (d) *Precipitates in β after annealing for 2 hours, (e) *Annealed for 5 hours, (f) Annealed for 6 hours, (g) Precipitates in α after annealing for 16 hours, (h) *Annealed for 19 hours. *by Hasan et al.[88] +by Culpan and Rose.[62]

- The fine particles present in the martensitic β-phase in the as-cast condition, returned into solution on heating, thus reconstituting the high temperature β phase. These particles (see Figs 13.21d, 13.21f and 13.21h), which were mentioned above (see Tables 13.3 and 13.4), are similar in composition to κ_{III} and are likewise based on NiAl. They re-precipitated during air cooling which accounts for the dark etching of the martensitic β-phase (see Figs 13.21b and 13.21e). Quenching after annealing suppresses these precipitates and results in a much lighter etched martensitic phase (not illustrated).

- The martensitic β phase is progressively eliminated with length of annealing – very little martensitic structure is left after 5 hours (see Figs. 13.21e and 13.21f) and none after 19 hours (see Fig. 13.21h). Prolonged annealing at 675°C therefore results in the complete transformation of the martensitic β-phase into a very fine dispersion of $\alpha + \kappa_{III}$. The progressive reduction in the proportion of martensitic β-phase in the microstructure did not markedly affect its composition. Hasan et al.[88] suggest that this is because the decomposition of the β-phase takes place at its interface with its matrix and that the aluminium concentration at the surface of β is reduced by diffusion into the surrounding α regions until it is low enough to permit β to decompose into $\alpha + \kappa_{III}$. The elimination of the martensitic β-phase is the most important effect of the heat treatment from the point of view of corrosion resistance.

- Increased density of κ_V precipitates in the α phase with length of annealing (see Figs. 13.21b, 13.21e, 13.21f and 13.21h). They were of two different types: one type was lath-shaped resembling a thin rectangular plate and the other consisted of a near-spherical shape at the centre of a thin rectangular plate with rounded corners. The different shapes of these precipitates can be seen in Fig. 13.21c. The presence of this finely dispersed cathodic κ_V phase within the α grains results in a much more general corrosion attack on the α-phase instead of the concentrated and penetrating attack on the α constituent of the $\alpha + \kappa_{III}$ eutectoid which may occur in as-cast material.

The lath-shaped particles, designated κ_V by Culpan and Rose,[62] had an ordered bcc structure based on NiAl and were composed of approximately 50% each of Ni and Al with traces of Cu and Fe. Culpan and Rose[62] report that, after annealing for 6 hours, their size was ~1μm×0.1μm, and after 16 hours they had grown to ~1μm×0.5μm (see Fig. 13.21g).

Hasan et al.[88] report that the spheroidal/lath particles were 'composite precipitates' in which the spheroidal centre had a different composition and structure to that of the lath-like extremities. The lath-like extremities had the same composition and structure as the lath-shaped κ_V particles, whereas the centre was similar to κ_{IV} with an ordered bcc structure based on Fe_3Al and were composed of approximately 75% Fe and 25% Al. The centre would have been as-cast κ_{IV} particles which nucleated the lath-like extremities. The analysis of the centre and extremities of these composite particles is given in Table 13.8 together with that of the κ_V lath-like particles.

- Prolonged annealing at 675°C resulted in the growth of the larger composite precipitates at the expense of the smaller lath-like (κ_V) particles which returned into solution, with the result that most particles had iron-rich centres even after 19 hours of annealing (see Fig. 13.21g).
- The presence of the κ precipitates increased the hardness of the alloy from 168 HV as-cast to a maximum of 179 HV after 2 hours annealing. It returned to the as-cast figure after prolonged annealing.

Table 13.8 Effect of on the composition of κ phases of annealing at 675°C for 2 and 6 hours.

Phases	Heat Treatment	wt % composition of κ phases				
		Al	Mn	Fe	Ni	Cu
Lath-like particles (κ_V)	*675°C for 2 hrs	28.5	1.9	14.6	42.1	12.9
	+675°C for 2 hrs	23±2	1±0.3	25±3	39±2	11±1
	+675°C for 6 hrs	27±4	1.5±0.3	27±4	35±3	10±2
'Composite' particles: Lath-like	*675°C for 2 hrs (ends)	30.5	1.9	16.8	36.1	14.7
precipitates formed on as-cast spheroidal κ_{IV} nuclei	*675°C for 2 hrs (centre)	13.6	2.3	71.8	5.0	4.0

*Typical composition only, by Hasan et al.[88]
+ Analysis by Culpan and Rose[62] using scanning transmission electron microscope (STEM)/ extraction replica, based on 20–30 specimens

Alloy compositions:	Al	Fe	Ni	Mn	Si	Cu
Hasan et al.[88]	9.40	4.40	4.90	1.20	0.07	bal
Culpan and Rose[62]	9.42	4.24	4.70	1.09	–	80.55

Culpan and Rose[62] report that, following heat treatment at 675°C, there was very little attack in the heat-affected zone of a welded casting, although the microstructure was similar to that of an as-cast structure which had shown severe selective phase attack.

Heat treatment above 675°C

Increasing the annealing temperature above 675°C had a similar effect to prolonged annealing at that temperature, except that, as reported by Culpan and Rose,[62] the lath-like κ_V particles became significantly larger as the annealing temperature rose, and assumed the form of large rods of approximate size 10μm × μ2μm at 840°C (see Fig. 13.22a). These particles were probably composite particles, but Culpan and Rose,[62] not being aware of this, analysed them as whole particles (see Table 13.9). The apparent increase in the iron content, and consequent proportional reduction in nickel and aluminium, with increases in time and temperature of annealing, would seem to suggest that most particles had a Fe_3Al (κ_{IV}) core.

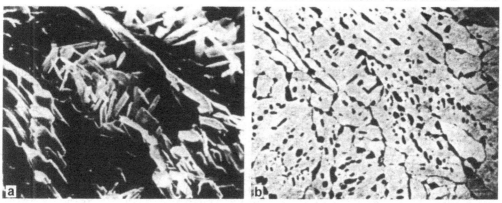

Fig. 13.22   Effect of annealing above 675°C on the microstructure of a nickel-aluminium bronze casting by Culpan and Rose.[62] (a) Precipitates in α after annealing at 840°C for 3 hours, (b) Annealed at 860°C for 72 hours.

At temperatures above 820–850°C, depending on its composition, the alloy transforms to α+β+κ which has the effect of partially or completely spheroidising the lamellar $κ_{III}$ phase (see Fig. 13.22b) and effectively dispersing the continuous grain boundary eutectoid. It seems that, after 72 hours exposure at 840°C, the various κ phases have re-dissolved and then precipitated on cooling into a single spheroidal κ particle. Culpan and Rose[62] thought that it might represent an equilibrium κ precipitate in the α grains which is a composite of both Fe_3Al and NiAl based compounds. This would seem to be a parallel development to the presence of composite particles which Hasan et al.[88] reported were already forming after annealing at 675°C.

Culpan and Rose[62] report that following heat treatment at 830°C of a welded casting, the resistance to corrosion in the heat-affected zone had much improved compared with that of cast material. The disadvantage of this higher temperature heat treatment is that significant amounts of β can be formed which, if not allowed time subsequently to decompose, can cause severe corrosion, particularly if it is continuous in the structure.

Quenching and tempering
S. Lu et al.[123] investigated the effect of quenching at 950°C followed by tempering for 2 hours at 400°C on a casting of the under-mentioned composition and of the following dimension: 20mm×12mm×120mm. As explained in Chapter 6, the object of this heat treatment is to improve hardness and/or strength, but it is at the expense of elongation. S. Lu et al.[123] point out, however, that by using the method of partial quenching, the excessive reduction in elongation can be avoided. The advantage of improving hardness is principally to improve wear properties (see below and Chapter 10).

Element	Al	Fe	Ni	Cu
Wt %	10.6	4.4	4.5	balance

Table 13.9 Effect of increasing the time and temperature of annealing on the chemical composition of intermetallic particles in a casting by Culpan and Rose.[62]

Phases	Heat Treatment	Composition wt %				
		Al	Mn	Fe	Ni	Cu
Lath-like particles:						
(probably κ_V)	675°C for 2 hrs	23±2	1±0.3	25±3	39±2	11±1
ditto	675°C for 6 hrs	27±4	1.5±0.3	27±4	35±3	10±2
(probably	675°C for 16 hrs	20±3	1.3±0.3	34±3	35±2	10±1
composite)	740°C	21±2	1.8±0.5	33±3	35±2	9±2
ditto	790°C	18±2	1.6±0.3	40±2	30±1	10±1
ditto	840°C	17±3	1.7±0.1	39±5	32±3	10±1
ditto and large						
Large spheroidal (thought to be in equilibrium condition)	840°C for 3 days	17±2	1.6±0.2	40±3	31±3	10±1

Analysis by scanning transmission electron microscope/extraction replica, based on 20–30 specimens each.

Fig. 13.23 shows the microstructure (a) as cast, (b) as quenched at 950°C and (c) as tempered at 400°C for 2 hours. The as-cast microstructure (Fig. 13.23a) is similar to that shown on Fig. 13.12c with its relatively large grains of α-phase. Some small and dendritic-shaped κ_{II} particles, which nucleated initially at the α/β boundary, have become enveloped by the growing α-phase and lamellar κ_{III} particles which have formed at the α/β grain boundary.

After quenching at 950°C (Fig. 13.23b), the α-phase has fragmented into evenly distributed slender needles with a Widmanstätten structure. The martensitic β-phase surrounds the α needles forming with them a 'Coarse Bainite' structure. Small particles of κ_{II} (not visible) are still present in the martensitic β-phase after quenching. After tempering at 400°C for 2 hours (Fig. 13.23c), the α-phase needles have become even smaller and more evenly distributed and more κ_{II} particles (not visible) have formed. If the tempering temperature is raised to 450°C, the κ_{II} particles become coarser.

Brezina[33] reports that quenching and tempering improves corrosion resistance due to the dispersed distribution of the κ-precipitates in the microstructure and their uniform composition.

Summary of effects of heat treatment of castings

1. The main reason for annealing a casting at 675°C for 2 to 6 hours followed by cooling in air, is to decompose any residual β-phase into α+κ_{III} in order to

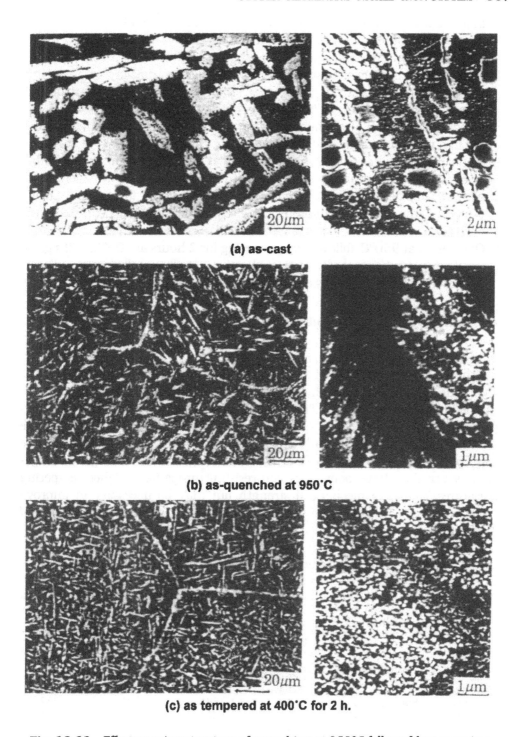

(a) as-cast

(b) as-quenched at 950˚C

(c) as tempered at 400˚C for 2 h.

Fig. 13.23   Effect on microstructure of quenching at 950°C followed by tempering at 400°C for 2 hours, by S Lu et al.[123]

improve corrosion resistance without detriment to mechanical properties. It, in fact, marginally improves mechanical properties. Annealing for 2 hours significantly improves hardness due to the presence of fine κ precipitates.

2. Increasing annealing time at 675°C will completely transform β-phase into α+κ$_{III}$ thereby improving corrosion resistance. It coarsens the structure, however, which is likely to improve ductility at the expense of hardness and tensile properties. It also refines and disperses the nickel-rich κ-phase, resulting in a more superficial and less penetrating corrosion attack.

3. Annealing at higher temperatures has similar effects to prolonged annealing at 675°C except that more high temperature β-phase will form which, if not allowed sufficient time to decompose, will transform to the martensitic β-phase. If the latter is continuous it will render the structure prone to severe corrosion.

4. Quenching at 950°C, followed by tempering for 2 hours at 400°C, will significantly improve hardness for wear applications. It also results in a fine dispersion of the κ precipitates resulting in good corrosion resistance.

Heat treatment of wrought products

Effect of tempering after cold working

One of the most common forms of heat treatment consists in tempering after hot or cold working in order to remove internal stresses and adjust properties to meet the requirements of the application. This consists in re-heating to a certain temperature, typically 400–540°C, for 1–2 hours and cooling in air. Figures given in Table 13.10 show that tempering a lightly drawn rod at the moderately low temperature of 500°C can result in exceptionally high proof strength values with good elongation. However they have poor corrosion-resisting properties. Higher tempering temperatures lower the mechanical strength but may be necessary to improve corrosion resistance by eliminating the martensitic β-phase (see above: 'Effects of heat treatment on cast structure').

Table 13.10 Effect of tempering on the mechanical properties of lightly drawn extruded rod in CuAl10Fe5Ni5 alloy.[127]

Form	Heat Treatment	Mechanical Properties			
	Tempering conditions	0.1% Proof Strength N mm^{-2}	Tensile Strength N mm^{-2}	Elongation %	Hardness HB
Extruded Rod	None	455	798	23	248
lightly drawn	500°C for 1 Hr	510	832	18	262
2.5% reduction	600°C for 1 Hr	515	821	21	281
	700°C for 1 Hr	441	753	24	229
	800°C for 1 Hr	377	742	28	197

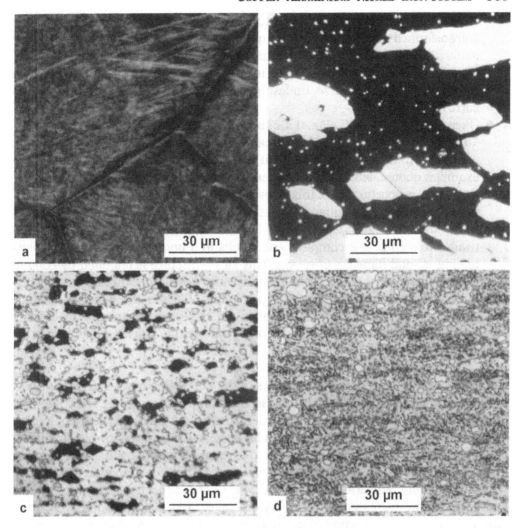

Fig. 13.24   Effect of hot work and quenching on the microstructure of a
CuAl10Fe5Ni5 alloy.[127] (a) Quenched from 1000°C: martensitic β with some α at
grain boundaries, (b) Hot worked and quenched from 900°C: martensitic β, some α
and κ, (c) Hot worked, held at 800°C and quenched: κ in α with some β, (d) Hot
worked, held at 750°C and quenched: κ in α.

Hot worked and quenched microstructure
As previously mentioned, the structure of Cu–Al–Fe–Ni alloy after hot working
consists of a β+κ matrix containing areas of α, which become elongated in the
direction of working (see Fig. 13.19).

Fig. 13.24 shows the effect on the microstructure of a 3/16 in. plate of a
CuAl10Fe5Ni5 alloy of hot working and quenching from various temperatures.

- Quenched from 1000°C (Fig. 13.24a). At this temperature the all-β structure is transformed to martensitic β on quenching in the same way as for binary alloys. In the example shown, transformation has not been completely prevented and a small proportion of α is evident at the grain boundaries. Such a heat treatment is, of course, undesirable from the point of view of resistance to corrosion.
- Quenched from 900°C (Fig. 13.24b), 800°C (Fig. 13.24c) and 750°C (Fig. 13.24d). This form of heat treatment gives structures approximately equivalent to those at equilibrium at the quenching temperature. In the three examples quoted, which involve quenching from progressively lower temperatures, the proportion of α increases as the temperature falls and is accompanied, in the case of quenching at 900°C and 800°C, by the formation of rounded particles of κ_{II}. In the case of quenching at 750°C (Fig. 13.24d), the transformation of β is complete and results in a homogeneous α matrix containing κ_{II} and lamellae of $\alpha+\kappa_{III}$. This ideal structure is obtained by cooling from the hot-working temperature sufficiently slowly to eliminate any β without excessive coarsening of the κ precipitate. It gives the most satisfactory combination of proof strength, ductility and corrosion resistance provided κ is in a finely divided state, as a coarse precipitate results in a severe drop in proof strength.

The effect on the mechanical properties of the above treatment is shown graphically in Fig. 13.25. It shows that the mechanical properties of a CuAl10Fe5Ni5 type of alloy are closely related to the microstructure. Material, which contains a large proportion of martensitic β, generally possesses high strength but low ductility, and as the proportion of martensitic β is reduced, with a consequent increase in the amount of α and κ, the strength is lowered and the elongation raised. The critical quenching temperature is around 800°C: above this temperature, strength and hardness are increased but ductility drastically reduced. Thus materials containing a very high proportion of martensitic β do not have satisfactory properties for most commercial applications in view of their low ductility. They also have poor corrosion resisting properties.

Annealing at 750°C for 1–2 hours followed by air cooling would give a similar microstructure to Fig. 13.24d and soaking at lower temperatures would give a structure of increasing fineness as the temperature decreases. On the other hand, annealing above 800°C, followed by air cooling, would result in a coarse κ precipitate with detrimental effect, particularly upon cold-working properties.

Hot worked, quenched and tempered microstructure
The previous heat treatment is likely to leave internal stresses in the material. It may therefore be desirable to temper after quenching.

The most common and readily controlled method of heat treatment involves quenching from a high temperature to obtain a martensitic structure, followed by

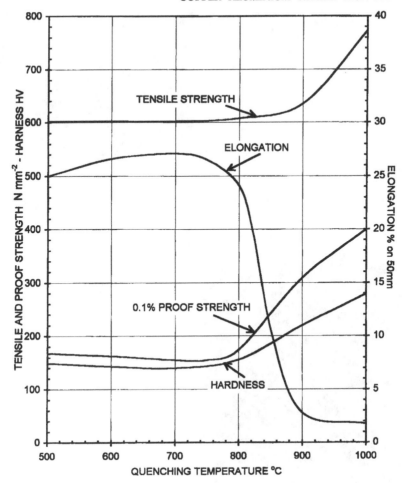

Fig. 13.25   Mechanical properties of a CuAl10Fe5Ni5 alloy after slow cooling from 1000°C to various temperatures and quenching.[47]

tempering at a lower temperature to give the required degree of decomposition of the martensitic β. This form of heat treatment is directly comparable to the tempering of steels.

Fig. 13.26 shows the effect of hot working, quenching from 1000°C and tempering at various temperature on the microstructure of a rolled plate of a CuAl10Fe5Ni5 alloy. Properties quoted in Table 13.11 resulting from the above heat treatment (quenched from 1000°C and tempering at different temperatures for short times) indicate that while high strength may be obtained by this method, elongation figures can remain exceedingly low. However, by extending the period of tempering to two hours, the degree of dissociation can be made more complete and a superior ductility obtained as may be seen from Fig. 13.27 and Table 13.11.

Fig. 13.26 Effect of hot working, quenching from 1000°C and tempering at various temperatures on the microstructure of a plate of a CuAl10Fe5Ni5 alloy.[127] (a) Tempered $1/_2$ hour at 600°C, (b) Tempered 10 minutes at 800°C, (c) Tempered 10 minutes at 900°C.

Alternatively, quenching from lower temperatures will give improved ductility as the proportion of martensitic β present on quenching is reduced. It is therefore frequently of advantage in commercial practice to quench from 900°C, when the structure contains a considerable proportion of the α phase. Subsequent tempering of the remaining β gives a much more favourable combination of proof strength and elongation values as shown in Table 13.11.

This treatment results in the dissociation of the martensitic β to form an extremely fine mass of α+κ$_{III}$ which, at the low temperature of 600°C (Fig. 13.26a), cannot be resolved clearly with the optical microscope. Tempering at progressively

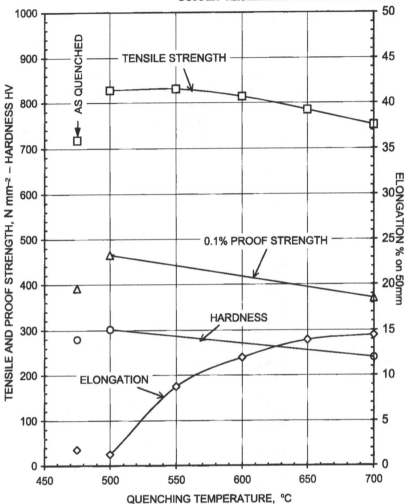

Fig. 13.27 Mechanical properties of CuAl10Fe5Ni5 alloy quenched at 1000°C and tempered for 2 hr various temperatures.[127]

higher temperatures allows a greater amount of diffusion and the precipitate becomes coarser. The effect of tempering for 10 minutes at 800°C is shown on Fig. 13.26b and at 900°C on Fig. 13.26c. Both these photomicrographs illustrate clearly the needle like structure of the $\alpha+\kappa_{III}$ eutectoid.

When quenched alloys are re-heated at moderate temperatures, a κ_{IV} phase comes out of solution in a finely divided form which results in a precipitation hardening effect and increases both proof strength and hardness. This is illustrated by the properties quoted in Table 13.12 for samples tempered at low temperatures.

Table 13.11 Properties of quenched and tempered rolled plates and extended rods.[127]

Form	Quench Temp.	Tempering conditions	0.1% Proof Strength N mm^{-2}	Tensile Strength N mm^{-2}	Elongation %	Hardness HV
Rolled plate	1000°C	600°C for ½ h	523	824	7	287
	1000°C	800°C for ½ h	303	773	9	221
	1000°C	900°C for ½ h	351	668	3	268
	1000°C	500°C for 2 h	464	866	2*	300*
	1000°C	600°C for 2 h	433	819	12*	260*
	1000°C	700°C for 2 h	340	758	14*	240*

*Values taken from published curves.   **HB**

Form	Quench Temp.	Tempering conditions	0.1% Proof Strength	Tensile Strength	Elongation %	Hardness HB
Extruded Rod	900°C	None	362	932	8	235
	900°C	500°C for 1 h	467	850	15	238
	900°C	600°C for 1 h	470	827	18	244
	900°C	700°C for 1 h	421	789	22	218
	900°C	800°C for 1 h	371	767	26	192

Quenching and tempering after hot-working to improve hardness
S. Lu et al.,[123] investigated the effect on hardness of hot-working, quenching and tempering a nickel aluminium bronze forging of the undermentioned composition.

Element	Al	Fe	Ni	Cu
wt %	10.6	4.4	4.5	balance

The advantage of improving hardness is principally to improve wear properties (see below and Chapter 10). Fig. 13.28 shows the microstructure (a) as forged at 950°C, (b) as forged at 980°C and immediately quenched and (c) as tempered at 400°C for 2 hours.

Table 13.12   Comparison of properties of a low nickel and iron alloy with other alloys before and after heat treatment.[127]

% Composition (bal Cu)			Condition and heat treatment	0.1% Proof N mm⁻²	Tensile Strength N mm⁻²	Elongation %	Izod Joules
Al	**Ni**	**Fe**					
9.3	2.0	2.7	As extruded	340	680	32	41
			Heat-treated*	340	711	33	53
9.0	–	1.8	As extruded	340	588	36	49
			Heat-treated*	216	588	48	76
9.6	5.0	5.0	As extruded	356	758	30	26
			Heat-treated*	325	696	32	35

*Quenched from 900°C and tempered at 600°C for 1 hour

The microstructure as forged at 950°C and cooled normally (Fig. 13.28a) is similar to the as-cast structure of the same alloy (Fig. 13.23a) except that the α-grains are smaller and more evenly distributed. As in the case of the as-cast structure, some small and dendritic-shaped κ_{II} particles, which nucleated initially at the α/β boundary, have become enveloped by the growing α-phase and lamellar κ_{III} particles have formed at the α/β grain boundary. The microstructure as forged at 980°C and quenched immediately when its temperature was ~900°C (Fig. 13.28b), is similar to that of the casting quenched at 950°C (Fig. 13.23b) but with a greater proportion of the α-phase. Tempering at 400°C for 2 hours (Fig. 13.28c) increases the proportion of κ-precipitates. The latter become denser and coarser if tempering is done at 450°C for the same period.

As was shown in Chapter 6 (Table 6.4), this heat treatment results in a higher hardness figure (38 HRC) than those obtained with the heat treatment described in Tables 13.10 and 13.11, bearing in mind that HRC numbers up to 40 are approximately 1/10th of corresponding Vickers or Brinell hardness numbers. This higher hardness figure is however obtained at the expense of elongation which is very low (1.96%).

Summary of effects of heat treatment of wrought alloys

1. Tempering at 700°C after cold-working is likely to give the best combination of strength and corrosion resistance.
2. Quenching at 1000°C after hot working will give the best combination of high strength and hardness but at the expense of ductility and corrosion resistance.
3. Quenching from 900°C after hot working will result in a better combination of ductility and strength but at the expense of hardening. Corrosion resistance will remain poor.

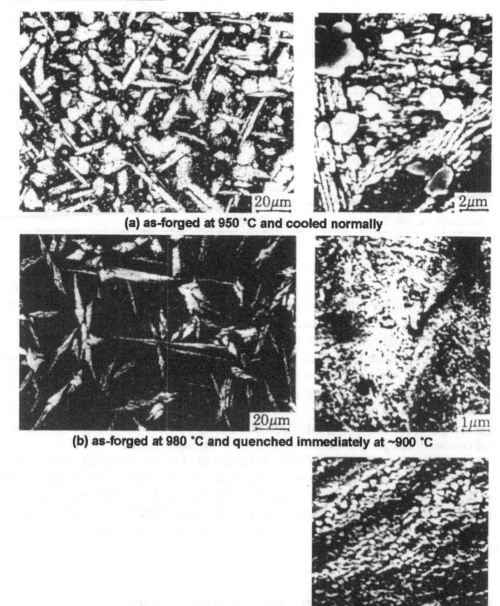

(a) as-forged at 950 °C and cooled normally

(b) as-forged at 980 °C and quenched immediately at ~900 °C

(c) as-tempered at 400 °C
for 2 h after (b)

Fig. 13.28   Effect on microstructure of various post-forging treatments, by S. Lu
et al.[123]

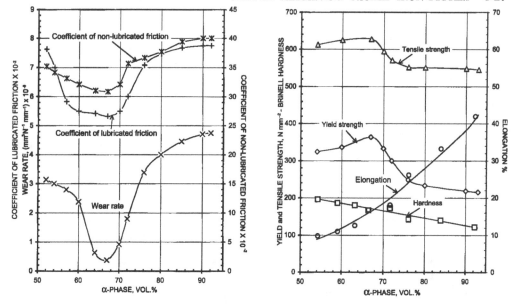

Fig. 13.29   Effect of percentage volume of α phase on wear and mechanical properties by Yuanyuan.[190] (a) Effect on wear rate and on lubricated and non-lubricated coefficients of friction, (b) Effect on hardness and mechanical properties.

4. Tempering for one hour at 700°C after quenching at 900°C will improve corrosion resistance without too much coarsening of the grain.
5. Slow cooling to 750°C after hot working followed by quenching is likely to give the best combination of strength, ductility and corrosion resistance.
6. As in the case of castings, quenching forgings at 950°C, followed by tempering for 2 hours at 400°C or 450°C, will significantly improve hardness for wear applications but at the expense of ductility. It also results in a fine dispersion of the κ precipitates resulting in good corrosion resistance.

E – Wear resistance

Effect of microstructure on wear performance

Yuanyuan Li et al.[190] have carried out wear tests on nickel–aluminium bronzes within the following ranges of wt % compositions:

Cu	Al	Fe	Ni	Mn
Bal	8–13	2–5	1–3	0.5–3

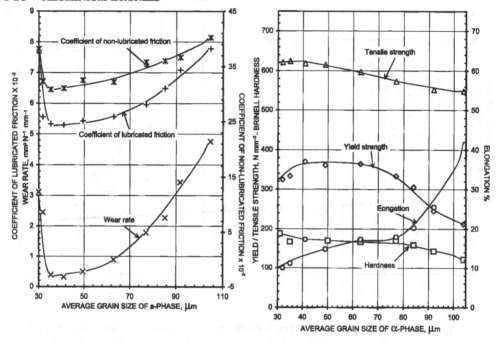

Fig. 13.30 Effect of average α grain size on wear and mechanical properties by Yuanyuan.[190] (a) Effect on wear rate and on lubricated and non-lubricated coefficients of friction, (b) Effect on hardness and mechanical properties.

They added unspecified quantities of titanium and boron as 'modifying elements' and of lead as an 'anti-friction component'.

The tests were carried out at only one combination of pressure (7.486 MPa) and sliding velocity (3.989 m s⁻¹) but under both non-lubricated and lubricated conditions. The lubricant used was WA elevator lubricating oil.

As shown on Fig. 13.29, they found that the proportion of α in the microstructure had a very significant effect on wear performance with the best results obtained when the α phase accounted for 67% of the structure.

They also found, as shown on Fig. 13.30, that average α grain size had similarly a very significant effect on wear performance with the best results obtained when the average α grain size lies between 33 and 46 µm.

The reason why these two types of variation in structure affect friction and wear in similar ways lie in their similar effect on the two opposite properties of plasticity and hardness. A high proportion of α (over 70%) or a large grain size (over 63 µm) render the structure soft, more plastic and more prone to adhesion. Consequently these structures result in a high coefficient of friction and wear rate. A high proportion of β and a small grain size, on the other hand, render the structure hard and brittle and cause it to be abrasive which leads to rapid deterio-

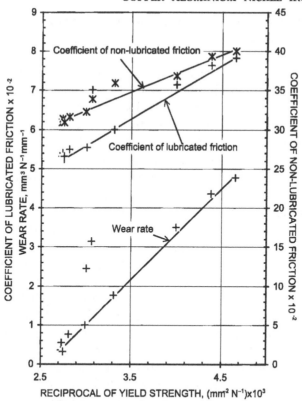

Fig. 13.31   Wear rate and coefficients of friction versus the reciprocal of yield strength by Yuanyuan.[190]

ration of at least one of the surfaces in contact. An other contributory factor to abrasion is the effect of the proportion of α and its grain size on the role played by the κ particles in wear resistance. As explained below, these κ particles can have a beneficial effect on wear resistance but a low proportion of α or a small grain size may result in these hard particles becoming dislodged from the α matrix and causing abrasion. The right balance of α and β or a medium grain size results in the lowest friction and wear rate as well as in the most favourable tensile and yield strength.

Unfortunately, the authors do not say at what precise composition or at what cooling rate both the 67% α proportion of the structure and the 35 μm average α grain size were obtained; nor do they say what was the average α grain size corresponding to the 67% α proportion of the structure nor what was the percentage α proportion of the structure when the average α grain size was 35 μm. The aluminium content is of course the main factor governing the percentage proportion of the α phase and the cooling rate and iron content the factors governing grain size.

They have also made the following valuable observations which can be deduced from Figs. 13.29 and 13.30:

- The wear rate and the coefficients of both lubricated and non-lubricated friction follow very similar trends. It should be noted that the wear rate relates to the applied pressure and velocity mentioned above. It will be seen from Figs. 13.29 and 13.30 that the trend of the yield strength is the inverse of that of the wear trend. In fact, both the coefficients of non-lubricated and lubricated friction and the wear rate are inversely proportional to the yield strength as shown in Fig. 13.31 except at extremely low α grain size associated with very small percentage volume of α. It will be noted that the peak of the tensile strength curve also corresponds to the lowest point on the wear curve. This relationship between wear and tensile properties may simply indicate that the type of structure that causes least wear is also the type of structure that results in highest yield strength. It does not necessarily mean therefore that yield strength has an influence on wear.
- Figs 13.29 and 13.30 show no apparent relationship between hardness and wear rate corresponding to the percentage volume of the α phase or to grain size. But, as explained above, the effects of structure on plasticity and hardness is known to affect friction and wear, even though it may not be possible to express the relationship by a mathematical formula.
- The hard κ particles embedded in the relatively soft α matrix are an ideal feature for wear performance. They reduce the tendency of the sliding pairs in a bearing adhering to one another, thereby reducing friction and wear. Furthermore, if the κ particles are harder than the material of the sliding counterpart, they can provide excellent resistance to abrasion. But, as explained above, if the percentage volume of the α matrix and its average grain size is not sufficient to retain the κ particles in the matrix, the peeling off of these hard particles can themselves cause severe abrasion. This would happen if the volume of α is less than 60% or if the average α grain size is less than 33 μm.

Alloys with high aluminium content
We have seen in Chapters 10 and 12 that Cu–Al–Fe alloys with aluminium contents of 14–15% have exceptional hardness and are used in dies for drawing steel sheets. A nickel content of 5–6% or, alternatively, ~1% nickel with a small addition of titanium of 0.3–0.45% resulted in the highest hardness figures. These alloys rely however for their hardness on the presence of the γ_2 phase which would be a problem in a corrosive environment. In an alloy with the required hardness for use as a die in sheet drawing, approximately 50% of the microstructure would consist of that phase.[154] On the other hand, we have seen above that, in the case of a Cu–Al–Fe–Ni alloy having more than 11% aluminium, and at

temperatures between 600°C and 575°C, for β to transform fully into the corrodible α+γ$_2$ eutectoid, the cooling rate must be less than 0.5°K min^{-1} whereas the transformation of β into α+κ$_{III}$ takes place even at cooling rates as high as 5°K min^{-1}. A Cu–Al–Fe–Ni alloy with aluminium content of 14–15% would therefore have appreciably less γ$_2$ phase than a Cu–Al–Fe alloy and would be less corrodible.

Roucka et al.[154] have shown that an alloy containing 14.9% Al, 4.9% Fe and 5.2% Ni, (in other words, the standard nickel–aluminium bronze) would have comparable hardness and tensile properties to those of an alloy containing 14.6% Al, 3.4% Fe and 1% Ni, when both cooled at 1.8°K min^{-1} from 960–650°C and at 1.0°K min^{-1} from 650–500°C (see Chapter 10). At such cooling rates, the nickel aluminium bronze would have little, if any, of the corrodible β and possibly none of the even more corrodible γ$_2$ phase.

Summary of effect of microstructure on wear rate

1. A high proportion of α (over 70%) or a large grain size (over 63 μm) render the structure soft, more plastic and more prone to adhesion. A high proportion of β and a small grain size, on the other hand, render the structure hard and brittle. The best wear performance is obtained at an in-between combination of a 67% proportion of α phase in the structure and an average α grain size that lies between 33 and 46 μm. The right balance of α and β and a medium grain size also results in the most favourable tensile and yield strength.

2. The aluminium content is the main factor governing the percentage proportion of the α phase and the cooling rate and the iron content the factors governing grain size (unfortunately the alloy composition and cooling rates to achieve the above combinations is not known).

3. Wear performance is not related to hardness alone but relies on the combination of hard κ particles, imbedded in a relatively soft α matrix. The above combination of percentage volume of α phase and range of grain size is ideal for wear performance.

4. For applications requiring exceptional hardness, such as in dies for drawing steel sheets, an alloy with aluminium contents of 14–15% and a nickel content of 5–6% (or ~1% nickel with a small addition of titanium of 0.3–0.45%) provides the highest hardness figures. Such an alloy relies however for its hardness on the presence of the γ$_2$ phase which would be a problem in a corrosive environment. It is also exceptionally brittle.

14

COPPER–MANGANESE–ALUMINIUM–IRON–NICKEL SYSTEM

Copper–manganese–aluminium–iron–nickel–alloys

Alloys that are rich in manganese have better castability due to their lower melting points and greater fluidity. They are sometimes preferred therefore for large marine propellers.[33] They contain typically 12% manganese, 8–9% aluminium and 3% each of nickel and iron.

Equilibrium diagram

Fig. 14.1a shows a section through the equilibrium diagram of a copper-manganese-aluminium system containing 12% manganese, 8% aluminium, 2.8% iron and 2% nickel. Bearing in mind that 6% manganese is equivalent to 1% aluminium, the equivalent aluminium content of this system is 2% higher than actual. Plotted against this equivalent aluminium content (Fig. 14.1b), the equilibrium diagram looks very similar to the binary diagram (see Chapter 11). The presence of iron and low percentage of nickel have little effect on the phase boundaries. Iron causes the precipitation of the κ phase in various forms below 850°C.

The solubility of manganese in the α phase above 650°C is 8%, whereas that in the β phase at 650°C is 26% and increases sharply with temperature.

The standard Cu–Mn–Al–Ni–Fe alloy, CuMn11Al8Fe3Ni3-C (formerly BS 1400 CMA1) has a range of aluminium content of 7.5% to 8.5% and, as may be seen from Fig. 14.1a, it has a freezing range of about 990–935°C. It solidifies into an all-β structure.

Iqbal, Hasan and Lorimer[101] have investigated the microstructure development of an alloy of the composition, given in Table 14.1, which complies with CuMn11Al8Fe3Ni3-C and ASTM C95700. The experiment consisted in heating a number of specimens to 880°C for 30 minutes, cooling them slowly at 5°K min⁻¹ and quenching them in succession at various temperatures.

Their findings were as follows (see Fig. 14.2a-g). The nature and composition of the various phases are given below:

- Quenched at 880°C, the alloy has the structure shown in Fig. 14.2a consisting of β plus non-dissolved dendritic shaped particles. Iqbal et al. report that these same particles were also observed in a sample which was heated to 950°C (close to the solidus) for two hours prior to quenching.
- Quenched at 825°C, the alloy has the structure shown in Fig. 14.2b. The α phase is the first phase to form from the β phase. It formed at the grain

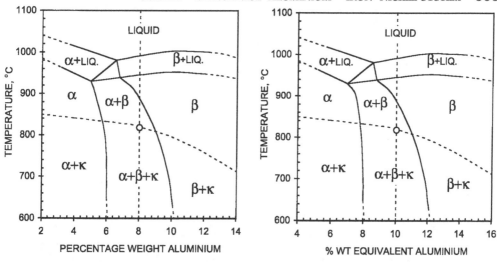

Fig. 14.1 Section through equilibrium diagram of Cu–Mn–Al–Ni–Fe alloy containing 12% manganese, 8% aluminium, 2.8% iron and 2% nickel.[101]. (a) Plotted against actual aluminium content, (b) Plotted against equivalent aluminium content.

boundary of the latter as well as around the existing dendritic-shaped particles.

- Quenched at 775°C, the alloy has the structure shown in Fig. 14.2c. Small dendritic-shaped particles have begun to appear in the β phase.
- Quenched at 730°C, the alloy has the structure shown in Fig. 14.2d and 14.2e. Some of the small dendritic-shaped particles have been enveloped in the growing α phase. In Fig. 14.2e, transmission electron microscopy reveals the formation of globular precipitates at the α/β grain boundary.
- Quenched at 670°C, the alloy has the structure shown in Fig. 14.2f. The growth of the α phase has moved the α/β grain boundaries and enveloped the globular precipitates but more of the latter formed at the displaced α/β boundaries. Some cuboid-shaped particles have also precipitated in the α grains
- Quenched at 500°C, the alloy has the structure shown in Fig. 14.2g which is similar to the as-cast structure Fig. 14.2h
- The as-cast structure shown in Fig. 14.2h reveals: (a) light etching α phase, (b) dark etching β phase, (c) large dendritic-shaped particles, (d) small dendritic-shaped particles, (e) globular precipitates and (f) cuboid precipitates.

In addition to the phases mentioned above, a needle-like phase is occasionally observed. It is sometimes referred to as 'sparkle-phase' (see below).

- If the alloy is allowed to cool very slowly (slower than in the case of the above as-cast structure), a further transformation occurs at 400°C: α+β converts to the α+γ_2 eutectoid, but this requires very long transformation time.

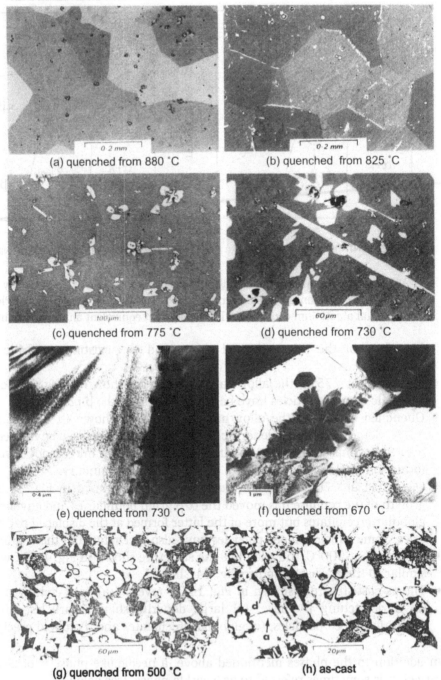

(a) quenched from 880 °C

(b) quenched from 825 °C

(c) quenched from 775 °C

(d) quenched from 730 °C

(e) quenched from 730 °C

(f) quenched from 670 °C

(g) quenched from 500 °C

(h) as-cast

Fig. 14.2 Microstructure of manganese–aluminium–bronze quenched at various temperatures as it cools slowly from 880°C, by Iqbal et al.[101]

Nature of phases

The composition of the various phases of manganese aluminium bronzes are given in Table 14.1 by Iqbal et al.[101]

The α phase

The light etching α phase is a copper-rich solid solution with a face-centred cubic (fcc) space lattice arrangement.

The β phase

The dark etching β phase is a copper-rich solid solution with an aluminium content approximately twice that of the α phase. It has an ordered bcc structure based on Cu_3Al.

The γ_2 phase

The γ_2 phase is part of the $\alpha+\gamma_2$ eutectoid which forms very slowly at 400°C. Its composition is given in Table 14.2 by Körster and Gödecke.[112] The figures given in this Table for the composition of α and β vary significantly from those given in Table 14.1 by Iqbal et al.

Table 14.1   Composition of the phases of a cast Cu–Mn–Al–Fe–Ni alloy based on 10 analyses of each phase.[101]

Phases	Method	Composition, wt %					
		Al	Si	Mn	Fe	Ni	Cu
α	bulk	5.9±0.22	0.2±0.1	12.1±0.2	2.4±0.5	1.4±0.1	78±1.0
β	bulk	12.5±1.1	0.3±0.1	13.5±0.5	1±0.2	2.2±0.3	70.5±10
Large dendritic-shaped particles (κ_I)	bulk	3.6±1	1.8±0.3	29.5±0.7	56.4±0.8	1.3±0.5	7.4±0.9
Small dendritic-shaped particles (κ_{II})	bulk	15.9±0.8	3±0.8	25.1±	47.1±3.5	1.2±0.4	7.8±2.5
Globular particles (κ_{III})	Thin foil	12.2±1	0.7±0.5	29.6±2.5	32.6±4	4.4±1.4	20.3±7.4
Cuboid particles (κ_{IV})	Thin foil	8.2±1.7	0.6±0.4	28.9±3	36.7±7	2.9±1.3	22.6±7.4
'Sparkle phase'*	Thin foil	1.3±1.2	0.6	28.3±1.7	61.0±5.3	0.4±0.2	8.4±4.1
Alloy composition		7.78	0.06	13.95	3.24	2.17	72.42

* Contains carbon also

Table 14.2   Composition of the phases of the Cu–Al–Mn system at
the eutectoid reaction temperature of 400°C.[112]

Phase	Composition: wt %		
	Cu	**Al**	**Mn**
β	80.5	12	7.5
α	87.5	9	3.5
γ_2	81	16	3

Inter-metallic κ particles

The inter-metallic particles in Cu–Mn–Al–Fe–Ni alloys are here designated as κ in line with the nomenclature used for nickel aluminium bronze although their compositions are different. As may be seen from Table 14.1, they are all rich in iron and manganese and low in nickel, whereas nickel aluminium bronze contains one phase, κ_{III}, which is nickel-rich. The four types of κ phases are shown on Fig. 14.3 taken from a Cu–Mn–Al–Fe–Ni alloy in the as-cast condition.

The κ_I particles

Fig. 14.3a shows a large dendritic-shaped particle, here referred to as κ_I. As may be seen from Table 14.1, it is an iron-rich particle with manganese as the other main constituent. The κ_I particles are located at the centre of the α grains (see Fig. 14.2h). They have been observed in the microstructure of samples quenched just below the solidus (950°C), leading one to suspect that they are formed in the melt. They are typically 20–40 μm in diameter, and are based on γ(Fe) with a fcc space lattice arrangement.

The κ_{II} particles

Fig. 14.3b shows a small dendritic-shaped particle, designated κ_{II}. The κ_{II} particles begin to appear in the β phase at about 750°C (see Fig. 14.2c) but become enveloped in the α phase as it grows with falling temperature (see Fig. 14.2d and e). They are much smaller than the preceding particles, being 5–10 μm in diameter. As may be seen from Table 14.1, they too are mainly composed of iron and manganese but with appreciably more aluminium than the large particles. They have an ordered bcc structure based on Fe_3Al but with manganese, copper and nickel substituting partially for iron, and silicon substituting partly for aluminium.

Gaillard and Weill[75] report a deep blue area in the interior of their grey-blue iron-rich κ_{II} precipitates. Microprobe analysis showed this to contain 50–80% Mn in alloys which varied in overall manganese content from 10.20–11.75%, suggesting this part of the precipitate may be based on MnSi.

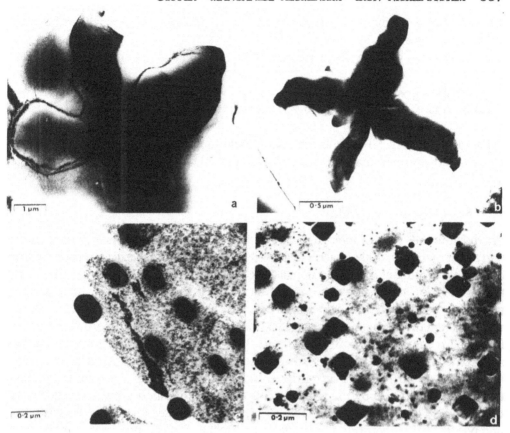

Fig. 14.3 Micrographs of an as-cast Cu–Mn–Al–Fe–Ni alloy showing the different types of intermetallic particles. (a) Large dendritic-shaped particles, (b) Small dendritic-shaped particles, (c) Globular-shaped particles, (d) Cuboid-shaped particles, by Iqbal et al.[101]

The κ_{III} particles

Fig. 14.3c shows globular particles, designated κ_{III}. They begin to appear in the α grains near the α/β boundary at about 730°C (see Fig. 14.2d and e) and, as the temperature falls, they get enveloped in the growing α phase as more of these particles form in the receding α/β boundary (see Fig. 14.2f). As may be seen from Table 14.1, they are composed of significant proportions of iron, manganese, copper and aluminium in that order of magnitude. They are small dendritic-shaped particles with an ordered bcc structure based on Fe_3Al.

The κ_{IV} particles

Fig. 14.3d shows cuboid-shaped particles, designated κ_{IV}, distributed throughout the α grains. They begin to appear between 730°C and 670°C in the α phase (see Fig. 14.2f). Their composition is similar to that of the globular particles but slightly richer in iron and with less aluminium (see Table 14.1). They are based on γ(Fe) with a fcc space lattice arrangement.

It will be noted that the above inter-metallic particles fall into two distinct categories:

(a) κ_I and κ_{IV} which are both based on γ(Fe) with a fcc structure and
(b) κ_{II} and κ_{III} which have both an ordered bcc structure based on Fe_3Al.

Langham and Webb[114] and Brezina[33] have reported the presence of nickel-rich particles (but without analyses) which Iqbal et al did not observe.

The 'sparkle-phase' particles

There is another type of inter-metallic particles which are needle-like in appearance and are known as 'sparkle-phase' particles. They are the result of excess carbon pick-up due to overheating of the melt and prolonged exposure in the molten state. They may render the alloy very brittle and unfit for use. Fig. 14.4 shows the microstructure of a casting containing these particles. As seen in Table 14.1, their composition is not unlike that of the large dendritic particles but with slightly more iron and less aluminium. The presence of carbon was detected by Iqbal et al. using electron energy loss spectrometry and have characterised the 'sparkle phase' as being based on cementite (Fe_3C) with an approximate composition of $(Fe_2Mn)C$ in which a substantial proportion of iron is replaced by manganese. This raises the question of the likely source of carbon which is most likely to be the waste gases in the case of oil or gas fired furnaces and possibly the carborundum crucibles. The likelihood of carbon contamination with induction furnaces could only arise from the carbon dioxide in the atmosphere and is therefore less likely to occur.

Experience has shown that repeated remelting of this alloy can have a cumulative effect in rendering it brittle. This is not surprising since carbon would be retained in remelted metal and added to, particularly if overheating occurs in remelting. Great care needs to be exercised therefore in not overheating the melt. Once the alloy is contaminated in this way, it cannot be 'de-contaminated' and is unfit for use. Other aluminium bronzes do not fortunately suffer in this way and can be repeatedly re-melted without detrimental effects (provided any loss of aluminium is compensated).

Effects of manganese

The main effect of manganese is as a substitute for aluminium, 6% manganese being equivalent to 1% aluminium. Consequently, like aluminium, it has a strengthening effect on the alloy.

A high concentration of manganese has a stabilising effect on β. Thus, whereas in nickel aluminium bronze the retained β has a martensitic structure, in manganese aluminium bronze the bcc β is retained at room temperature.

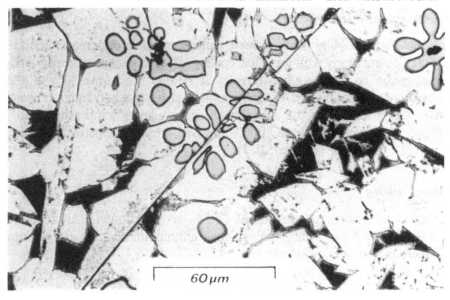

Fig. 14.4 Micrograph of an as-cast Cu–Mn–Al–Fe–Ni alloy showing the 'sparkle phase', by Iqbal et al.[101]

The presence of a high concentration of manganese also stabilises the fcc form of iron, with the result that two fcc precipitates are formed in the metal as it cools: κ_I (large dendritic particles and κ_{IV} (cuboid shaped particles).

Iqbal et al. also report that the solubility of iron in copper is substantially reduced by a high concentration of manganese and that it reduces with temperature in both the α and β phases.

Manganese also lengthens considerably the time of transformation of β into the $\alpha+\gamma_2$ eutectoid and lowers to 400°C the temperature at which it takes place.

Corrosion resistance

Manganese-aluminium bronze alloys experience only limited corrosive attack which is concentrated in certain more anodic phases. Lorimer, Hasan, Iqbal and Ridley[122] have established that the more vulnerable phases are the ordered β phase and the κ_I and κ_{IV} particles in the α phase. Both these particles which are based on $\gamma(Fe)$ are anodic to α whereas κ_{II} and κ_{III} which are based on Fe_3Al are cathodic to α as are the Fe_3Al-based κ particles (κ_I, κ_{II}, κ_{IV} and κ_V)in nickel-aluminium bronze. The rate of corrosion is influenced by the proportion of anodic to cathodic phases, and to their morphologies and distribution in the microstructure.

Magnetic properties

The magnetic permeability of manganese-aluminium bronze increases considerably below 500°C, particularly with very slow cooling. Thus the magnetic permeability of a sand casting can be anything between 2 and 10 and the permeability of a slowly cooled casting can be as high as 15. The reason for this is not known but is likely to be related to the reduction in the solubility of iron at lower temperature and to its morphology. By quenching above 500°C, however, a magnetic permeability figure in the region of 1.03 can be obtained.

Standard alloys

There are two standard Cu–Mn–Al–Fe–Ni alloys: **CuMn13Al8Fe3Ni3** and **CuMn13Al9Fe3Ni3** which differ only in their aluminium content. The difference in their properties is shown in Chapter 3.

APPENDIX 1
STANDARD AMERICAN SPECIFICATIONS

Cast alloys

Table A1.1 Composition and mechanical properties of cast aluminium bronze alloys to American ASTM standards.

ASTM ALUMINIUM BRONZE ALLOYS									
DESIGNATION		**COMPOSITION (%)**							
Current American ASTM designation	Nearest European CEN/TC 133 equivalent	Al	Fe	Ni	Mn	Si	Total other elements	Cu	
C 95200	CuAl10Fe2-C	8.5–9.5	2.5–4.0	–	–	–	1.0	86.0 min	
C 95300	–	9.0–11.0	0.8–1.5	–	–	–	1.0	86.0 min	
C 95400	–	10.0–11.5	3.0–5.0	2.5 max	0.5 max	–	0.5	83.0 min	
C 95500	–	10.0–11.5	3.0–5.5	3.0–5.5	3.5 max.	–	0.5	78.0 min	
C 95600	Br. Nav. NES 834 Pt3	6.0–8.0	–	0.25 max	–	1.8–3.2	1.0	88.0 min	
C 95700*	CuMn11Al8Fe3Ni3-C	7.0–8.5	2.0–4.0	1.5–3.0	11.0–14.0	0.10 max	0.5	71.0 min	
C 95800+	CuAl10Fe5Ni5-C	8.5–9.5	3.0–5.5	4.0–5.5	3.5 max	0.10 max	0.5	78.0 min	

* 0.03% max lead
+ 0.02% max lead and iron content shall not exceed nickel content.

DESIGNATION		MECHANICAL PROPERTIES OF CAST ALLOYS					
Current American ASTM designation	Nearest European CEN/TC 133 equivalent	Modes of casting	Condition	Tensile Strength *kg/mm² min	0.5% Proof Strength *kg/mm² min	Elongation % min	Hardness Brinell 3000 kg (Typical)
C 95200	CuAl10Fe2-C	Sand Permanent mould Centrifugal Continuous Plaster	sand-cast	45.7	17.6	20	125
C 95300	–	ditto	sand-cast	45.7	17.6	20	140
			heat-treated	56.2	28.1	12	174
C 95400	–	ditto	sand-cast	52.7	21.1	12	170
			heat-treated	63.3	31.6	6	195
C 95500	–	ditto	sand-cast	63.3	28.1	6	195
			heat-treated	77.3	42.2	5	230
C 95600	Br. Nav. NES 834 Pt3	ditto	sand-cast	42.2	19.7	10	140
C 95700*	CuMn11Al8Fe3Ni3-C	ditto	sand-cast	63.3	28.1	20	180
C 95800+	CuAl10Fe5Ni5-C	ditto	sand-cast	59.8	24.6	18	159

* To convert to N mm⁻², multiply by 9.807 (g)

Wrought alloys

Table A1.2   Composition of wrought aluminium bronze alloys to American ASTM standards.

ASTM WROUGHT ALUMINIUM BRONZE ALLOYS							
DESIGNATION		**COMPOSITION (%)**					
American ASTM designation	Nominal Composition	Al	Fe	Elements with max and min specified limits	Elements with max limits only		
				Cu + Elements with specific limits = 99.5 min			
C 60800	CuAl5	5.0–6.5	< 0.10	As: 0.20 – 0.35	Pb <0.10		
C 61000	CuAl8	6.0–8.5	< 0.50		Pb <.02	Zn <.20	Si <.10
C 61300	CuAl7Sn0.3	6.0–8.0	< 3.5	Sn: 0.2 – 0.5	Mn < 0.5	Ni < 0.5	
C 61400	CuAl7Fe2	6.0–8.0	1.5–3.5		Pb <0.01 P <.015	Zn <0.2	Mn <1.0
C 61800	CuAl10Fe1	8.5–11.0	0.5–1.5		Pb <0.2	Zn <0.02	Si <0.10
C 61900	CuAl9.5Fe4	8.5–10.0	3.0–4.5		Pb <0.02	Sn <0.6	Zn <0.8
C 62300	CuAl10Fe3	8.5–11.0	2.0–4.0		Sn <0.6 Ni <1.0	Mn <0.5	Si <0.25
C 62400	CuAl11Fe3	10.0–11.5	2.0–4.5		Sn <0.2	Mn <0.3	Si <0.25
C 62500	CuAl13Fe4.3	12.5–13.5	3.5–5.0		Mn <2.0		
C 63000	CuAl10Fe3Ni5	9.0–11.0	2.0–4.0	Ni: 4.0–5.5	Sn <0.20 Si <0.25	Zn <0.30	Mn <1.5
C 63200	CuAl9Fe4Ni5	8.5–9.5	3.0–5.0	Ni: 4.0–5.0	Pb <0.02	Mn <3.5	Si <0.10
C 63800	CuAl2.8Si1.8Co0.4	2.5–3.1	<0.05	Si: 1.5–2.1 Co: .25 – .55	Pb <0.05 Ni <.10	Zn <0.50	Mn<0.10
C 64200	CuAl7Si1.8	6.3–7.6	<0.30	Si: 1.5–2.2	Pb <0.05 As <0.15	Sn <0.20 Mn <0.10	Zn <0.50 Ni <0.25

Table A1.3   Mechanical properties of wrought aluminium bronze alloys to American ASTM standards.

MECHANICAL PROPERTIES OF ASTM WROUGHT ALUMINIUM BRONZE ALLOYS

ASTM designation	Form	Size Section (in.)	Temper	Tensile strength (kg/mm²)	Yield strength (kg/mm²) 0.5% Ext. under load	Yield strength (kg/mm²) 0.2% offset	Elongation in 2 in %	Rockwell Hardness F	Rockwell Hardness B	Rockwell Hardness 30T	Shear strength (kg/mm²)	Fatigue strength (kg/mm²)	10⁶ Cycles
C 60800	Tube	1.0 OD × 0.065	0.025 mm	60.0	27.0	–	55	77	–	–	–	–	–
C 61000	Rod	1.0 dia	Soft anneal	70.0	30.0	–	65	–	60	–	–	–	–
C 61300	Flat Products	0.125	Hard (40%)	80.0	55.0	–	25	–	85	–	52.0	28.0	100
		0.500	Soft anneal	80.0	40.0	–	40	–	82	–	50.0	26.0	100
		1.0	Soft anneal	78.0	35.0	–	42	–	81	–	45.0	25.0	100
		3.0	Soft anneal	76.0	33.0	–	42	–	79	–	42.0	21.0	100
	Rod	0.500	Hard (25%)	70.0	30.0	–	40	–	78	–	48.0	–	–
		1.0	Hard (25%)	85.0	58.0	–	35	–	91	–	45.0	–	–
		2.0	Hard (25%)	82.0	55.0	–	35	–	90	–	40.0	30.0	100
C 61400	Flat Products	0.125	Soft	80.0	48.0	–	35	–	88	–	45.0	–	–
		0.312	Hard	82.0	45.0	–	40	–	84	–	42.0	–	–
		0.500	Soft	89.0	60.0	–	32	–	87	–	–	28.0	100
		1.0	Hard	80.0	40.0	–	40	–	83	–	40.0	–	–
		0.500	Soft	85.0	58.0	–	35	–	86	–	–	26.0	100
		1.0	Hard	78.0	35.0	–	42	–	82	–	40.0	–	–
	Rod	2.0	Soft	80.0	54.0	–	38	–	85	–	–	25.0	100
			Hard	76.0	33.0	–	45	–	81	–	–	–	–
			Hard	78.0	45.0	–	40	–	84	–	–	–	–
C 61800	Rod	0.500	Hard	85.0	45.0	–	35	–	91	–	48.0	28.0	100
		1.0	Hard	82.0	40.0	–	35	–	90	–	45.0	28.0	100
		2.0	Hard	80.0	35.0	–	35	–	88	–	40.0	26.0	100
		1.0	1/2 Hard (15%)	85.0	42.5	–	23	–	89	–	47.0	–	–
		2.0	1/2 Hard (15%)	82.5	39.0	–	25	–	88	–	45.0	–	–
		3.0	1/2 Hard (15%)	80.0	39.0	–	28	–	88	–	43.0	–	–
C 61900	Flat Products	0.040	Annealed	92.0	49.0	60.0	30	–	–	72	–	–	–
			Half Hard	104.0	**	80.0	21	–	93	78	–	–	–
			Hard	121.0	**	102.0	8	–	98	83	–	–	–
			Hard*	135.0	**	120.0	2	–	–	–	–	–	–
			Spring	130.0	**	108.0	4	–	100	83	–	33.5	100
			Spring*	145.0	**	135.0	1.5	–	–	–	–	38.5	100
			Extra Spring	137.0	**	117.0	3	–	100	84	–	–	–
			Extra Spring*	152.0	**	145.0	1	–	–	–	–	–	–

* After heat treatment for one hour at 450F   ** Not applicable because the 1/2% extension under load yield strength is below the elastic limit.

ASTM designation	Form	Size Section (in.)	Temper	Tensile strength (kg/mm²)	Yield strength (kg/mm²) 0.5% Ext. under load	Yield strength (kg/mm²) 0.2% offset	Elongation in 2 in %	Rockwell Hardness F	Rockwell Hardness B	Rockwell Hardness 30T	Shear strength (kg/mm²)	Fatigue strength (kg/mm²)	10⁶ Cycles
C 62300	Bar	3.0	As Extruded	75.0	35.0	–	35	–	80	–	–	–	–
		1.0	Half Hard (15%)	90.0	45.0	–	30	–	85	–	–	–	–
		2.0	Half Hard (15%)	85.0	40.0	–	30	–	82	–	–	–	–
	Rod	4.0	As Extruded	75.0	35.0	–	35	–	80	–	35.0	25.0	100
		0.5	Half Hard (15%)	98.0	52.0	–	22	–	89	–	52.0	31.0	100
		1.0	Half Hard (15%)	95.0	50.0	–	25	–	88	–	50.0	30.0	100
		2.0	Half Hard (15%)	92.0	48.0	–	28	–	87	–	48.0	30.0	100
		3.0	Half Hard (15%)	87	45.0	–	35	–	85	–	46.0	29.0	100

Alloy	Form	Size	Temper									
C 62400	Bar	1.0	Light Anneal	100.0	50.0	—	14	92	—	—	—	—
		2.0	Light Anneal	93.0	45.0	—	14	91	—	—	—	—
		3.0	As Extruded	90.0	40.0	—	18	87	—	—	—	—
	Rod	1.0	Half Hard (10%)	105.0	52.0	—	14	92	—	65.0	36.0	100
		2.0	Half Hard (10%)	95.0	48.0	—	14	92	—	62.0	35.0	100
		3.0	Half Hard (10%)	95.0	47.0	—	14	91	—	60.0	34.0	100
		4.0	As Extruded	90.0	40.0	—	18	87	—	58.0	32.0	100
C 62500	Rod and Bar	All	As Extruded	100.0	55.0	—	1	29	—	60.0	67.0	100
C 63000	Bar	1.0	Half Hard (10%)	110.0	62.0	—	15	97	—	—	—	—
		2.0	Half Hard (10%)	100.0	60.0	—	15	96	—	—	—	—
		3.0	As Extruded	90.0	50.0	—	15	96	—	—	—	—
	Rod	1.0	Half Hard (10%)	118.0	75.0	—	15	98	—	70.0	38.0	100
		2.0	Half Hard (10%)	115.0	65.0	—	18	96	—	69.0	37.0	100
		3.0	Half Hard (10%)	112.0	62.0	—	20	96	—	65.0	37.0	100
		4.0	As Extruded	100.0	60.0	—	15	96	—	62.0	36.0	100
C 63200	Plate	0.250	Light Anneal	100.0	50.0	—	20	92	—	—	—	—
		0.500	Light Anneal	95.0	47.0	—	22	89	—	—	—	—
		1.0	Light Anneal	90.0	45.0	—	25	85	—	—	—	—
	Rod	1.0	Light Anneal	105.0	53.0	—	22	96	—	—	—	—
		2.0	Light Anneal	103.0	52.0	—	24	94	—	—	—	—
		3.0	Light Anneal	102.0	51.0	—	24	93	—	—	—	—
C 63800	Forging	5.0	Light Anneal	100.0	45.0	—	22	92	—	—	—	—
	Flat Products	0.040	Soft Anneal	82.0	53.0	54.0	36	86	—	—	—	—
			Quarter Hard	96.0	—	77.0	23	93	73	—	—	—
			Half Hard	106.0	—	88.0	15	96	78	—	—	—
			Hard	120.0	—	101.0	7	99	79	—	—	—
			Extra Hard	124.0	—	107.0	5	100	82	—	—	—
			Spring	128.0	—	110.0	4	101	82	—	—	—
			Extra Spring	130.0	—	114.0	—	—	83	—	—	—
			As Extruded	75.0	35.0	—	32	77	—	—	—	—
C 64200	Bar	0.250	As Extruded	75.0	35.0	—	32	77	—	—	35.0	100
	Rod	0.250	Light Anneal	92.0	58.0	—	22	98	—	—	—	—
		0.500	Light Anneal	90.0	55.0	—	28	89	—	—	—	—
		0.750	Hard (15%)	102.0	68.0	—	22	94	—	59.0	50.0	100
		1.500	Hard (10%)	93.0	60.0	—	26	90	—	—	—	—
	Forging	2.0	As Forged	79.0	38.0	—	30	78	—	—	—	—

APPENDIX 2

Elements and symbols

Elements	Symbol	Elements	Symbol	Elements	Symbol
Aluminium	Al	Helium	He	Rhodium	Rh
Antimony	Sb	Hydrogen	H	Selenium	Se
Argon	Ar	Iridium	Ir	Silicon	Si
Arsenic	As	Iron	Fe	Silver	Ag
Barium	Ba	Lead	Pb	Sodium	Na
Beryllium	Be	Magnesium	Mg	Strontium	Sr
Bismuth	Bi	Manganese	Mn	Sulphur	S
Boron	B	Mercury	Hg	Tantalum	Ta
Cadmium	Cd	Molybdenum	Mo	Tellurium	Te
Calcium	Ca	Nickel	Ni	Thallium	Tl
Carbon	C	Niobium	Nb	Thorium	Th
Cerium	Ce	Nitrogen	N	Tin	Sn
Chlorine	Cl	Osmium	Os	Titanium	Ti
Chromium	Cr	Oxygen	O	Tungsten	W
Cobalt	Co	Palladium	Pd	Uranium	U
Columbium	Cb	Phosphorus	P	Vanadium	V
Copper	Cu	Platinum	Pt	Zinc	Zn
Gold	Au	Potassium	K	Zirconium	Zr

APPENDIX 3

Comparison of nickel aluminium bronze with competing ferrous alloys in sea water applications[139]

Competing ferrous alloys

The following are the main ferrous alloys that compete with nickel aluminium bronze in sea water applications:

Duplex Stainless Steels (wrought and cast)
Superduplex Stainless Steels (wrought and cast)
Type 316 Austenitic Stainless Steel (wrought and cast)
High Molybdenum Superaustenitic Stainless Steels (wrought and cast)
Ni-Resist Cast Iron.

Factors affecting choice of alloy

Many factors affect the choice of alloy for a given application:

- mechanical properties,
- physical properties, notably heat and electrical transfer properties,
- nature of the environment,
- operating conditions,
- required life span,
- corrosion/erosion resistance in relation to environment, operating conditions and required life span,
- cost in relation to life span and reliability,
- ease of weld fabrication and/or weld repair,
- wear resistance,
- weight in relation to strength.
- castability
- wrought forms available
- lead time

Choosing the most suitable alloy for a particular application is made difficult by the fact that each alloy is 'best' in a number of respects but less desirable in other respects.

Alloy compositions (wt %)

The composition of nickel aluminium bronzes and typical compositions of competing ferrous alloys for sea water applications are given in Table A3.1. Most stainless steels are produced under proprietary names which have slightly different compositions to those given in this table.

366

Table A3.1 Compositions (wt %) of nickel-aluminium bronzes and competing ferrous alloys for sea water environment, by Oldfield and Masters.[139]

NICKEL ALUMINIUM BRONZE						
Alloy	**Cu**	**Al**	**Fe**	**Ni**	**Mn**	**Impurities**
BS 1400 AB2	Bal	8.8–10.0	4.0–5.5	4.0–5.5	3.0 max	0.20 excl.Zn
NES 747 Pt. 2†	Bal	8.8–9.5	4.0–5.0	4.5–5.5*	0.75–1.3	0.25 total
ASTM 958	Bal	8.5–9.5	3.0–5.5	4.0–5.5*	3.5 max	Pb: 0.02 max
						Si: 0.1 max
						Other: 0.5 max

* Ni must exceed Fe
†British Naval Specification – Castings annealed at 675°C for to 2–6 hours and cooled in air – Best nickel aluminium bronze for corrosion resistance.

STAINLESS STEELS							
Grade	**Form**	**Cr**	**Ni**	**Mo**	**N**	**C (max)**	**Other**
Duplex							
UNS S31803	wrought	21–23	4.5–6.5	2.3–3.5	0.08–0.2	0.03	Cu: 1.0 max
UNS J92205	cast	21–23.5	4.5–6.5	2.5–3.5	0.1–0.3	0.03	
Superduplex							W: 0.5–1.0
UNS S32760	wrought	24–26	6–8	3–4	0.2–0.3	0.03	Cu: 0.5–1.0
UNS S32750	wrought	24–26	6–8	3–4	0.24–0.32	0.03	
UNS J93380	cast	24–26	6.5–8.5	3–4	0.2–0.3	0.03	W: 0.51–1.0
							Cu: 0.5–1.0
Type 316							
UNS S31600	wrought	16–18	10–14	2–3	–	0.08	
UNS S31603	wrought	16–18	10–14	2–3	–	0.03	
UNS J92900 (CF-8M)	cast	18–21	9–12	2–3	–	0.08	
UNS J92800 (CF-3M)	cast	17–21	9–13	2–3	–	0.03	
Superaustenitic							
UNS N08367	wrought	20–22	23.5–25.5	6–7	0.18–0.25	0.03	Cu: 0.75 max
UNS S31254	wrought	19.5–20.5	17.5–18.5	6–6.5	0.18–0.22	0.02	Cu: 0.5–1.0
UNS N08926	wrought	20–21	24.5–25.5	6–6.8	0.18–0.20	0.02	Cu: 0.8–1.0
UNS J93254	cast	19.5–20.5	17.5–19.7	6–7	0.18–0.24	0.025	Cu: 0.5–1.0

NI-RESIST							
	Fe	**Ni**	**Cr**	**Si**	**C**	**Mo**	**Other**
D2-W**	bal	18–22	1.5–2.2	1.5–2.4	3.0 max	0.5–1.5	Nb: 0.12–0.2
							P: 0.05 max
							Mn: 0.5–1.5
							Cu: 0.5 max

** A modified version of D2 with controlled close composition for improved welding.

368   ALUMINIUM BRONZES

Table A3.2 Mechanical properties of cast nickel-aluminium bronze at temperatures compared with those of competing cast ferrous alloys, by Oldfield and Masters.[139]

	Nickel* alumin. bronze	Duplex SS	Super-duplex SS	Type** 316 SS	Super-austenitic SS	Ni-Resist D2-W
0.2% Proof strength N mm^2	250 min	415 min	450 min	205 min	250 min (290 typic)	241 (D2)
Tensile strength N mm^2	620 min	620 min	700 min	485 min	550 min (630 typic)	407 (D2)
Elongation %	15–20	25	25	40	35–50	7
Young's modulus kN mm^2	124–130	200	180–200	200	200	113–128 (D-2)
Hardness (HB)	140–180	240	285 max	156	155 typic.	160–200

MECHANICAL PROPERTIES AT AMBIENT TEMPERATURE

* Mechanical properties given are for (heat treated) cast nickel aluminium bronze to Brit. Nav. Spec.NES 747 Part 2
** Cast designation: CF-8M or CF-3M

Mechanical properties

Table A3.2 gives the mechanical properties of nickel aluminium bronze at ambient and at elevated temperatures and compares them with those of ferrous alloys that compete with nickel aluminium bronze for sea water environments.

The mechanical properties of cast stainless steels may be adversely affected by the problem of segregation which may occur with heavier sections or as a result of welding.

It will be seen that nickel aluminium bronze has comparable proof strength at ambient temperature to Ni-Resist and to austenitic and superaustenitic stainless steels, but has significantly lower proof strength than duplex and superduplex stainless steels. It is often not possible, however, to take advantage of the higher proof strength of an alloy by reducing cast section thickness because of the limitations of castability. Although the proof strength of nickel–aluminium bronze can be raised to around 440 N mm^{-2} by quenching from 925°C, this is not recommended for sea water conditions as it would adversely affect the corrosion resisting properties of the alloy, due to the presence of the martensitic beta phase (see Chapter 13).

Nickel aluminium bronze has elongation properties significantly better than those of Ni-Resist and only slightly less than those of superduplex stainless steel. Austenitic and superaustenitic stainless steel have remarkably high elongation properties.

The greater rigidity of stainless steels by comparison with nickel aluminium bronze and Ni-Resist is evident from their higher moduli of elasticity.

Nickel aluminium bronze has hardness figures comparable to Ni-Resist and to most stainless steel with the exception of duplex and superduplex stainless steels which have outstanding hardness properties.

Table A3.3   Physical properties of nickel aluminium bronze compared
with those of competing ferrous alloys, by Oldfield and Masters.[139]

	Nickel alumin. bronze	Duplex SS	Super-duplex SS	Type 316 SS	Super-austenitic SS	Spheroidal Graphite Ni-Resist D2-W
Density g/cm^3	7.5	7.9	7.8	8.0	8.0	7.4
Thermal conductivity J s^{-1}m^{-1}K^{-1} at 20°C	33–46	14	12.90	15	13.5	13.4
Specific heat capacity J kg^{-1}K^{-1}	419	450	460–500	470–500	500	460–500*
Coefficient of thermal expansion per K $\times$ 10^{-6}	16.2	13	13	16.5–18.5	16.5	18.7
Electrical resistivity 10^{-7} Ω m^{-1}	1.9–2.5	8	9.16	7.5	8.5	10.2

* These figures are for Flake Graphite Ni-Resist but are likely to be similar to those for Speroidal Graphite Ni-Resist

Physical properties

Table A3.3 shows a comparison between the physical properties of cast nickel aluminium bronze and those of competing ferrous alloys.

The densities of nickel aluminium bronzes, Ni-Resist and duplex stainless steels are very similar, with austenitic stainless steels being slightly heavier. If weight saving is a factor in the choice of alloy, it can only be achieved significantly by taking advantage of the difference in mechanical strength, provided castability permits.

Being a copper alloy, nickel aluminium bronze has much greater thermal conductivity than stainless steels or Ni-Resist. For the same reason, the electrical resistivity of nickel aluminium bronze is much lower than that of stainless steels and of Ni-Resist.

Corrosion resistance

The various types of corrosive attack affecting alloys used in sea water applications are explained in Chapter 8.

There is little available data on the corrosion resisting properties of cast stainless steels as opposed to wrought material. Generally speaking, the corrosion resisting properties of cast stainless steels varies from alloy to alloy and is unlikely to be better than that of corresponding wrought alloys. It may be adversely affected by the problem of segregation which may occur with heavier cast sections or as a result of welding.

Other factors, such as mechanical strength, physical properties, castability, weldability and cost, are taken into account in choosing the appropriate alloy and are often the deciding factors where the corrosion performance of competing alloys are comparable.

General corrosion

Uniform or general corrosion is that which succeeds in permeating to some extent the protective oxide film of an alloy under galvanic action or which directly attacks this film chemically. As explained in Chapter 8, direct chemical attack only occurs in polluted sea water containing hydrogen sulphide. If this condition is sustained and severe, it is considered as a special type of corrosion and not as general corrosion.

Galvanic action takes place in an electrolyte such as sea water either between components of different alloys, due to their different electro-chemical potential, or between different phases of the same alloy, or due to differential aeration.

Table A3.4 Electro-chemical potentials of various alloys in ambient temperature sea water (see Chapter 8, Fig. 8.3).

Alloy	mV
Anodic or most vulnerable	
Ni-Resist	−220 to −450
Nickel aluminium bronze	−80 to −250
Type 316 austenitic stainless steel (passive)	100 to 300
Duplex and superduplex stainless steels (passive)	+250 to +350
Superaustenitic stainless steel (passive)	+250 to +350
Cathodic or most 'noble'	

The electro-chemical or galvanic series is given in Chapter 8, Fig. 8.3. Metals with the lowest potential are anodic to metals with higher potential which are said to be more 'noble'. A more anodic metal will tend to corrode in the presence of a more noble metal and the more noble metal is thereby 'protected' by the more anodic metal. Thus if stainless steel is connected to ordinary steel in sea water, the stainless steel accelerates the corrosion of the steel and the latter gives protection to the former. Table A3.4 gives figures for electrochemical potentials in ambient temperature sea water of the alloys under comparison. The austenitic stainless steels are the most noble whereas Ni-Resist is the most active. Nickel aluminium bronze is more active than the austenitic alloys and marginally more so than the duplex alloys. It should therefore not cause any problem in an application that is not of concern to duplex alloys. Linked to super-duplex or super-austenitic stainless steels, it corrodes heavily in the kappa3 phase in natural sea water. These facts need to be taken into consideration in any mixed alloy sea water system.

Except in special cases of localised corrosion mentioned below, the alloys under comparison are virtually unaffected by electro-chemical action in sea water, as long as the oxide protection is not undermined. General corrosion in non-sulphide polluted sea water is therefore not significant with any of these alloys.

Inter-phase corrosion in nickel aluminium bronze can be prevented by control of composition and by heat treatment (see Chapter 13). Nickel aluminium bronze, free of continuous $gamma_2$ and beta phases, corrodes only at about 0.1 mm/year.[6]

The general rate of corrosion of Ni-Resist is less in de-aerated water (0.02 mm/y) but higher in aerated water (0.2 mm/year). The average rate of corrosion in vertical pipe pumps in the North Sea over a period of ten years was found to be 0.08 mm/year which is very close to that of nickel aluminium bronze.

Stainless steel are not subject to general corrosion.

Differential aeration is only a problem in the case of pitting and crevice corrosion (see below).

Table A3.5   Pitting Resistance Equivalent (PREN) for Stainless Steels.[74]

	Duplex SS	Superduplex SS	Type 316 SS	Superaustenitic SS
$PREN_1$	31 – 36	40 – 44	24.5 – 31	42 – 47
$PREN_2$	32.5 – 38.5	40 – 49	24.5 – 31	45 – 51

Pitting corrosion

As explained in Chapter 8, pitting results from localised damage to the protective oxide film which forms a recess or 'pit' on the metal surface. This recess is inaccessible to oxygen and a galvanic couple is created due to differential aeration between the inside of the 'pit' and the remaining surface of the component. Nickel aluminium bronze and Ni-Resist are hardly affected at all by this form of attack in sea water.

Stainless steels, on the other hand, rely for their resistance to corrosion on a thin protective 'passive' oxide film which can be easily damaged. As explained in Chapter 8, the passive oxide film renders the metal more cathodic or 'noble' and therefore less corrodible. If this film is damaged however, it exposes the more anodic parent metal which then corrodes and the corrosive effect is further aggravated by differential aeration. The smaller the ratio of the damaged area to the undamaged area, the more severe the rate of pitting corrosion. The pitting resistance of stainless steel is a function of its contents of Chrome, Molybdenum and Nitrogen. There are various formulae for calculating the Pitting Resistance Number (PREN) of stainless steels, two of which are as follows:

$$PREN_1 = (\%Cr) + (3.3 \times \%Mo) + (16 \times \%N)$$
$$PREN_2 = (\%Cr) + (3.3 \times \%Mo) + (30 \times \%N)$$

Table A3.5 gives both sets of PREN values for the stainless steels under consideration. A value of 40 or above is usually considered to indicate a satisfactory resistance to pitting corrosion in sea water. On this basis, superduplex and superaustenitic stainless steels have good resistance to pitting corrosion at ambient temperature and the latter is the more resistant. The PREN number is an indication of resistance to the *initiation* of pitting corrosion. If the latter does start, however, the PREN number gives no indication on the rate of propagation which is sometimes significantly greater for duplex than for austenitic alloys, depending on temperature and potential.

Crevice corrosion

As explained in Chapter 8, a crevice is a 'shielded area' where two components or parts of the same component are in close contact with one another although a thin film of water can penetrate between them: between flanges, within fasteners etc. A shielded area can also be created by marine growth (biofouling) or other undisturbed deposits on the surface of the component. The shielded area is starved of oxygen and crevice corrosion is another example of the effect of differential aeration. Practically all metals and alloys suffer accelerated local corrosion either within or just outside a crevice.

Ni-Resist is not susceptible to crevice corrosion. Nickel aluminium bronze experiences some selective phase attack resulting in 'de-aluminification'. The depth of attack is however minimal provided the alloy is free of continuous beta phase or gamma $_2$ phase (see Chapter 13). Being a copper alloy, nickel aluminium bronze is moderately resistant to biofouling, a common cause of crevice corrosion.

The more highly alloyed stainless steels which, as we have seen, are more resistant to pitting corrosion, are also more resistant to crevice corrosion, whereas type 316 stainless steel and type 2205 duplex are susceptible to attack. Very small differences in the depth and width of the crevice gap makes a big difference on the degree of crevice corrosion attack and renders comparisons difficult between sets of data. Thus in the case of a threaded joint, which leaves a very small gap, crevice corrosion of the more highly alloyed stainless steels can occur. Other environmental conditions may also lead to crevice corrosion of these alloys: for example if the temperature is increased and/or chlorination has been carried out.

Chlorination, which is effective in preventing biofouling, is detrimental to most stainless steels as it causes deep crevice attack of very small cross section if the chlorine content is high.

Erosion corrosion

Under conditions of service involving exposure to liquids flowing at high speed or with a high degree of local turbulence in the stream or containing abrasive particles such as grit, the flow generates a shear stress which is liable to damage the protective oxide film, locally exposing unprotected bare metal. Nickel aluminium bronze is vulnerable to such attack at flow speed in excess of 4.3m s^{-1} with clean water. J. P. Ault[17] found that the annual corrosion/erosion rate of nickel aluminium bronze in fresh unfiltered sea water varied logarithmically with velocity and therefore becomes rapidly unacceptable at higher velocities. Nickel aluminium bronze is even more vulnerable if grit is present in the water.

Ni-Resist is more vulnerable to corrosion/erosion than nickel aluminium bronze, particularly in aerated water. Thus at 14.5 ft/sec (4.4 m s^{-1}) it has been found to corrode at a rate of 0.27 mm/year in aerated sea water but only at 0.02 mm/year in de-aerated sea water.

The protective film on stainless steels, on the other hand, although very thin, is resistant to such attack and can withstand flow velocities even as high as 40 m s^{-1}

Cavitation erosion
Rapid changes of pressure in a water system, as may occur with rotating components such as propellers and pump impellers, cause small vapour bubbles to form when the pressure is lowest. As the pressure suddenly increases, the bubbles collapse violently on the surface of the metal, generating stresses which may erode the surface of the metal or even tear out small fragments by fatigue. The soundness of the casting is of critical importance in resisting cavitation erosion since any subsurface porosity may give way under the hammering effect of cavitation.

Table A3.6   Cavitation erosion rates in fresh water by I. S. Pearshall.[41]

Material	Cavitation Erosion Rate mm^3 h^{-1}
Nickel aluminium bronze	0.6
Austenitic stainless steel 316	1.7
Ni-resist cast iron	4.4

Table A3.7   Cavitation erosion rates at 40m/sec in natural sea water by P. A. Lush.[125]

Material	Cavitation Erosion Rate mm^3 h^{-1}
Nickel aluminium bronze	0.9 – 1.1
Nickel alloys	0.35 – 1.7
Titanium alloys	0.35 – 0.8
Austenitic stainless steel 316	0.3 – 0.45
Duplex stainless steel	0.20 – 0.22

Table A3.6 gives comparative cavitation erosion rates in fresh water for cast nickel aluminium bronze, type 316 austenitic stainless steel and Ni-Resist. Table A3.7 compares cavitation erosion rates at 40 m s^{-1} in natural sea water of nickel-aluminium bronze with various alloys.

Chloride stress corrosion cracking
Stress corrosion is a highly localised attack occurring under the simultaneous action of internal tensile stresses in a component and a particular type of corrosive

environment. Thus stainless steels are vulnerable to stress corrosion in warm chloride solutions (sea water) whereas aluminium bronzes are not affected. Although the total amount of corrosion may be small, cracking occurs in a direction perpendicular to that of the applied stress and may cause rapid failure. Internal stresses due to welding or cold work may be minimised by a stress relief heat treatment.

There are significant differences in vulnerability to chloride stress corrosion cracking between the various stainless steels. Type 316 is the most vulnerable and is likely to be affected above 80–100°C. Superaustenitic alloys are affected above 100–150°C, Type 2205 duplex alloys above 120–150°C and superduplex alloys above 150–200°C.

Ni-Resist, on the other hand, is susceptible to stress corrosion cracking in sea water at ambient temperature and must therefore be stress relieved.

Sulphide pollution
Hydrogen sulphide is generated by decaying organic matter and is a common form of polluted seawater. It attacks chemically all the alloys under consideration with the exception of Ni-Resist. Copper based alloys such as aluminium bronzes are particularly vulnerable as was explained in Chapter 8. The oxide film is reduced by the hydrogen sulphide and replaced by copper sulphide which is porous and does not adhere to the metal surface. This effect is aggravated by flow velocities which remove the corrosion products resulting in severe pitting corrosion.

In the case of stainless steels subjected to very high sulphide concentrations, the chemical reaction results in a protective iron sulphide film. With lower concentrations however, the hydrogen sulphides reduces the oxide film and makes the alloy more vulnerable to localised attack such as pitting and crevice corrosion. It is these lower concentrations of sulphides which are more generally relevant. In these conditions, the 316 stainless steels and 2205 duplex alloy are susceptible to attack and the corrosion resistance of the superaustenitic and superduplex alloys is somewhat reduced, although they are not susceptible to significant corrosion under these conditions.

Effect of segregation on corrosion resistance
In the case of superduplex and 2205 duplex stainless steels, there is a danger of segregation occurring with thicker section castings with detrimental effects on corrosion resistance.

Summary comparison of corrosion resistance
Table A3.8 compares the corrosion resistance of nickel aluminium bronze in sea water with that of competing ferrous alloys by means of arbitrary corrosion resistance ratings estimated out of 10 for each type of corrosion. The table also shows the total rating of each alloy for four different sets of service conditions to illustrate how a given alloy is more suited to a particular set of service conditions. Thus nickel

Table A3.8 Comparison of corrosion properties of nickel aluminium bronze in sea water with those of competing ferrous alloys.[139,74]

		Nickel alumin. bronze	Duplex SS	Super-duplex SS	Type 316 SS	Super-austenitic SS	Ni-Resist D2-W
a	General corrosion	9	10	10	10	10	8
b	Pitting corrosion	10	6	10	4	10	10
c	Crevice corrosion	9	4	8	3	8	10
d	Erosion/corrosion	8	10	10	10	10	6
e	Cavitation	8	8	9	7	8	4
f	Stress corrosion	10	10	10	7	10	5
g	Corrosion fatigue	9	9	9	6	6	6
h	Sulphide polluted	1	5	9	6	6	6

In the above arbitrary values, 10 ranks highest in corrosion resistance.
E = Estimated

Condition		Total ratings for each service conditions below					
A	= a+b+c	28	20	28	17	28	28
B	= a+b+e+f+g	46	43	48	34	44	33
C	= a+b+d+e	35	34	39	31	38	28
D	= b+d+e+h	27	29	38	27	34	26

Service conditions

Condition A: Static condition or low velocity water flow, exposed to general corrosion, to shielded areas (crevice corrosion), to local surface damage (pitting), but to no significant pollution: (e.g.) valve parts, offshore vertical fire pumps, etc.

Condition B: Rotating or moving parts, flow velocity less than 4.3 m s^{-1}, exposed to general corrosion, to little or no abrasive particles, to local surface damage (pitting), to cavitation, to fluctuating stresses, but to no significant pollution: (e.g.) propellers and some turbines.

Condition C: Fast rotating parts, flow velocity in excess of 4.3 m s^{-1}., exposed to general corrosion, to abrasive particles, to local surface damage, to cavitation, but to no significant pollution: (e.g.) some pump impellers.

Condition D: Rotating and moving part, exposed to sustained sulphides pollution and to cavitation: (e.g.) pump impellers in sulphide polluted location.

aluminium bronze is a good choice for conditions A and B but is not suited to conditions C and D.

Fabrication properties

As explained in Chapter 7, nickel aluminium bronze is readily welded by various techniques. Castings can therefore be repaired and can be incorporated in part fabrications. Worn components can also be built up during overhaul. As explained in Chapter 7 and in greater details in Chapter 13, welding will render a nickel aluminium bronze casting more vulnerable to corrosion but this can be remedied by heat treatment.

Low carbon versions of type 316 stainless steel have been developed to facilitate welding of castings in this alloy. The high austenitic alloys are weldable using similar material to the nickel based alloy 625. This results in welds that have better corrosion resisting properties than the parent metal.

Welding of 2205 duplex and superduplex wrought alloys has been developed in recent years and produces good results, provided welding procedures are closely followed.

SG Ni-Resist type 2, subject to control of composition and of Nb addition in grade D-2W, can be repair welded. Most welding is done to reclaim defective castings, usually by manual arc welding, using flux-coated electrodes. Oxyacetylene welding is occasionally used.

Comparison of casting costs

It is impossible to make reliable comparisons between the casting costs of nickel aluminium bronze with that of competing ferrous alloys in sea water applications, because of the conflicting effects of the following factors:

- the cost of raw materials,
- the size and complexity of castings,
- the consequent difficulty of producing a sound casting in each alloy,
- the way the foundry allocates its overheads and its general pricing policy,
- the demand and competition for the alloy,
- the pattern cost which has to suit the running system of each alloy,
- the machining costs etc.

The following gives therefore only an approximate idea of price ranking of the various alloys under consideration. Each case has nevertheless to be treated on its own merits and various quotations obtained, if cost is likely to be a determining factor in the choice of alloy.

Most expensive	Superaustenitic stainless steel
	Super duplex stainless steel
	2205 duplex stainless steel
	Nickel aluminium bronze
	316 stainless steel
Least expensive	Ni-Resist

Difference in costs can be very significant. A relatively simple valve body casting may cost twice as much in superaustenitic stainless steel than in nickel-aluminium bronze and the difference in cost of fully machined castings in these two alloys can be in the ratio of 5:1 or more respectively.

Summary of comparison

The following are the main attractive features and drawbacks of the alloys compared above:

Nickel aluminium bronze

Attractive features	**Drawbacks**
Good mechanical properties	Vulnerable to polluted water corrosion
Good general corrosion resistance	
Excellent pitting resistance	
Good cavitation resistance	
Good corrosion fatigue resistance	
Best anti-fouling properties	
Good resistance to crevice corrosion	
Will withstand clean water flows of up to 4.3m/sec without significant erosion	
Immune to chloride stress corrosion cracking	
Best heat and electrical conductivity	
Excellent wear properties	
Excellent shock properties	
Weldable	
Easily machined	
Medium cost	

Superaustenitic and superduplex stainless steels

Attractive features	**Drawbacks**
Best mechanical properties	Poor corrosion fatigue resistance (superaustenitic)
Excellent general corrosion resistance	Susceptible to bio-fouling
Good pitting resistance	High cost
Good cavitation resistance	
Good corrosion fatigue resistance (superduplex)	
Excellent fluid flow erosion resistance	
High stress corrosion cracking resistance	
Best resistance to polluted water corrosion	
Excellent wear properties	
Medium heat and electrical conductivity	

Duplex and type 316 stainless steels

Attractive features	Drawbacks
Good mechanical properties	Poor pitting resistance (type 316)
Excellent general corrosion resistance	Vulnerable to polluted water corrosion
Good cavitation resistance	Low corrosion fatigue resistance (type
Good corrosion fatigue resistance	316)
(duplex)	Vulnerable to stress corrosion cracking
Excellent fluid flow erosion resistance	in warm sea water (type 316)
Good stress corrosion cracking	Susceptible to bio-fouling
resistance (duplex)	Susceptible to crevice corrosion
Excellent wear properties	
Medium heat and electrical conductivity	
Medium cost	

Ni-Resist

Attractive features	Drawbacks
Good mechanical properties	Low corrosion fatigue resistance
Good general corrosion resistance	Low cavitation resistance
Excellent pitting resistance	Low fluid flow erosion resistance
Best resistance to crevice corrosion	Low stress corrosion cracking resistance
Good wear properties	Vulnerable to polluted water attack
Medium heat and electrical conductivity	Susceptible to bio-fouling
Lowest cost	

APPENDIX 4
MACHINING OF ALUMINIUM BRONZES

Introduction

Because of their growing popularity as a high strength and excellent abrasion and corrosion resisting material, aluminium bronzes are increasingly being machined in most large and small engineering companies. It will be appreciated that, to ensure the most economical production, materials of this calibre require correct machining methods. Though many machine shops have developed their own standard practice to suit their particular requirements, these notes will serve as a general guide for machining aluminium bronzes. Aluminium bronzes must not be confused with free machining brass, but treated as a bronze with mechanical properties similar to those of high grade steel.

The handling of aluminium bronzes need present no difficulty to the average machine shop. They can readily be machined using modern tools and the correct workshop techniques. It is not possible to specify precise values for maximum feeds, speeds and depth of cut since these are influenced by several factors: the equipment being used, the operator, and his experience in handling the material. The recommendations given below may be taken as representing a reliable average, offering maximum production output for reasonable tool life and efficiency. Whilst some machine shops may fail to achieve the recommended values, others will exceed them.

Little distortion normally occurs on machining but, in cases where dimensions are critical, it may be found useful to carry out a stress-relief heat treatment of one hour at 350°C prior to final machining.

The scrap value of aluminium bronze swarf is relatively high. This can help offset machining costs and should be considered when costing component manufacture.

The information contained in this Appendix is derived from CDA Publication No 83, 'Aluminium Bronze Alloys for Industry'.[48] For further information on the machining of copper and its alloys, see CDA Technical Note TN44, 'Cost-effective Manufacturing-Machining Brass, Copper and its Alloys'.

Turning

The use of tungsten carbide tipped tools is considered desirable for turning aluminium bronzes. It is most important that the work should be held rigidly and that tools should be properly supported, with minimum overhang from the tool post. To obtain the best results, plant must be kept in good condition: excessively worn headstock bearings and slides will give rise to tool shatter and rapid tool breakdown. The first roughing cut on a casting should be deep enough to penetrate the

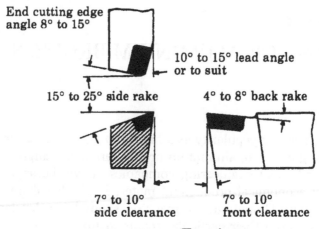

End cutting edge
angle 8° to 15°

10° to 15° lead angle
or to suit

15° to 25° side rake

4° to 8° back rake

7° to 10°
side clearance

7° to 10°
front clearance

Turning

Use full rake angle.
Do not flatten cutting edge.

Fig. A4.1 Details of carbide-tipped tools for turning

skin, and a steady flow of soluble oil is essential for both roughing and finishing cuts. The work must be kept cool during precision machining. If is allowed to heat up, difficulty will be experienced in maintaining accuracy.

Suitable designs for tungsten carbide roughing and finishing tools are illustrated in Fig A4.1 and depth of cut, speeds and feeds recommended for use with these tools are given in Table A4.1. High efficiency with carbide-tipped tools is achieved by using a light feed, a moderately heavy depth of cut and the highest cutting speed consistent with satisfactory tool life.

Table A4.1 Turning speeds and feed rates for aluminium bronzes.

		Roughing	Finishing
Depth of cut			
	mm	3–6	0.12–0.25
	in	1/8–1/4	0.005–0.010
Speed			
	m min^{-1}	30–60	120–180
	ft min^{-1}	100–200	400–600
Feed			
	mm/rev	0.25	0.12
	in/rev	0.010	0.005

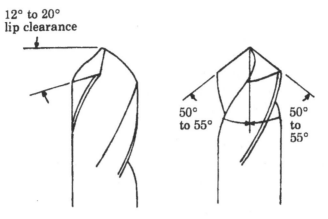

Fig. A4.2   Drill point and clearance angles

Drilling

Since aluminium bronze is hard, close-grained and free from the 'stringy' characteristic of copper, a fine drilled finish is obtainable. Fig A4.2 shows drill point and clearance angles. The best results are obtained with high-speed steel drills ground with negative rake at an included angle of 110° to 120°. Straight fluted drills will give a fine surface finish. Binding in the hole can be overcome by grinding the drill very slightly 'off-centre', thereby providing additional clearance.

Where counter-sinking is required, a counter-boring tool will give the best results. If a counter-boring tool is not available, it may be found preferable to carry out counter-sinking before drilling.

A coolant must be used, especially with the harder grades of aluminium bronzes and overheating must be avoided. Medium speeds and moderate feeds give the best results:

Speed: 15–40 m min^{-1}
       50–130 ft/min
Feed: 0.075–0.5 mm/rev
       0.003–0.02 in/rev

Reaming

Excellent results can be obtained with aluminium bronzes, but normal reaming practice is not suitable. It has been found that a simple 'D' bit, made up with a tungsten carbide insert, will maintain the closest limits and give a highly finished bore. Approximately 0.12mm (.005in) of metal should be removed. Adjustable type reamers with carbide inserts can also be used and it will be found that chatter is eliminated if a reamer, having an odd number of inserts, is chosen. If hand reaming is carried out, a left-hand spiral type is to be preferred. Avoid undue heating and use coolant.

Tapping

The principal reason for torn threads or broken taps may be selecting a tap drill which is either too small or to close to the size of the root diameter. In the majority of cases, where a specified thread fit is not needed and where the depth of hole is at least equal to the diameter of the tap, a 75% to 80% depth of thread is sufficient. A 100% thread is only 5% stronger than a 75% thread, yet it needs more than twice the power to tap. It also presents problems of chip ejection and requires the tap to be specially designed for the particular alloy.

For hand tapping, where the quantity of work or nature of the part does not permit the use of a tapping machine, regular commercial two-flute and three-flute high-speed steel taps should prove satisfactory. The rake should be correct for the metal being cut and the chamfer should be relatively short so that work hardening or excess stresses do not result from too many threads being cut at the same time.

High-speed steel taps with ground threads are used in machine tapping. In instances where the threads tend to tear as the tap is being backed out, a rake angle should be ground on both sides of the flute.

In the case of aluminium bronzes which produce tough and stringy chips, spiral-pointed taps (see Fig A4.3), with two or three flutes, are preferred for tapping through holes or blind holes drilled sufficiently deep for chip clearance. These taps produce long and curling chips, which are forced ahead of the tap.

Spiral-fluted bottoming taps can be used for machine (and hand) tapping of blind holes and wherever adequate chip relief is a problem.

Rake angle should be 8°–15° (see Fig A4.3), modified for the particular conditions of the job and used at speeds of 10–20 m/min (30–60 ft/min). The speeds indicated are based on the use of taps to produce fine to moderate pitch threads and speeds should be reduced by about 50% if carbon steel taps are used.

If the work is allowed to overheat, a re-tapping operation may be necessary. The use of tapping compound, having a high tallow content, will prevent binding in the case of softer grades of aluminium bronzes, and will prevent cracking of the work in the case of harder grades.

Milling

Undue heating must be avoided and a coolant should be used. Good results can be achieved using standard steel practice. It is recommended that the cutting edge of teeth should be on a radial line from the centre of the cutter; this applies to end-mills as well as to standard milling cutters. Speeds and feeds will depend upon the job and machining conditions, but the work must not be 'forced' or tearing and chipping may result.

Grinding

All grades of aluminium bronzes can readily be given an excellent ground finish. Even the softer grades will not clog the grinding wheel. Again, a coolant must be

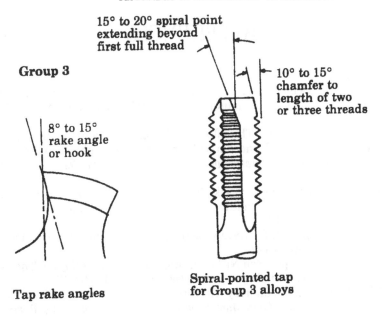

Group 3

15° to 20° spiral point
extending beyond
first full thread

10° to 15°
chamfer to
length of two
or three threads

8° to 15°
rake angle
or hook

Tap rake angles

**Spiral-pointed tap
for Group 3 alloys**

Fig. A4.3   Tap rake angles and spiral-pointed taps

used and overheating must be avoided. A bauxite type wheel gives satisfactory results and the grades recommended for particular operations are as follows: 30 grit for roughing, 46 grit for general purposes and 60 grit for fine-finish work. Since aluminium bronze is non-magnetic, it cannot be finished using a magnetic chuck.

REFERENCES

1. P. Aaltonen, K. Klemetti and H. Hanninen, 'Effect of tempering on corrosion and mechanical properties of cast aluminium bronzes', *Scand. J. Metall*, 1985, **14**, 233–242.
2. Z. Ahmad and P. Dvami, 'The effect of alloying additions on the optimisation of corrosion resistance and mechanical properties of alpha and beta aluminium bronzes', Paper from *6th International Congress on Metallic Corrosion*, ???Books, Sydney, 1975.
3. S. Alam, R. I. Marshall and S. Sasaki, 'Metallurgical and Tribological Investigations of Aluminium Bronze bushes made by a novel centrifugal casting technique', *Tribol. Int.*, 1996, **29**, No. 6, 487–492.
4. S. Alan and R. I. Marshall, 'Testing performance of various bronze bushes', *J. Appl. Phys.* 1992. **25**, 1340–1344.
5. W. O. Alexander, 'Copper–rich Nickel–Aluminium–Copper Alloys II. The Constitution of the Copper–Nickel rich alloys', *J. Inst. Met.*, 1938, **63**, 163–83.
6. A. Al–Hashem and H. M. Shalaby, *The role of Microstructure on the Cavitation Corrosion Behaviour of cast Nickel–Aluminium Bronze*, Nat Ass. of Corr. Eng. Report No. 283, March 1995.
7. A. Al-Hashem, P. G. Caceres, W. T. Riad and H. M. Shalaby, 'Cavitation corrosion behaviour of cast nickel aluminium bronze in sea water', *Corrosion* (US), 1995, **51** (5), 331–342.
8. Anonymous, 'The making and casting of aluminium bronze', *Brass World*, 1911, **7**, 100–102.
9. Anonymous, 'New wear resistance data for aluminium bronze', *Mater. Eng.*, Apr. 1972, **62** (4), P. 50.
10. D. Arnaud, 'Thermal analysis of copper alloys', *Trans AFS*, 1970, **78**, 25–32.
11. D. Arnaud, *The elevated temperature properties of cast copper alloys*, Incra Rep., Apr. 1972, Project No. 182, pp 36.
12. D. Arnaud, 'La désaluminisation des cupro–aluminiums, Importance de la composition, Sa justification electro–chimique', *Fonderie* 312, May 1972, p159–167.
13. D. Arnaud, 'Etude des caractéristiques mécaniques et de la limite d'endurance des cupro–aluminium' (Study of the mechanical characteristics and of the endurance limit of cupro–aluminium alloys), *Fonderie*, Oct. 1976, **13**, No. 5, 681–693.
14. N. C. Ashton, 'The production technology of aluminium bronze plate, bar, fasteners and welding', Papers presented at a symposium on 'Copper and copper alloy semi products', Bombay, Indian Copper Infom. Ctr. , Calcutta, Apr. 1980.
15. D. J. Astley and J. C. Rowlands, 'Modelling of bi–metallic corrosion in sea water systems', *Brit. Corr. J.*, 1985, **20**, (2), 90–94.
16. B. G. Ateya, E. A. Ashour and S. M. Sayed, 'Stress corrosion behaviour of a–aluminium bronze in saline water', *Corrosion*, January 1994.
17. J. P. Ault, *Erosion Corrosion of Nickel–Aluminium Bronze in Flowing Seawater*, Nat Ass. of Corr. Eng. Report No. 281, March 1995.
18. P. De Baets, 'Comparison of the wear behaviour of six bearing materials for a heavily loaded sliding system in Sea water', *Wear*, Jan 1995, **180** (1–2), 61–72.
19. J. Bailey, *Fundamentals of Engineering Metallurgy and Materials*, Cassel, 1973.
20. J. T. Barnby, E. A. Culpan, A. E. Morris, L. J. Hussey, M. S. Atherton and D. M. Rae, 'The fracture resistance of nickel aluminium bronze tubes', *Int. J. Fract.*, Oct. 1977, **13** (5), 681–693.

21. N. Battistelli and H. Lietveaux, 'Résistance à la corrosion des cupro–aluminiums binaires', *Fonderie, Fondeur d'aujourd'hui*, October 1993, **128**.

22. J. A. Beavers, G. H. Y. Koch and W. E. Berry, *Corrosion of metals in marine environments, a state–of–the–art report,* ???Reports, July 1986, Report No. MCIC–86–50, pp 8–1/8–79.

23. E. Belkin, 'Cast Ni–containing Aluminium Bronze Properties and Microstructure', *Modern Castings,* August 1961, **40**, 87–97.

24. N. V. Belyaev, 'Fatigue strength of welded joints in complexly alloyed aluminium bronzes', *Weld. Intl.,* 1989, **3** (5), 386–388.

25. R. M. Bentley and D. J. Duquette, *The effect of environment and applied current and potential on the corrosion fatigue properties of an as–cast duplex aluminium bronze alloy,* INCR Report, Project No. 241, Nov. 1979.

26. R. E. Berry, 'Contribution to discussion on the control of quality on the production of wrought non–ferrous metals and alloys', *J. Inst. Met.* 1954–55, **83**, 357–58.

27. A. W. Blackwood and NS. Stoloff, 'The mechanism of stress corrosion cracking in Cu–Al alloys', *ASM. Trans. Q,* 1969, **62** (3), 677–689.

28. A. J. Bradley *et al, J. Inst. S. I,* 1940, **141**, 99.

29. J. NBradley, 'Recent developments in copper–base alloys for naval marine applications', *Int. Met. Rev.* 1972, **17**, 81–99.

30. P. Brezina, *Giessereiforschung,* 1970,???(2) 81.

31. P. Brezina, *Sulzer Tech Rev,* 1972, Research No. 1, 29–36.

32. P. Brezina, *Proc. Conf. 'Aluminium Bronzes, a state of the art',* Leoben, Austria, April 1977.

33. P. Brezina, 'Heat treatment of complex aluminium bronzes', *Internat. Met. Reviews,* 1982, **27** (2).

34. British Defence Standard 01/2, *Guide to Engineering Alloys Used in Naval Service: Data Sheets.*

35. S. C. Britton, 'The Corrosion of Copper and Some Copper Alloys in Atmospheres Highly Polluted with Coal Smoke', *J. Inst. Metals,* 1941, **67**, 119–33.

36. C. V. Brouillette, 'Corrosion Rates in Port Hueneme Harbour.', *Corrosion,* August 1958, **14**, 352t–6t.

37. H. F. Brown, *The Application of Corrosion-Resisting Materials to Railroad Electrical Construction,* Assoc. Amer. Railroads, 1950.

38. L. Brown, *Joining of Copper and Copper Alloys,* CDA (UK) Publication No. 98, September 1994.

39. C. L. Bulow, 'Corrosion and Biofouling of Copper–Base Alloys in Sea Water', *Trans. Electrochem. Soc.,* 1945, **87**, 319–52.

40. B. Bhushan and B. K. Gupta, *Handbook of Tribology: Materials, Coatings and Surface Treatments,* McGraw–Hilll, New York, 1991.

41. H. S. Campbell, *Aluminium Bronze Alloys Corrosion Resistance Guide,* CDA (UK) Publication No. 80, July 1981.

42. H. S. Campbell, Private communication, Feb. 1998.

43. R. J. T. Caney, 'Corrosion Resistant Aluminium Bronze', *Australasian Engineer,* June 1954, **46**, 54–69.

44. H. C. H. Carpenter and C. A. Edwards of the National Physical Laboratory, *Teddington, England, Eighth Report to the Alloys Research Committee: on the properties of some alloys of aluminium and copper,* The Institution of Mechanical Engineers, Jan. 1907.

45. P. Collins and D. J. Duquette, 'Corrosion fatigue behavior of a duplex aluminium bronze alloy'. *Corrosion (US),* Apr. 1978, **34** (4), 119–124.

46. J. F. G. Conde and J. C. Rowlands, 'Copper base alloys in ship and marine applications', Papers presented at Copper '83, The Metals Soc., London, Nov. 1983, No. 34, pp 34.1–34.13.

47. M. Cook, W. P. Fentiman and E. Davis, 'Observations on the Structure and properties of Wrought Copper–Aluminium–Nickel–Iron Alloys', *J. Inst. Met.*, 1951–52, **80**, 419–29.

48. CDA (UK), *Aluminium Bronze Alloys for Industry*, CDA (UK) Publication No. 83, March 1986.

49. CDA (UK), *Welding of Aluminium Bronzes*, CDA (UK) Publication No. 85, 1988.

50. CDA (UK), *Cost–effective Manufacturing: Copper Alloy Bearings*, CDA (UK) Publication No. TN45, December 1992.

51. CDA (UK), *Copper and Copper Alloys: Compositions, Applications and Properties*, CDA (UK) Publication No. 120, 1998.

52. D. J. H. Corderoy and NOno, 'The work hardening of copper–aluminium alloys', *Proceedings of the 6th intern. confer.*, Melbourne, Aug. 1982, Pergamon, 1983, 941–946.

53. W. M. Corse and G. F. Comstock, 'Some copper–aluminium–iron alloys', *Transaction Am. Inst. Metals X*, 119, 1916.

54. L. P. Costas, 'Atmospheric corrosion of copper alloys exposed for 15 to 20 years', *Atmospheric Corrosion of Metals*, Report No. ASTM. STP–767, Jun. 1982, 106–115.

55. B. Cotton and B. P. Downing, 'Corrosion Resistance of Titanium to Sea Water', *Trans. Inst. Marine Eng.*, 1957, **69**, August. 311–19.

56. A. Couture, M. Sahoo, B. Dogan and J. D. Boyd, 'Effect of heat treatment on the properties of Mn–Ni–Al Bronze Alloys', *Trans. AFS*, 1987, **95**, 537–552.

57. R. A. Cresswell, 'Development of the Tungsten-Arc Cutting Process', *Brit. Welding J.*, August 1958, **5**, 346–55.

58. W. I. J. Crofts, D. W. Townsend and A. P. Bates, 'Soundness and Reproducibility of Properties of Sand–Cast Complex Aluminium Bronzes', *Brit. Foundryman*, 1964, **57**, 89–103; disc., *Brit. Foundryman*, 1964, 502–3.

59. W. I. J. Crofts, 'The influence of aluminium, iron, nickel and manganese on the tensile properties of complex aluminium bronzes', *Austr. Inst. Metals* 1965, **10** (2), 49.

60. J. Cronin and J. Warburton,, 'Amplitude-Controlled Transitions in Fretting: the Comparative Behaviour of Six Materials', *Tribol. Inst.* 21, 309–315 Dec. 1988.

61. E. A. Culpan and J. T. Barnby, 'The metallography of fracture in cast nickel aluminium bronze', *J. Mater. Sci.*, Feb. 1978, **13** (2), 323–328.

62. E. A. Culpan and G. Rose, *Microstructural characterization of cast nickel aluminium bronze*, Chapman and Hal, 1978 (See Ref. 193 also).

63. E. A. Culpan and A. G. Foley, 'The detection of selective phase corrosion in cast nickel aluminium bronze by acoustic emission techniques', *J. Mater. Sci.*, Apr. 1982, **17** (4), 953–964.

64. Delta Encon, *The Encon continuous casting process*, Delta (Manganese Bronze) Ltd brochure.

65. C. P. Dillon and Associates, 'Copper Alloys in Alkaline Environments', *USA Mater Perform.*, Feb 1996, **35** (2), 97.

66. A. G. Dowson, 'A new intermediate phase in the aluminium–copper system', *J. Inst. Met.*,1937, **61**, 197.

67. K. Drefahl, M. Kleinau and W. Steinkamp, 'Creep behaviour of copper and copper alloys as design criteria in pressure vessel manufacture', *J. Test Eval.* , Sept. 1985, **13**, No. 5, 329–343.

68. B. Dubois, G. Ocampo, G. Demiraj and G. Bouquet, 'Mechanical properties during the tempering of martensitic copper–aluminium alloys', *Proceedings of the 6th intern. confer., Melbourne*, Aug. 1982, Pergamon Press,1983, 281–287.

69. J. O. Edwards and D. A. Whittaker, 'Aluminium Bronzes containing Manganese, Nickel and Iron: Chemical Composition, Effect on Structure and Properties', *Trans. A.F.S.*, 1961, **69**, 862–72.

70. E. A. Feest and I. A. Cook, 'Pre–Primary Phase Formation in Solidification of Nickel–Aluminium Bronze', (Retroactive Coverage), *Met. Technol.* Apr. 1983, **10** (4), 121–124.

71. R. J. Ferrara and T. E. Caton, 'Review of de–alloying of cast aluminium bronze and nickel aluminium bronze alloys', *Mater. Performance*, Feb 1982, **21** (2), 30–34.

72. Louis Figuier, 'L'aluminium et le bronze d'aluminium', *'Les merveilles de la science'* vol. 4, Furne, Jouvet et Cie, Paris, 1877.

73. R. Francis and C. R. Maselkowski, *The effects of chlorine on materials for sea water cooling systems*, BNF Metals Technology Centre. Wantage. Oxfordshire, May 1985.

74. R. Francis, Private communication, 1997.

75. F. Gaillard and P. Weill, 'Identification des phases dans le cupro–aluminium allié', *Me. Sci. Rev. Met.*, 1965, **62**, 591.

76. W. A. Glaeser, 'Wear properties of heavy loaded copper–base bearing alloys', *JOM*, Oct 1983, **35** (10), 50–55.

77. W. A. Glaeser, *Materials for tribology*, Elsevier, Tribology series 20, 1992.

78. S. Goldspiel, J. Kershner and G. Wacker, 'Heat Treatment, Microstructure and Corrosion Resistance of Aluminium Bronze Castings', *Modern Castings*, Jan 1965, **47**, 818–823.

79. E. Gozlan and M. Bamberger, 'Phase transformations in permanent–mould cast Aluminium Bronze', *J. Mater. Sci*, 1988, **23** (9), 3558–3562.

80. E. Gozlan and M. Bamberger, Private communication Jan. 1998.

81. Gronostajski, Jerzy; Ziemba, Henryk, 'The Effect of High-Temperature Deformation of BA1032 Aluminium Bronze on its structure', *Rudy Met. Niezelaz.* May 1978, **23** (5), 218–222.

82. L. Guillet, 'Etude théorique et industrielle des alliages de Cuivre et d'Aluminium', *Revue de Métallurgie*, 1905, 568–571.

83. L. Guillet, 'Les bronzes d'aluminium au silicium', *Revue de Métallurgie*, 1923, **XX**, 771.

84. L. Habraken, C. Rogister, A. Davin and D. Coutsouradis, 'Eine eue anlaufbeftandigere Kupferlegienrung', *Metal*, 1969, **23**, 1148–1155.

85. A. P. C. Hallowes and E. Voce, 'Attack of Various Superheated Steam Atmospheres upon Aluminium Bronze Alloys', *Metallurgia*, 1946, **34**, 119–22.

86. M. Hansen and K. Anderko, *Constitution of Binary Alloys*, McGraw–Hill, New York, 1958.

87. F. Hasan, A. Jahanafrooz, G. W. Lorimer and N. Ridley, 'The Morphology, Crystallography and Chemistry of Phases in as–cast Nickel–Aluminium Bronze', *Met. Trans. A*, Aug 1982, **13A**, 137.

88. F. Hasan, G. W. Lorimer and N. Ridley, 'Tempering of Cast Nickel–Aluminium Bronze', *Met. Sci.*, June 1983, **17** (6), 289–295.

89. F. Hasan, J. Iqbal, and N. Ridley, 'Microstructure of As–Cast Aluminium Bronze Containing Iron', *Mater. Sci. Technol.* Apr. 1985, **1** (4), 312–314.

90. F. Hasan, G. W. Lorimer and N. Ridley, *Phase Transformations During Welding of Nickel–Aluminium Bronze*, The Institute of Metals, 1988 131–134.

91. J. L. Heuze, B. Masson, A. Karimi, 'Comportement à l'erosion de cavitation des cupro-aluminiums complexes', Association Technique Maritime et Aeronautique, Session 1991.

92. R. A. Higgins, *Engineering Metallurgy, Part I, Applied Physical Metallurgy*, Hodder and Stoughton 1980.

93. O. Hindbeck, 'Resistance to Wear of Copper and Copper Alloys', Paper read at 4th Internat. Mech. Eng. Congress, Stockholm, June 1952.

94. W. Hirst and J. K. Lancaster, 'Surface Film Formation and Metallic Wear', *J. Appl. Phys.*, September 1956, **27**, 1057–65.

95. C. Hisatsune, 'On the Equilibrium Diagram of the copper–aluminium alloy system', *Mem. Coll. Eng. Kyoto Imp. Univ. ???*1034, **8** (2), 74.

96. F. Hudson, 'High Duty Brasses and Bronzes Available to the Engineer', *Metallurgia*, 1937, **16**, 195–8; 1936, **17**, 61–64.

97. I. M. Hutchings, *Tribology: Friction and Wear of Engineering Materials*, Edward Arnold, 1992.

98. I. M. Hutchings, Private communication, December 1997.

99. International Copper Research Association. Research programme sponsored at Battelle and Stanford Research Insts., 1962.

100. J. Iqbal, F. Hasan and G. W. Lorimer, 'The characterisation of phases and the development of microstructure in an as–cast silicon–aluminium bronze', Unpublished, University of Manchester/UMIST (*circa* 1985).

101. J. Iqbal, F. Hasan and G. W. Lorimer, 'The development of microstructure and characterisation of phases in an as–cast copper–manganese–aluminium alloy', Unpublished, University of Manchester/UMIST (*circa* 1985).

102. A. Jahanafrooz, F. Hasan, G. W. Lorimer and G. W. Ridley, 'Microstructural development in complex nickel aluminium bronzes', *Metall Trans. A*, Oct 1983, **14**, (10), 1951–1956.

103. J. Jellison and E. P. Klier, 'The Cooling Transformations in the Beta Eutectoid Alloys of the Cu–Al System' *Trans. A I M. E*, 1965, **233**, 1694–1702.

104. R. L. Jones and J. C. Rowlands, *Recent research and development on copper alloys at Adm, Res, Estab.*, CDA (UK) Papers, 1985, Report No. 17.

105. G. Joseph, R. Perret and M. Ossio, 'Corrosion resistance of copper and copper aluminium bronzes', *Copper 87*, Vol. 1, Physical Metallurgy of Copper, Conference held in Vina, Chile, Dec. 1987, 325–338.

106. M. Kanamori, S. Ueda, S. Matsuo and H. Sakaguchi, 'On the Properties of Cu–Al–Ni–Fe Alloys in Slowly-Cooled Castings', *Nippon Kinzoku Gakkai–Si*, 1960, **24** (5), 265–8.

107. M. Kanamori, S. Ueda, and S. Matsuo, 'On the Corrosion Behaviour of Copper–Aluminium–Nickel–Iron Cast Alloys in Sea Water', *Nippon Kinzoku Gakkai*, 1960, **24** (5), 268–71.

108. M. Kanamori, S. Ueda and S. Matsuo, 'Nickel Aluminium Bronze for Marine Propeller Alloys, I. Structure, II. Mechanical Properties, III. Effect of other Alloying Elements', *Nippon Kinzoku Gakkai–Si*, 1960, **24** (4), 201–5; 205–9; 209–13 (In Japanese).

109. A. H. Kasverg and D. J. Mack, 'Isothermal Transformation and properties of a Commercial Aluminium Bronze', *Trans. AIME*, 1951, **191**, 903–8.

110. J. McKeown, D. NMends, E. S. Bale and A. D. Michael, 'The creep and fatigue properties of some wrought complex Aluminium Bronzes'', *J. Inst. Met.*, 1954–5, **83**, 69–79.

111. J. F. Klement, R. E. Maersch and P. A. Tully, 'Use of Alloy additions to Prevent Intergranular Stress–Corrosion Cracking in Aluminium Bronze', *Corrosion*, October 1960, **16**, 519*t*–22*t*.

112. W. Körster and T. Gödecke, 'Das dreistoffsystem kupfer–mangan–aluminium', Z. *Metllkd.*, 1966, **57**, 889–901.

113. E. J. Kubel, 'Curbing corrosion in marine environments', *Adv. Mater. Process*, Nov. 1988, **134** (5), 29–32; 41.

114. J.M. Langham and A.W.O. Webb, 'The New High Strength Copper–Manganese–Aluminium Alloys, Their Development, Properties and Applications', *British Foundryman*, June 1962, **55**, 246–62.

115. T. S. Lee, 'Preventing galvanic corrosion in marine environments', *USA Chem. Eng.*, (New York), April 1985, **92**, No. 7, 89–92.

116. H. Leidheiser, *Stress corrosion corrosion cracking of copper alloys*, International Copper Research Association, December 1985, 18.

117. R. E. Lismer, 'The properties of some metals and alloys at low temperatures', *J. Inst Met*, May 1960–61, **89**, 145–61.

118. B. J. Little and F. B. Mansfeld, 'The corrosion behaviour of Stainless Steel and Copper Alloys exposed to Natural Sea water', *Werks Korros*, July 1991, **42** (7), 331–340.

119. B. Little and P. Wagner, 'Micro–biological influenced corrosion in offshore oil and gas systems', *Proceedings of New Orleans Offshore Corrosion Conference*, Dec. 1993, Report No. NRL/PP/7333–93–0020, AD–A275 195.

120. D. M. Lloyd, G. W. Lorimer and N. Ridley, 'Characterisation of Phases in a Nickel–Aluminium Bronze', *Met. Technol.* Mar. 1980, **7** (3), 114–119.

121. G. W. Lorimer, F. Hasan, J. Igbal and N. Ridley, 'Observations of microstructure and corrosion behaviour of some aluminium bronzes', CDA Conference 'Copper alloys in marine environments', Paper 1, Birmingham, Mar. 1985.

122. G. W. Lorimer, F. Hasan, J. Iqbal and N. Ridley, 'Observation of Microstructure and Corrosion Behaviour of Some Aluminium Bronzes', *Br. Corros. J.*, 1986, **21** (4), 244–248.

123. S. Lu, A. Wu and S. Zhou, 'The hardening of complex aluminium bronze by heat treatment', *Copper 91*, Vol. 1, *Economics and Applications of Copper, Intern. Symposium*, Ottawa, Aug 1991, Pergamon, 211–220.

124. Lucey, 'Mechanism of pitting corrosion of copper in supply waters', *Br. Corros. J.* September 1967, **2**.

125. P. A. Lush, *Chartered Mechanical Engineer*, Oct. 1987, 31.

126. C. McCombe, 'Aluminium Bronze: a Victorian Wonder Metal', *Foundry Trade Journal*, March 1986.

127. P. J. Macken and A. A. Smith, *The Aluminium Bronzes*, CDA (UK) Publication No. 31 1966.

128. F. Mansfeld, G. Liu, H. Xia, C. H. Tsai and B. J. Little, 'The corrosion behaviour of Copper Alloys, Stainless Steels and Titanium in Sea water.', *Corros. Sci*, Dec. 1994, **36** (12), 2063–2095.

129. D. Medley, Scotforge, Spring Grove, USA, Private communication (and indirectly from Baltimore Speciality Steels), *June 1997*.

130. C. H. Meigh, *The Practical Applications of Aluminium Bronze*, McGraw–Hill, 1941.

131. H. J. Meigh, 'Designing aluminium–bronze castings', *Engineering*, August 1983, Technical File No. 116.

132. J. R. Moon and M. D. Garwood, 'Transformation during continuous cooling of the b phase in copper–aluminium alloys', *J. I M*, 1968, **96**, 17.

133. J. S. Msahana, O. Vosikovsky and M. Sahoo, 'Corrosion fatigue behaviour of nickel aluminium bronze alloys', *Canada Met. Q.* , Jan–Mar. 1984, **23** (1), 7–15.

134. J. A. Mullendore and D. J. Mack, 'The Iron-Rich Phase in Aluminium Bronze' *Trans. AIME*, 1958, **212**, 232–53.

135. G. S. Murgatroyd, 'Effects of Composition and process variables on cast structures of aluminium bronzes', Unpublished thesis for Master of Science Degree, University of Birmingham, 1975.

136. J. Murphy, 'Contribution to discussion on The Resistance of Some Special Bronzes to Fatigue and Corrosion Fatigue', *J. Inst. Met.*, 1937, **60**, 154–7.

137. I. Musatti and L. Dainelli, 'Influence of Heat Treatment on the Fatigue and Corrosion Resistance of Aluminium Bronze', *Alluminio*, 1935, **4**, 51–63.

138. T. M. Mustaleski, R. L. McCaw and J. E. Sims, 'Electron beam welding of nickel–aluminium bronze', *Weld J.*, July 1988, **67** (7), 53–59.

139. J. W. Oldfield and G. L. Masters, *Collation of data comparing properties of Aluminium Bronze with cast Stainless Steels and Ni–resist in offshore sea water environments*, CDA (UK) Publication No. 115, June 1996.

140. H. R. Oswald and P. Sury, 'On the corrosion behaviour of individual phases present in aluminium bronzes', *Corros. Sci. (UK)*, Jan 1972, **12** (1), 77–90.

141. M. A. Phoplonker, J. Byrne, TV. Duggan, R. D. Scheffel and P. Barnes, 'Near–threshold fatigue crack growth in and aluminium bronze alloy', *Fatigue of Engineering Materials and Structures*, Vol. 1, 1986, Report no. C277/86, 137–144.

142. R. A. Poggie, J. J. Wert and L. A. Harris, 'The effects of surface oxidation and segregation on the adhesional wear behaviour of aluminium bronze alloys', *J. Adhes. Sci, Technol.*, 1994, **8** (1), 11–28.

143. W. Pool and J. L. Sullivan, 'The wear of aluminium bronze on steel in the presence of aviation fuel', *ASLE Transactions*, 1977, **22**, 154–161.

144. W. Pool and J. L. Sullivan, 'The role of aluminium segregation in the wear of aluminium bronze–steel interfaces under conditions of boundary lubrication', *ASLE Transactions*, 1980, **23**, 401–408.

145. B. J. Popham, 'Copper–nickel and aluminium bronze alloys: forming and fabrication', *Métalurgie*, Apr. 1983, **50** (4), 168–171.

146. L. E. Price and G. J. Thomas, 'Oxidation resistance in copper alloys', *J. Inst. Met.* 1938, **63**, 21–28.

147. E. Rabald, *Corrosion Guide*, Second Edition, Elsevier, 1965.

148. A. A. Read and R. H. Greaves, 'The influence of nickel on some copper–aluminium alloys', *J. Inst. of Metals*, 1914, **XI**, 169.

149. J. V. Reid and J. A. Schey, 'Adhesion of copper alloys', *Wear*, 1985, **104**, 1–20.

150. J. V. Reid and J. A. Schey, 'The effect of surface hardness on friction', *Wear*, 1987, **118**, 113.

151. M. Richard, Centre Technique des Industries de la Fonderie, Paris, Private communication May 1998.

152. H. Rininsland and E. Wachtel, 'Uber den aufbau und die magnetischen eigenschaften von festen und flussigen kupfer–aluminium–legierungen mit zusatzen von ubergangsmetallen', *Giesserei–forschung*, 1970, **3**, 129–142.

153. W. Rosenhain and C. A. H. Lantsberry of the National Physical Laboratory, Teddington, England, *Ninth Report to the Alloys Research Committee: on the properties of alloys of copper, aluminium and manganese*, The Institution of Mechanical Engineers, 21 Jan. 1910.

154. J. Roucka, I. Macásek, K. Rusin and J. Svejcar, 'Possibilities of applying aluminium bronze in the production of cast tools for sheet drawing', *Solidification technology in the foundry and casthouse*, The Metals Society, 1983, 392–397.

155. J. C. Rowlands, 'Studies of the preferential phase corrosion of cast nickel aluminium bronze in sea water', *Metal Corr.*, 1981, **2**, 1346–1351.

156. K. Rutkowski, 'Cu–Al–Mn and Cu–Mn–Al Casting Alloys with Ni and Fe Additions' *Modern Castings*, February 1962, **41**, 99–116.

157. M. Sadayappan, F. A. Fasoyinu, R. Orr, R. Zavadil and M. Sahoo, *Impurity limits in aluminium bronzes*, Progress Report, Year 1, Materials Technology Laboratory, CANMET, March 1998.

158. S. Sarker and A. P. Bates, 'Impact resistance of sand–cast aluminium bronze BS. 1400 AB2', *British Foundryman* 1967, **60**, 30.

159. W. J. Schumacker, 'Wear and galling can knock out equipment', *Chemical Engineering* (New York), May 1977.

160. W. J. Schumacker, Private communication to E. E. Denhart, Manager, Stainless Steel Research and Technology, Armco, Feb. 1980.

161. A. Schüssler and H. E. Exner, 'The corrosion of the nickel–aluminium bronzes in sea water, I. Protective layer formation and the passivation mechanism, II. The corrosion mechanism in the presence of sulphide pollution', *Corros. Sci.*, **34** (11), 1993, 1793–1815.

162. H. M. Shalaby, A. Al-Hashem, H. Al–Mazeedi and A. Abdullah, 'Field and Laboratory Study of Cavitation Corrosion Behaviour of cast Nickel–Aluminium Bronze', *Br. Corr. J.*, 1995, **30** (1), 63–70.

163. Z. Shi, Y. Sun, A. Bloyce and T. Bell, 'Un-lubricated rolling–sliding wear mechanisms of complex aluminium bronze against steel., *Wear*, May 1996, **193** (2), 235–241 ISSN:0043–1648. SINTEF Corrosion Center, 'Corrosion Handbook'.*???*

164. C. S. Smith and W. E. Lindlief, 'A micrographic study of the decomposition of the beta phase in the copper–aluminium system', *Trans. Amer. Inst. Min. Met. Eng.*, 1033, 104, 69. *???*

165. C. Soubrier and M. Richard, 'Soudures de pièces en cupro-aluminiums: étude électro-chimique du choix du recuit', *Fonderie, Fondeur d'aujourd'hui*, January 1991, **101**.

166. E. S. Sperry, 'Aluminium bronze and what can be done with it; its good and bad qualities', *Brass World*, Jan. 1910, **6** (1), 3–7.

167. M. S. Stamford, 'Copper alloys for diecasting', *Br. Foundryman*, May 1980, **73** (5), 146–148.

168. J. L. Sullivan and L. W. Wong, 'Wear of aluminium bronze on steel under conditions of boundary lubrication', *Tribol. Int.*, 1985, **18**, 275–281.

169. J. L. Sullivan, 'Boundary lubrication and oxidational wear', *J. Physics, D*, 1986, **19**, 1999.

170. H. A. Sundquist, A. Mathews and D. G. Teer, 'Ion-plated aluminium bronze coatings for sheet metal forming dies', *Thin solid films*, 17 Nov. 1980, **73** (2), 309–314.

171. P. J. Le Thomas, D. Arnaud and A. Lethuiller, 'L'enrouillement des cupro-aluminium au fer', *Mém. Sci. Rev.Métall.* 1960, **57**, 313–323.

172. R. Thomson, 'Charpy Impact Properties of Bronze Propeller Alloys', *Modern Castings*, April 1968, **53**, 189–199.

173. C. H. Thornton, *Aluminium Bronze Alloys Technical Data*, CDA (UK) Publication No. 82, January 1986.

174. W. Thury and W. Meyer, *The properties of cast cobalt–bearing aluminium bronzes*, INCRA Report No. PE–13, Sept. 1971, 29 pages.

175. D. F. Toner, 'The nature of b phases in Copper–Aluminium System', *Trans. AIME*, 1959, **215**, 223–25.

176. W. Tracy, 'Resistance of Copper Alloys to Atmospheric Corrosion', *A.S.T.M. Symposium on Atmospheric Exposure Tests on Non–Ferrous Alloys*, February, 1946.

177. A. W. Tracy, *Effect of Natural Atmospheres on Copper Alloys: 20–Year Test*, A.S.T.M. Spec. Tech. Pub., 175, 1955.

178. B. W. Turnbull, 'The effects of heat treatment on the mechanical properties and corrosion resistance of cast aluminium bronze', *Corros. Australas.*, Oct. 1983, **8** (5), 4–7.

179. A. H. Tuthill, 'Guidelines for the use of Copper Alloys in Sea water', *Mater. Perform.* 1987, **26** (9), 12–22.

180. H. H. Uhlig, *The Corrosion Handbook*, sponsored by the Electrochemical Society, New York, John Wiley and Sons Inc., *circa* 1947.

181. E. Uusitalo, 'Galvanic Corrosion of Copper Alloys and Stainless Steels in Cellulose Wash Waters.' *Teknillisen Kemian Aikakauslehti*, May 1959, 273–81. (In English.).

182. C. Vickers, 'Manganese in aluminium bronze', *Brass World*, Jul. 1918, **14** (7), 202–203.

183. P. Weill–Couly and D. Arnaud, 'Influence de la composition et de la structure des cupro–aluminium sur leur comportement en service', *Fonderie*, April 1973, **322**, 123.

184. P. Weill–Couly, 'Welding aluminium bronze castings', *Proc Int Conf on welding of castings*, Welding Institute, Cambridge, 1977, vol. 1, 253–266.

185. P. Weill–Couly, Private communication with D. Meigh 1986.
186. D. Weinstein, *Thermo–mechanical processing and transformation–induced plasticity*, Stanford Research Institute, Report No. 7428/6, 1970, to International Copper Research Institute.
187. P. Wenschot, 'The properties of Nickel Aluminium Bronze sand cast ship propellers in relation to section thickness', *Nav. Eng. J.*, Sept 1986, **98** (5), 58–69.
188. Williams, 'Aluminium Bronzes for Marine Applications', *J. Amer. Soc. Naval Engineers, August* 1957, **69** (3), 453–61.
189. J. H. Woodside, *Report of Investigations to Improve the Wear Resistance of Manganese and Naval Bronze Castings for Naval Service*, US. Navy Bur. Ships Welding Engrs. Conference, Naval Res. Lab, April 30, May 4, 1956, 47–58.
190. Yuanyuan li, T. L. Ngai, 'Grain refinement and microstructural effects on mechanical and tribological behaviours of Ti and B. modified aluminium bronze', *Journal of Materials Science*, 1996, **31**, 5333–5338.
191. A. Yutaka, 'The Equilibrium Diagram of Iron–bearing Aluminium Bronze' *Nippon. Kinzoky. Gakkai–Si*, 1941, **5** (4), 136–55.
192. R. N. Singh, S. K. Tiwari and W. R. Singh, *Effects of Ta, La, and Nd additions on the corrosion behaviour of aluminium bronze in mineral acids*, Chapman and Hall, 1992.
193. E. A. Culpan and G. Rose, 'Corrosion behaviour of cast nickel aluminium bronze in sea water', *Br. Corros. J.* **14** (3), 160–166.
194. Y. S. Sun, G. W. Lorimer and N. Ridley, 'Microstructure and its Development in Cu–Al–Ni Alloys', *Metallurgical Transactions A*, March 1990, **21A**,.

Current CDA Publications can be obtained from Copper Development Association, Verulam Industrial Estate, 224 London Road, St Albans, Herts AL1 1AQ, Great Britain.

Publications by INCRA (International Copper Research Association) may be obtained from European Copper Institute, Avenue de Tervueren 168 b 10, B–1150 Brussels, Belgium.

INDEX